SALINITY TOLERANCE IN PLANTS

SALINITY TOLERANCE IN PLANTS

Strategies for Crop Improvement

Edited by

RICHARD C. STAPLES

Boyce Thompson Institute
Cornell University

GARY H. TOENNIESSEN

The Rockefeller Foundation

A Wiley-Interscience Publication

JOHN WILEY & SONS
New York   Chichester   Brisbane   Toronto   Singapore

Library of Congress Cataloging in Publication Data:
Main entry under title:

Salinity tolerance in plants.

(Environmental sciences and technology, ISSN 0194-
0287)
Papers presented at an international conference
held in Bellagio, Italy, sponsored by Rockefeller
Foundation.
Includes indexes.
1. Plants, Effect of salt on—Congresses.
2. Salt-tolerant crops—Physiology—Congresses.
3. Salt-tolerant crops—Breeding—Congresses.
I. Staples, Richard C. II. Toeniessen, Gary H.
III. Rockefeller Foundation. IV. Series.

QK753.S3S24 1984   581.19′24   83-14759
ISBN 0-471-89674-8

CONTRIBUTORS

AYADI, ABDELKADER, Physiologie Végétale, Faculté des Sciences, Campus Universitaire, Tunis El Menzah, Tunis, Tunisia

FLOWERS, T. J., Biology Building, The University of Sussex, Falmer-Brighton, Sussex, United Kingdom

GALE, J., Blaustein Institute of Desert Research, Ben-Gurion University of the Negev, Sde Boger, Israel

GORHAM, J., Department of Biochemistry and Soil Science, University College of North Wales, Bangor, Wales, United Kingdom

HAMZA, M., Physiologie Végétale, Faculté des Sciences, Campus Universitaire, Tunis El Menzah, Tunis, Tunisia

HANSON, MAUREEN R., Department of Biology, University of Virginia, Charlottesville, Virginia

JANA, S., Department of Crop Science and Plant Ecology, University of Saskatchewan, Saskatoon, Canada

JEFFERIES, R. L., Department of Botany, University of Toronto, Toronto, Ontario, Canada

JESCHKE, W. DIETER, Lehrstuhl Botanik 1, Universitat Würzburg, Würzburg, FRG

KRAMER, DETLEF, Institut für Botanik, Technische Hochschule Darmstadt, Darmstadt, FRG

KUIPER, PIETER J. C., Biologisch Centrum, Afd. Plantenfysiologie, Rijksuniversiteit Groningen, The Netherlands

LÄUCHLI, ANDRÉ, Department of Land, Air and Water Resources, University of California, Davis, California

LEOPOLD, A. C., Boyce Thompson Institute for Plant Research, Cornell University, Ithaca, New York

LÜTTGE, ULRICH, Institut für Botanik, Technische Hochschule Darmstadt, Darmstadt, FRG

MCDONNELL, E., Department of Biochemistry and Soil Science, University College of North Wales, Bangor, Wales, United Kingdom

MOORE, CHARLES V., National Economics Division, USDA, Department of Agricultural Economics, University of California, Davis, California

O'LEARY, JAMES W., Environmental Research Laboratory, Tucson International Airport, The University of Arizona, Tucson, Arizona

PITMAN, M. G., School of Biological Sciences, The University of Sydney, Sydney, Australia

PONNAMPERUMA, F. N., International Rice Research Institute, Manila, Philippines

RAINS, D. W., Plant Growth Laboratory, Department of Agronomy and Range Science, University of California, Davis, California

RUDMIK, T., Department of Botany, University of Toronto, Toronto, Ontario, Canada

SACHER, ROBERT F., Boyce Thompson Institute for Plant Research, Cornell University, Ithaca, New York

SHANNON, MICHAEL C., U.S. Salinity Laboratory, U.S. Department of Agriculture, Riverside, California

SMITH, J. ANDREW C., Institut für Botanik, Technische Hochschule Darmstadt, Darmstadt, FRG

SRIVASTAVA, J. P., Cereal Improvement Program, International Center' for Agricultural Research in Dry Areas, Aleppo, Syria

STAPLES, RICHARD C., Boyce Thompson Institute for Plant Research, Cornell University, Ithaca, New York

STAVAREK, S. J., Plant Growth Laboratory, Department of Agronomy and Range Science, University of California, Davis, California

TAL, MOSHE, Department of Biology, Ben-Gurion University of the Negev, Beer Sheva, Israel

TOENNIESSEN, GARY H., Agricultural Sciences, The Rockefeller Foundation, New York, New York

WILLING, R. P., Biological Sciences Department, State University of New York, Albany, New York

WYN-JONES, R. GARETH, Department of Biochemistry and Soil Science, University College of North Wales, Bangor, Wales, United Kingdom

YEO, ANTHONY R., Biology Building, The University of Sussex, Falmer-Brighton, Sussex, United Kingdom

ZERONI, M., Blaustein Institute for Desert Research, Ben-Gurion University of the Negev, Sde Boger, Israel

To John J. McKelvey, Jr.,
a friend and colleague, who stimulates scientists from throughout the world to address important problems of agriculture.

SERIES PREFACE

Environmental Science and Technology

The Environmental Science and Technology Series of Monographs, Textbooks, and Advances is devoted to the study of the quality of the environment and to the technology of its conservation. Environmental science therefore relates to the chemical, physical, and biological changes in the environment through contamination or modification, to the physical nature and biological behavior of air, water, soil, food, and waste as they are affected by man's agricultural, industrial, and social activities, and to the application of science and technology to the control and improvement of environmental quality.

The deterioration of environmental quality, which began when man first collected into villages and utilized fire, has existed as a serious problem under the ever-increasing impacts of exponentially increasing population and of industrializing society. Environmental contamination of air, water, soil, and food has become a threat to the continued existence of many plant and animal communities of the ecosystem and may ultimately threaten the very survival of the human race.

It seems clear that if we are to preserve for future generations some semblance of the biological order of the world of the past and hope to improve on the deteriorating standards of urban public health, environmental science and technology must quickly come to play a dominant role in designing our social and industrial structure for tomorrow. Scientifically rigorous criteria of environmental quality must be developed. Based in part on these criteria, realistic standards must be established and our technological progress must be tailored to meet them. It is obvious that civilization will continue to require increasing amounts of fuel, transportation, industrial chemicals, fertilizers, pesticides, and countless other products; and that it will continue to produce waste products of all descriptions. What is urgently needed is a total systems approach to modern civilization through which the pooled talents of scientists and engineers, in cooperation with social scientists and the medical profession, can be focused on the development of order and equilibrium in the presently disparate segments of the human environment. Most of the skills and tools that are needed are already in existence. We surely have a right to hope a technology that has created such manifold environmental problems is also capable of solving them. It is our hope that this series in Environmental Science and Technology will not only serve to make this challenge more explicit to the established professionals, but that it also will help to stimulate the student toward the career opportunities in this vital area.

Robert L. Metcalf
Werner Stumm

PREFACE

The scientific manipulation of the genetic composition of plants in order to produce improved crop varieties has contributed significantly to increased food production. Packaging scientific and technological know-how in the form of an improved seed has been an effective means of reaching and benefiting millions of poor farmers in developing countries as well as serving commercial agriculture.

Many of the high-yielding varieties that contributed to production increases over the past few decades were deliberately developed to maximize yield under favorable environmental conditions. However, where it is not feasible to modify the environment to suit the plant, scientists are now being challenged to modify the plant to suit adverse environments while maintaining reasonable and reliable yields. To accomplish this requires knowledge of the plant's physiological and biochemical response to stress conditions, a germ plasm resource of plants in which response mechanisms vary significantly, the ability to select plants having a desired response to stress, and techniques for shuffling or transferring germ plasm in ways that produce varieties having new combinations of desired characteristics. As the genetic manipulation techniques become more precise, a thorough understanding of the biochemical and physiological mechanisms involved becomes more important.

Salinity in soil or water presents a stress condition for crop plants that is of increasing importance in agriculture. It is estimated that a third of the world's irrigated area is already affected to some degree by excess salinity, primarily caused by inadequate drainage. In addition, the need to produce more food is continually pushing agriculture farther onto marginal lands often characterized by soils and waters with a high degree of natural salinity. Salinity tolerance would therefore be a highly desirable characteristic to introduce into crop plants if yield potential could also be maintained.

This volume is based on papers presented at an international conference designed to review the current state of knowledge concerning the physiological and biochemical mechanisms used by plants to accommodate high saline conditions. We wanted to identify opportunities where such knowledge could help plant breeders and cell biologists in their efforts to select plants having good tolerance to salinity, and to assess strategies that might be used today or in the future to produce tolerant crop varieties.

The chapters are presented in three parts. The first on "Mechanisms of Salt Tolerance" deals with cytology and physiology. These chapters indicate that salt tolerance is primarily a whole plant characteristic resulting from cells functioning as tissues, rather than as individuals. The second part on "Crop Selection and Improvement" reviews the successes and difficulties that plant breeders and cell biologists have encountered in trying to select for and use salt-tolerant plants in breeding programs. Chapters are also included on the potential use of halophytic plants and on genetic engineering and cell culture as promising future plant improvement technologies. The third part entitled "Controlled Environments and Economic Analyses" has a chapter that considers the use of controlled environment agriculture as an alternative strategy for using saline waters. Two other chapters present an economic analysis of plant improvement strategies, and a review of the world food situation including the role of salt-tolerant plants.

We thank The Rockefeller Foundation for hosting the conference at its Study and Conference Center in Bellagio, Italy. We also thank Dr. Ulrich Lüttge for assisting with the organization of the conference.

RICHARD C. STAPLES
GARY H. TOENNIESSEN

New York, New York
January, 1984

CONTENTS

SALINITY
TOLERANCE
IN PLANTS

PART ONE

MECHANISMS OF
SALT TOLERANCE

1

CYTOLOGICAL ASPECTS OF SALT TOLERANCE IN HIGHER PLANTS

Detlef Kramer

Institut für Botanik
Technische Hochschule Darmstadt
Darmstadt, Federal Republic of Germany

Natural salinity is a widespread phenomenon on earth, and the evolution of living organisms has resulted in numerous species that show special adaptive mechanisms to growth in saline environments. The majority of plants are relatively salt sensitive. In particular, almost all crop plants are unable to tolerate permanently saline conditions in the soil. Nevertheless, specialists have developed in many families that are able to live in such habitats.

In principle, two different mechanisms of salt tolerance are found in the plant kingdom. Halophilic bacteria have a highly salt-resistant cytoplasm, which means that enzymes and membrane systems function in the presence of high concentrations of NaCl [1]. In contrast, such mechanisms have not been established for any eukaryotes. Most enzymes that have been isolated from salt-tolerant higher plants show no salt resistance in excess of that found in salt-sensitive species [2-4]. Therefore, it must be assumed that salt-tolerant eukaryotes are able to maintain a reasonably low salt concentration in the cytosol. For a single cell this can theoretically be achieved by exclusion of NaCl or sequestration within cellular compartments such as the vacuole and accumulation of K^+ and organic osmotica in the cytosol. It appears that marine algae probably maintain their physiological activity by a combination of these mechanisms. Because no cytological features have been found in marine algae that can be correlated with these transport mechanisms, it must be concluded that the transport properties at the plasmalemma and tonoplast have adapted to the increasing salt concentration of the seawater during evolution. A special characteristic of marine organisms is that each cell is bathed in the same solution and is therefore more or less independent in terms of nutrient supply. Furthermore, seawater contains high concentrations of the ions (potassium, calcium, magnesium) that are important for the physiological processes in plant cells and these concentrations are constant at a given site.

Terrestrial plants that grow on saline soils are confronted with more complex problems. In the rhizosphere the concentration of salt in the soil solution fluctuates because of changes in water supply, drainage, and evaporation and transpiration. Salinity is caused not only by $NaCl$ but also by Na_2CO_3, $NaHCO_3$, and Na_2SO_4 and the relations of these salts to each other as well as to other nutrients like K^+, Ca^{2+}, and Mg^{2+} are important and may differ greatly at different sites.

Another important aspect of plant adaptation to different environments is the differentiation of the plant body. The specific problems of life on land have been resolved by the development of a division of labor within the plant body. Each organ of a terrestrial plant has particular functions but only few, specialized cells are in direct contact with the soil solution and responsible for nutrient uptake. A sophisticated system assures transport and distribution of water and nutrients to the other plant organs. This heterogeneity may result in concentration gradients in different cell types and demands high transport "efficiency" in certain tissues such as the absorbing part of the root or salt glands.

In higher plants that are able to tolerate salinity, several mechanisms have been recognized. Roughly speaking one can distinguish between salt excluders and salt includers (5). Salt excluders possess mechanisms that ensure that salt reaches the shoot only in very small amounts. This might be due to a very efficient selectivity toward K^+ during absorption, as in many grasses (6). Another possibility is that Na^+ is absorbed in significant amounts but is reabsorbed from the xylem sap in proximal parts of the root (7) or in the shoot (8, 9) and is then either stored or retranslocated to the soil (10–12).

In contrast, salt includers absorb salt and store it at high amounts in the stem and leaves. Here, sequestration of salt from the cytosol is the problem, as described for marine algae. Many salt includers are succulent, probably because of the accumulation of salt in large vacuoles of the mesophyll cells (Chapter 7). Other salt-including species have special glands on the leaf surface that excrete salt at a high concentration (13).

It is evident that, whatever the strategy by which a plant is able to adapt to salinity, transport phenomena play a significant role. It is therefore valid to ask whether modifications or specializations at the cellular level are found which might be correlated with these transport processes. It is an important question whether the knowledge of the mechanisms of salt tolerance will be useful to plant breeders. At least we can agree that a knowledge of the structural features that are correlated with salt tolerance is essential for an understanding of the mechanisms of salt tolerance. An attempt is therefore made in this chapter to summarize cytological observations on salt-tolerant plants and to discuss how far they contribute to the adaptation to salinity.

1. ABSORPTION OF NUTRIENTS AND LATERAL TRANSPORT ACROSS THE ROOT

Along the longitudinal axis of a root, different developmental stages can be distinguished. For the uptake of ions from the surrounding soil solution most probably the so-called *absorption zone* has an important role. This segment of the root is the first functional stage directly above the differentiation zone. It has characteristic anatomical features: the radial walls of the innermost cortical cells, the endodermis, are suberized (the casparian strips) and thus impermeable to water and ions; at least the protoxylem vessels have lost their living content, and half-bordered pits have developed between them and the surrounding xylem parenchyma cells; and sieve tubes and companion cells are differentiated. Epidermal cells are often developed as root hairs in this zone.

The zone above this primary stage of development is characterized by secondary thickening in dicotyledons and lignification of the endodermis and sclerification of the xylem parenchyma in monocotyledons. The length of the absorption zone is variable: it may be only a few millimeters in slowly growing halophytic plants or up to several centimeters in rapidly growing roots. Although this zone has often been termed the absorption zone because of its structural pecularities, many experimental data suggest that ion uptake may also occur in more proximal parts of the root (14). Certainly further experimental work needs to be undertaken to clarify this issue.

Four transport steps occurring during lateral transport to the vessels are of importance in regard to salt tolerance:

1. Absorption by the symplast through the plasmalemma of epidermal and cortical cells: this step at least is highly selective and may be actively driven by a proton pump (15). Under conditions of salt stress active excretion of Na^+ out of the symplast into the rhizosphere may occur at the same site (15).

2. Centripetal transport through the symplast: once the ions have entered the symplast they are transported laterally within the symplast via plasmodes-mata. Since the concentration of K^+ within the symplast may be about 50- to 200-fold higher than in the apoplast and the soil solution, leakage must be prevented under conditions of low transpiration (e.g., during the night). This is assured by the low permeability of the plasmalemma of the root cortex cells which also assure selectivity.

3. Sequestration of ions in the vacuole: Na^+ and Cl^- may be absorbed by the tonoplast of root cells and thus be sequestered in the vacuole (Chapter 3).

4. Entry into the xylem: there is still discrepancy in the literature as to whether ions enter the vessels purely passively (16) or whether they are actively transported (17–19). Some authors even believe that ions may reach the vessels entirely symplastically (i.e., before maturation of the vessels) and that they are released during degradation of the cytoplasm (20,21).

For the salt-excluder type of salt tolerance, absorption and lateral transport of nutrients are the steps that limit the degree of adaptation. But even for salt includers these steps are of great importance because these plants also accumulate large amounts of K^+ against the electrochemical gradient (6). The few cytological investigations of this problem indicate that adaptations on the cellular level may in fact be found which can be correlated with the different transport steps as described above. X-ray microanalysis of roots of *Atriplex hastata* (= *A. triangularis*?) (22) and of *Avicennia marina* (23) has revealed clearly that the first step (the entry into the symplast at the plasmalemma of an epidermal or cortical cell) accounts for a great deal of the observed exclusion of Na^+ and accumulation of K^+. However, so far no significant ultrastructural alterations have been reported that could account for this high selectivity. Kramer et al. found that under saline conditions the roots of *Atriplex hastata* develop epidermal transfer cells (Fig. 1.1) (22). However, this has turned out to be not a specific response to salinity but rather an indirect effect that can be overcome by increasing the iron concentration in the nutrient solution relative to that of the salt (24). In other words, salt added to the nutrient solution seems to cause iron deficiency. The formation of transfer cells in the epidermis of roots in response of iron deficiency has been demonstrated in *Helianthus annuus* (25) and many other species (26). Their physiological significance is discussed elsewhere (25, 27). It is only necessary to point out here that salinity apparently interferes

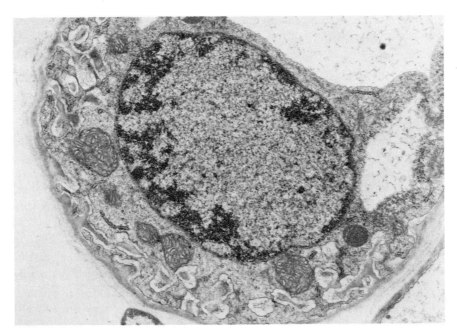

Figure 1.1. Transfer cell in the epidermis of an *Atriplex hastata* root, grown in a nutrient solution with 300 mM NaCl.

with the mechanism of iron uptake. The cause of this interference is still unknown.

Lawton et al. (28) have made comparative investigations on two mangrove species, differing in the degree of salt exclusion, namely, *Avicennia marina*, a salt includer with glands, and *Bruguiera gymnorhiza*, a salt exluder. By using lanthanum as a tracer of the apoplast, they observed that the higher influx of Na^+ into the xylem by the salt includer is apparently due to a leak in the apoplastic pathway. They argue that lanthanum (or Na^+) diffuses into the stele in a segment directly behind the tip, where the casparian strip is not yet formed and thus the apoplast open. The better capability of *Bruguiera* to exclude La^{3+}/Na^+ is, according to their interpretation, due to the root cap that, in this species, protects a much longer part of the root so that there is no gap between root cap and developed casparian strips as it is in *Avicennia*. However, since in neither of the species are xylem vessels differentiated at this stage, it is not clear how La^{3+} enters the premature vessels since La^{3+} cannot cross the plasmalemma, assuming there is no longitudinal diffusion via the cell walls up to the mature part of the stele. Another uncertain point in this interpretation is the protective role of the mucilage of the cap. It may be questioned whether these pectic substances really offer as much resistance to the monovalent ions as they may do to the trivalent La^{3+}. In any case, this work highlights the issue of how the apoplastic pathway in the root tip is blocked in the longitudinal direction. If it were really open, the suberin lamellae of the endodermis would be unnecessary.

It is possible that even in the cortex there is some differentiation of cell types. Stelzer and Läuchli (29) have made the interesting observation that, in the very efficient Na^+ exluding grass *Puccinellia peisonis*, the cortical cells that face the passage cells differ significantly from other cells in the tissue. These specialized inner cortical cells are smaller in volume but contain large proportions of cytoplasm with numerous mitochondria. The authors stress that, in contrast to the outer cortex, no intracellular spaces are formed between inner cortical and endodermal cells. Hence, the contact between both cell types appears to be close, and there are many plasmodesmata perforating the common cell walls. It is thought that these inner cortical cells may add to the passage cells and to the xylem parenchyma cells as important elements in the control of the lateral symplastic transport. This, however, would imply that the symplastic transport itself can be "controlled," an assumption for which experimental evidence is lacking. There are, on the other hand, data from X-ray microanalyses which strongly suggest that cells may differ significantly in their ion ratios even though they are symplastically connected (23, 30).

Casparian strips may also vary according to the ecotype. Poljakoff-Mayber (31) has calculated from data of Ginsburg that the ratio of the width of the casparian strip to the width of the whole tangential wall is highest in halophytes and smallest in mesophytes.

As discussed earlier, an important part in the selectivity of lateral transport to the stele may be attributed to the xylem parenchyma cells. Under conditions of

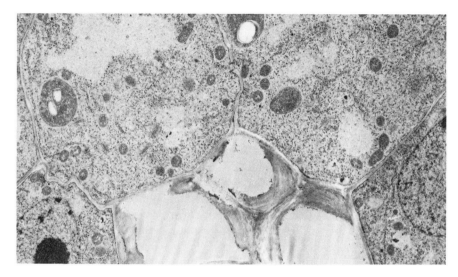

Figure 1.2. Vessels and surrounding xylem parenchyma cells in the absorbing part of a root of *Atriplex hastata*, treated with 400 m*M* NaCl. Xylem parenchyma cells are very rich in cytoplasm under these conditions.

salt stress these cells always have a very active appearance, with a large amount of cytoplasm and only small vacuoles. Mitochondria and cisternae of rough endoplasmic reticulum are the predominent structures (Fig. 1.2). This is an indication of energy-consuming transport steps at the symplast/xylem apoplast border. It has in fact been demonstrated by X-ray microanalysis that in *Atriplex hastata* the K^+/Na^+ ratio in the vessel is much higher than in the adjacent xylem parenchyma cells (22). The possible role of the endoplasmic reticulum in this context will be discussed later.

The observations that have been reviewed in this chapter are far from providing a complete and detailed survey of the structural basis of selective uptake and lateral transport of ions across the root. It is rather a summary of individual observations that points out the problems that remain to be resolved. It is necessary to put these observations into a broader context by incorporating more species in order to avoid generalizations of individual results. It will be necessary to intensify localization techniques, particularly X-ray microanalysis and laser microprobe techniques, to obtain better correlations of structure and function.

2. SECONDARY CONTROL OF THE XYLEM SAP

Transport of ions to the shoot can also be regulated during the upward transport in the vessel. Jacoby (8), Jacoby and Ratner (7), and Rains (9) found that in proximal parts of the root and the lower parts of the stem of *Phaseolus* Na^+ is removed from the xylem sap and replaced by K^+. The same phenomenon was observed by Yeo et al. (32) in proximal parts of the maize root. By means of X-ray

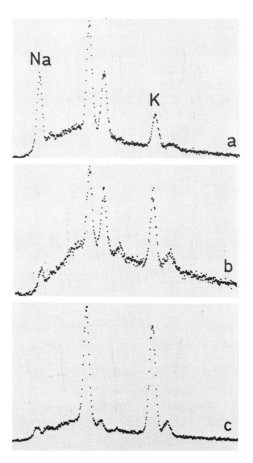

Figure 1.3. X-ray spectra from unfixed, frozen root specimens of *Phaseolus coccineus*. Na peaks have to be corrected by multiplication with 2.5 to give accurate values. Plants were grown in the presence of 50 mN Na$_2$SO$_4$ · (*a*) Xylem parenchyma cell from proximal root. (*b*) Adjacent vessel. (*c*) Xylem parenchyma cell, 10 cm behind the tip.

microanalysis on freeze-fractured tissue this mechanism has shown in both cases to be located at the border between vessel and xylem parenchyma cells (Fig. 1.3). In *Phaseolus coccineus*, xylem parenchyma cells are significantly different from those of other parts of the root in that they are differentiated as transfer cells (33). The characteristic wall labyrinth always occurs on the plasma side of the half-bordered pits. The cells are rich in cytoplasm, as are all transfer cells, but a notable feature is the high proportion of rough endoplasmic reticulum (RER) (Fig. 1.4). Cisternae of RER often appear to be connected with the plasmalemma by small fibrillar bridges; sometimes they surround the protuberances of the wall labyrinth inside the cell. In *Zea mays* roots xylem parenchyma cells are not differentiated as transfer cells but show a peculiar development of the cell wall, which in the half-bordered pits is swollen and composed of loosely packed

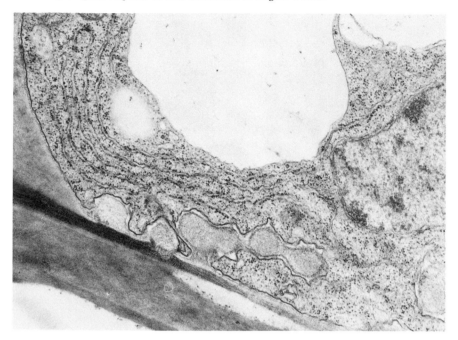

Figure 1.4. Xylem parenchyma cell from the proximal part of a root of *Phaseolus coccineus* grown in the presence of 100 m*M* NaCl. Note the increased number of RER cisternae.

fibrillar material (32). An explanation for this feature has not yet been found. However, it is important to note that high proportions of RER cisternae also occupy the cytoplasm of xylem parenchyma cells.

It is not known yet whether Na$^+$ is stored in the xylem parenchyma and cortical cells or whether it is excreted into the soil. Marschner and Ossenberg-Neuhaus (10) have shown that in *Phaseolus* Na$^+$ is retranslocated from the shoot to the proximal parts of the root and is here excreted into the external medium. Winter has performed careful reinvestigations of this mechanism by combining tracer experiments, X-ray microanalysis, and conventional electron microscopy (11, 12). She used the moderately salt-tolerant *Trifolium alexandrinum*. Her experiments show convincingly that upon application of NaCl to the root, Na$^+$ accumulates in the companion cells of petioles and leaf laminae and is then retranslocated downward to the root (12). At the sites of the transfer of Na$^+$ from the xylem to the phloem in the leaves, companion cells and (sometimes) xylem parenchyma cells were differentiated as transfer cells (11). Thus, there is clear evidence of a correlation between intraveinal recycling and the occurrence of transfer cells. It is, however, not clear whether the active uptake of Na$^+$ is really an important function of these companion cells, because companion cells are differentiated as transfer cells in leaves of many plants. At least in the leaf lamina companion cells are believed to accumulate assimilates actively from the apoplast by a proton/sugar symport (34). Further, K$^+$ is also accumulated during this step, and it could well be that Na^+ competes with sugar or potassium (or both). Moreover, according to Winter's observations phloem parenchyma cells are damaged after several days of salt treatment, with the result that sugars cannot be exported from the leaves. It is thus not clear whether this mechanism has any adaptive significance for the plant. Only short-time salt treatments can be tolerated as long as the recycling capacity of stem and petioles is sufficient to prevent Na^+ from reaching the leaf lamina. The important question of how and through which tissues the Na^+ is exported to the soil remains open.

Many salt includers excrete excess salt through salt glands at the leaf surface or sequester it in hairlike structures as does *Atriplex spongiosa* (35, 36). Of all ultrastructural adaptations to salinity this is by far the most elaborate one. There are several excellent reviews on this subject available (13, 36) so that a further description in this chapter is unnecessary. Only some aspects will be touched on here that might be important for the present discussion. The salt glands of many species are in fact transfer cells, though other species have salt glands without any wall labyrinth. How far this is correlated with different types of function is unknown (13). At least in some species there is clear evidence of an increased quantity of internal membrances, such as small vesicles and ER cisternae. A common feature of all salt glands is the high number of mitochondria, suggesting high rates of energy turnover.

3. CONCLUSIONS

Having described those structures that might be correlated with the adaptation to salinity and to the transport processes that are involved, two aspects should be focused on: the importance of transfer cells and the possible role of endoplasmic reticulum.

Transfer cells may occur at many sites in the plant, and it is widely held that their occcurrence is correlated with enhanced transport activities between apoplast and symplast. Their most characteristic feature is a wall labyrinth and, although direct experimental evidence is still lacking, the most favored hypothesis holds that the protuberances lead to a large increase of the plasmalemma surface (37). Other hypotheses maintain that in the vicinity of the protuberances a standing osmotic gradient is built up that could support transport processes against a concentration gradient (38). In this chapter, several types of transfer cells have been described that might have a function in the adaptation of plants to salinity. Epidermal transfer cells in the root, which can be induced by salt stress in *Atriplex*, are probably not involved directly in the Na^+/K^+ transport or the Cl^- exclusion as has been suggested previously (22) but appear to be a response to salinity-induced iron deficiency (24).

Xylem parenchyma transfer cells in *Phaseolus* roots and stems are certainly responsible for the observed K^+/Na^+ exchange. However, it is not clear whether these transfer cells have really evolved as an adaptation to temporary salt stress; very little is known about the ecology of *Phaseolus* species. Intraveinal recycling is described by Winter (11) occurs via transfer cells in xylem and phloem, but again it must be doubted whether this is the main function of these cells. Although salt glands are adaptations to overcome salinity stress in the leaf tissue, they are not necessarily differentiated as transfer cells. The conclusion can only be drawn that the occurrence of transfer cells might be an important adaptive feature. Further research is required on their possible significance in the regulation of ion relations.

Another general aspect of the adaptation of higher plants to salinity is the compartmentation of the cytoplasm, because the cytosol itself is apparently not salt resistant at all. As has been pointed out several times, in cells that are involved in salt transport the RER occupies an unusually high proportion of the cytoplasm. It is possible that the cisternae of RER provide the compartment within the cytoplasm in which salt may be sequestered. The ER constitutes a compartment through which substances might be transported symplastically via the desmotubule, the narrow central tubule of plasmodesmata, without passing the cytosol (40), and the RER cisternae may then fuse with the tonoplast, releasing their contents into the vacuole. This can be illustrated by observing the

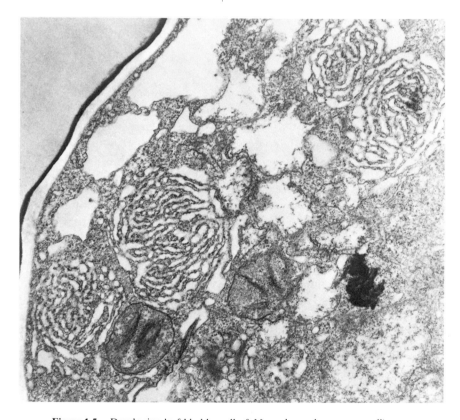

Figure 1.5. Developing leaf bladder cell of *Mesembryanthemum crystallinum.*

development of the huge vacuole of epidermal bladder cells of the halophyte *Mesembryanthemum crystallinum* (Fig. 1.5) (39). In the cytoplasm of the very young bladder cells many concentric stacks of ER are formed, which seem then to give rise to the vacuole.

Plant cytology is certainly not able to offer an explanation of how a salt-tolerant plant functions, and it is not possible to identify one particular structure that would explain the phenomenon of salt tolerance. It can only, in combination with other disciplines of plant research, contribute to our understanding of how plants adapt to saline environments.

REFERENCES

1. A. D. Brown. Halophilic prokaryotes. In O. L. Lange, P. S. Nobel, C. B. Osmond and H. Ziegler, eds. *Encyclopedia of Plant Physiol.*, New Series Vol. 12C, Physiol. Plant Ecol. III, Springer, Berlin, Heidelberg, New York, 1983.

2. H. Greenway and C. B. Osmond. Salt response of enzymes from species differing in salt tolerance. *Plant Physiol.* **49**, 256–259 (1972).

3. T. J. Flowers. Salt tolerance in *Suaeda maritima* L. (Dum). The effect of sodium chloride on growth, respiration, and soluble enzymes in a comparative study with Pisum. *J. Exp. Bot.* **23**, 310–321 (1972).

4. T. J. Flowers. The effect of chloride on enzyme activities from four species of Chenopodiaceae. *Phytochemistry* **11**, 1881–1886 (1972).

5. R. Collander. Selective absorption of cations by higher plants. *Plant Physiol.* **16**, 691–721 (1941).

6. R. Albert and M. Popp. Chemical composition of halophytes from the Neusiedler Lake region in Austria, *Oecologia (Berlin)* **27**, 157–170 (1977).

7. B. Jacoby and A. Ratner. Mechanism of sodium exclusion in bean and corn plants—A reevaluation. In J. Wehrmann, ed., *Plant Analysis and Fertilizer Problems*. German Society of Plant Nutrition, Hannover, 1973, pp. 175–184.

8. B. Jacoby. Sodium retention in excised bean stems. *Physiol. Plant.* **18**, 730–739 (1965).

9. D. W. Rains. Sodium and potassium absorption by stem tissue of bean and cotton. *Plant Physiol.* **44**, 547–554 (1969).

10. H. Marschner and H. Ossenberg-Neuhaus. Langstreckentransport von Natrium in Bohnenpflanzen. *Z. Pflanzenernähr. Düng. Bodenkd.* **193**, 129–142 (1976).

11. E. Winter. Salt tolerance of *Trifolium alexandrinum* L. III: Effects of salt on untrastructure of phloem and xylem transfer cells in petioles and leaves. *Aust. J. Plant Physiol.* **9**, 239–250 (1982).

12. E. Winter and J. Preston. Salt tolerance of *Trifolium alexandrinum* L. IV. Ion measurements by X-ray microanalysis in unfixed, frozen hydrated leaf cells at various stages of salt treatment. *Aust. J. Plant Physiol.* **9**, 251–259 (1982).

13. W. W. Thomson. Salt glands. In A. Poljakoff-Mayber and J. Gale, eds., *Plants in Saline Environments*. Springer-Verlag, Berlin, 1975.

14. D. T. Clarkson and J. B. Hanson. The mineral nutrition of higher plants. *Annu. Rev. Plant Physiol.* **31**, 239–298 (1980).

15. W. D. Jeschke, Roots: Cation selectivity and compartmentation, involvement of protons and regulation. In R. M. Spanswick, W. J. Lucas, and J. Dainty, eds., *Plant Membrane Transport: Current Conceptual Issues,* Elsevier/North Holland, Amsterdam, 1980.

16. J. Dunlop and D. J. F. Bowling. The movement of ions into the xylem exudate of maize roots. I. Profiles of membrane potential and vacuolar potassium activity across the root. *J. Exp. Bot.* **22,** 434–444 (1971).

17. A. Läuchli, A. R. Spurr, and E. Epstein. Lateral transport of ions into the xylem of corn roots. II. Evaluation of a stelar pump. *Plant Physiol.* **48,** 118–124 (1971).

18. M. G. Pitman. Uptake and transport of ions in barley seedlings. II. Evidence for two active stages in transport to the shoot. *Aust. J. Biol. Sci.* **25,** 243–257 (1972).

19. M. G. Pitman. Ion transport into the xylem. *Annu. Rev. Plant Physiol.* **28,** 71–88 (1977).

20. R. F. Davis and N. Higinbotham. Electrochemical gradients and K^+ and Cl^- fluxes in excised corn roots. *Plant Physiol.* **57,** 129–136 (1976).

21. M. F. Danilova and E. Y. Stamboltsyan. The ultrastructure of differentiating primary xylem cells of the root in relation to hypothesis of transfer of solutes into xylem elements. *Bot. Zh. (Leningrad)* **60,** 913–926 (1975).

22. D. Kramer, W. P. Anderson and J. Preston. Transfer cells in the root epidermis of *Atriplex hastata* L. as response to salinity: A comparative cytological and X-ray microprobe investigation. *Aust. J. Plant Physiol.* **5,** 739–747 (1978).

23. D. Kramer and J. Preston. A modified method for X-ray microanalysis of bulk-frozen plant tissue and its application to the problem of salt exclusion in mangrove roots. In P. Echlin and R. Kaufmann, eds., *Microsc. Acta Suppl.* 2. Microprobe analysis in biology and medicine. Hirzel, Stuttgart, 1978.

24. D. Kramer. Structure and function in absorption and transport of nutrients. In R. Brouwer, O. Gasparikova, J. Kolek, and B. C. Loughman, eds., *Structure and Function of Roots.* Martinus Nijhoff/Junk, The Hague, 1981.

25. D. Kramer, V. Römheld, E. Landsberg, and H. Marschner. Induction of transfer cell formation by iron deficiency in the root epidermis of *Helianthus annuus* L. *Planta* **147,** 325–339 (1980).

26. D. Kramer. Genetically determined adaptations in roots to nutritional stress: Correlation of structure and function. *Plant and Soil* **72,** 167–173 (1983).

27. E. Ch. Landsberg. Transfer cell formation in the root epidermis: A prerequisite for Fe-efficiency? *J. Plant Nutrition* **5,** 415–432 (1982).

28. J. R. Lawton, A. Todd, and D. K. Naidoo. Preliminary investigations into the structure of the roots of the mangroves *Avicennia marina and Bruguiera gymnorrhiza,* in relation to ion uptake. *New Phytol.* **88,** 713–722 (1981).

29. R. Stelzer and A. Läuchli. Salz- und Überflutungstoleranz von *Puccinellia peisonis.* II. Strukturelle Differenzierung der Wurzel in Beziehung zur Funktion. *Z Pflanzenphysiol.* **84,** 95–108 (1977).

30. R. F. M. Van Steveninck, M. E. Van Steveninck, R. Stelzer, and A. Läuchli. Electron probe X-ray microanalysis of ion distribution in *Lupinus luteus* L. seedlings exposed to salinity stress. In. R. M. Spanswick, W. J. Lucas, and J. Dainty, eds., *Plant Membrane Transport: Current Conceptional Issues.* Elsevier/North-Holland, Amsterdam, 1980.

31. A. Poljakoff-Mayber. Morphological and anatomical changes in plants as a response to salinity stress. In A. Poljakoff-Mayber and J. Gale, eds., *Plants in Saline Environments.* Springer-Verlag, Berlin, 1975.

32. A. R. Yeo, D. Kramer, A. Läuchli, and J. Gullasch. Ion distribution in salt stressed mature *Zea mays* roots in relation to ultrastructure and retention of sodium. *J. Exp. Bot.* **28,** 17–29 (1977).

33. D. Kramer, A. Läuchli, A. R. Yeo, and J. Gullasch. Transfer cells in roots of *Phaseolus coccineus:* Ultrastructure and possible function in exclusion of sodium from the shoot. *Ann. Bot.* **41,** 1031–1040 (1977).

34. R. Giaquinta. Sucrose/proton cotransport during phloem loading and its possible control by internal sucrose concentration. In R. M. Spanswick, W. J. Lucas, and J. Dainty, eds., *Plant Membrance Transport: Current Conceptual Issues*, Elsevier/ North-Holland, Amsterdam, 1980.

35. C. B. Osmond, U. Lüttge, K. R. West, C. K. Pallaghy, and B. Schachar-Hill. Ion absorption in *Atriplex* leaf tissue. II. Secretion of ions to epidermal bladders. *Aust. J. Biol. Sci.* **22,** 797–814 (1969).

36. U. Lüttge. Salt glands. In D. A. Baker and J. L. Hall, eds., Ion transport in plant cells and tissues. Elsevier/ North-Holland, Amsterdam, 1975.

37. J. S. Pate and B. E. S. Gunning. Transfer cells. *Annu. Rev. Plant Physiol.* **23,** 173–196 (1972).

38. A. Läuchli. Apoplasmic transport in tissues. In U. Lüttge and M. G. Pitman, eds., *Encyclopedia of Plant Physiol.* New Series, Vol. 2, Transport in Plants II, Part B, Springer-Verlag, Berlin, 1976.

39. D. Kramer. Ultrastructural observations on developing leaf bladder cells of *Mesembryanthemum crystallinum* L. *Flora* **168,** 193–204 (1979).

40. J. Burgess. Observations on the structure and differentiation of plasmodesmata. *Protoplasma* **73,** 83–95 (1971).

2

CHEMICAL MICROSCOPY FOR STUDY OF PLANTS IN SALINE ENVIRONMENTS

Robert F. Sacher and Richard C. Staples

Boyce Thompson Institute for Plant Research
Cornell University
Ithaca, New York

The impact of salinity on agriculture is now being felt in irrigated areas in which soil- and water-borne salts are becoming concentrated during repeated cycles of water reuse. Summer fallowed areas in which salts seep upward with rising groundwater are also impacted, as are marginal lands of high natural salinity. These problems will grow as increasing populations require more intensive use of land and water and as presently unused or marginally used resources are pressed into service. The entire physical and biological system involved in saline agriculture must be understood and carefully managed if increased production is to be achieved without amplifying the existing problems.

Future directions for use of saline resources include conventional agriculture as well as new systems specifically engineered for plant, animal, and microorganism production (1, 2). One aspect under the immediate control of plant scientists is the development of new cultivars or species capable of productive growth in saline situations. Candidate crops for such improvement are those that presently exhibit some degree of salt tolerance and those with natural sources of salt tolerance germplasm. Tomato (*Lycopersicon esculentum* Mill.) is one such crop. It has moderate salt tolerance relative to other crops and its abundant wild relatives possess many stress-tolerance traits (3). The collection, systematization, and genetic study of wild tomato relatives have been done by Rick (4) and others. Breeding programs have made progress using conventional methods and cultivars (5), and the introgression of wild genes (6–9). Biotechnology may eventually be of use as well (10, 11).

Improvement of crop performance often is more efficiently carried out if specific selection goals exist. Understanding mechanisms underlying valuable traits allows the establishment of specific selection parameters. For example, we have previously determined that regulation of sodium partitioning within the whole tomato plant is, statistically, the most important factor in salt tolerance. This information is valuable as it stands but could be used much more effectively if enough more were known to make possible direct, mechanistic manipulation of the relevant systems. Chemical microscopy of freeze-substituted tissue offers the possibility of examining intracellular partitioning of sodium and other mineral ions in images of entire tissue sections. Chemical microscopical techniques have been applied to the examination of intracellular sodium partitioning with varying degrees of success and are now continuing. We will discuss several aspects of microscopical techniques for chemical measurement and will particularly discuss our recent work with the Cameca ion microscope and the effects of salt on tomato.

1. SAMPLE PREPARATION TECHNIQUES

1.1. Unfixed Tissue

Biological chemical microscopy requires the preservation of both the structural and chemical distributions in a state as close as possible to the living condition. In some cases, it is possible to detect chemical distributions in living tissues without incurring the trauma of excision. Translocation of dyes and fluors or the natural fluorescence of certain cell components can be monitored. Such studies may be of interest as the tissue is affected by salt, but they will not be considered further in this discussion. Detection of mineral elements, which are of primary interest in saline research, requires either chemical derivatization or other rather elaborate preparation. For this discussion, we will classify sample preparation methods based on the characteristics of the initial fixation step. These are not exclusive categories.

1.2. Excised Living Tissue

Ideally, microscopists would directly observe living cells but this is not possible in most cases. Small tissue pieces can often be excised and observed microscopically. Correlated functions of the tissue with the whole plant are of course lost but individual cells remain viable. Artifacts may arise from mounting living tissue for microscopy because of the processes of excision and handling. In certain cases, a small contact stimulus can result in large ion fluxes between or within cells (12). If live mounts are used, it would seem worth a preliminary investigation of the effects of physical stimuli such as excision on cation status. A limitation to the use of live tissues which is of more consequence is the need to use sections at least as thick as the average cell diameter which prohibits high resolution microscopy. Visualization of the elements in a cell without upsetting its metabolism is virtually impossible except in certain situations. A notable exception is the use of fluorescence probes for membrane-bound calcium and magnesium (13, 14). This will be discussed in Section 2.1.

1.3. Aqueous Fixatives

Aqueous fixation media rely on precipitation of cellular contents into stable, insoluble forms (15–17). Acids, alcohols, and aldehydes and other chemicals are used for these fixation media. Osmium tetroxide is often included both for fixation and for staining of electron microscopical specimens. One function of these fixatives is to rapidly penetrate cells and stabilize the structure. Membranes rupture in the process and allow copious leaching of smaller, soluble molecules. To halt leaching of salt ions, precipitating reagents are included with the fixatives. Silver nitrate will form insoluble silver chloride. Anions, including citrate, tartarate, and antimonate, will precipitate the metal cations. If their chemical affinities are understood and accounted for they can contribute to useful quantitative procedures (18).

1.4. Rapid Freezing

Fixation may also be achieved by rapidly freezing the tissue. The success of freeze fixation methods depends heavily on achieving a fast heat transfer out of the specimen (quench) in order to minimize ice crystal size. Several quenching media are now in use. CO_2 jets provide temperatures low enough for routine histology but not for higher resolution microscopy. The heat transfer properties of several liquified gases used as quenching media have recently been evaluated (19, 20). Liquid N_2 is widely available and easily handled. However, liquid N_2 is normally at its boiling point in equilibrium with room temperature air, and an insulating gas envelope forms around a specimen almost immediately upon plunging into liquid N_2. To reduce the incidence of such an envelope, the temperature of liquid N_2 can be lowered by repeatedly imposing a high vacuum. Rapid evaporation lowers the temperature to the freezing point with formation of a slush. After release of the vacuum, the temperature remains below the boiling point long enough for quenching. Liquid hydrocarbons and fluorocarbons are prepared in liquid N_2 cooled vessels. They have higher heat transfer rates than liquid N_2 slush and can be supercooled for exceptional resistence to gassing. Mechanical considerations in the creation and handling of these quench media also need to be considered. Specimen characteristics, especially size, can be more limiting to heat transfer than are quench medium characteristics.

Once frozen, the specimen can be sectioned and imaged directly, it can be freeze-dried, or it can be infiltrated with an organic solvent followed by epoxy

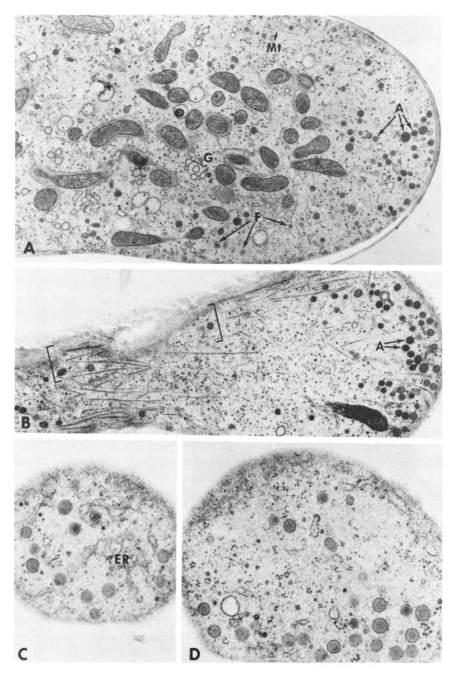

Figure 2.1. Bean rust spore germ tube prepared by freeze substitution for electron microscopy. Adapted from ref. 23. Used by permission.

resin for freeze substitution. Frozen hydrated specimens require imaging by an instrument equipped with a cold stage (21). Freeze-dried specimens are often physically distorted, though chemical retention is good (22). Freeze-drying is sometimes used for woody specimens. Freeze-substituted specimens are usable in any method that allows epoxy embedment. Chemical retention by freeze substitution can be virtually complete and, if proper specimens are used, structural preservation can be excellent. Very small specimens quench quickly for ideal preservation (23, 24) (Fig. 2.1). With thicker, porous specimens such as leaves, the preservation, while quite acceptable, is usually less than perfect (Fig. 2.2).

In recent studies we have developed relatively rapid procedures for freeze-substitution processing. We have used these methods as preparation for ion microscopy, which allows the imaging of monovalent and divalent ions with greater than 90% cation retention and maintenance of cell structure and organization (26).

The freeze-substitution procedure recently employed in our laboratory is as

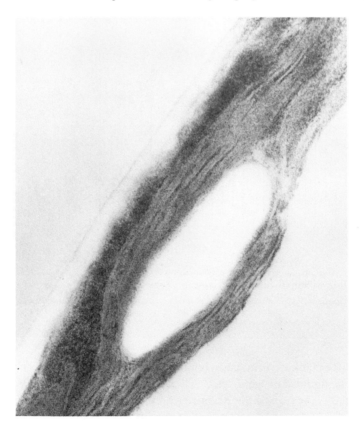

Figure 2.2.   *Suaeda* leaf tissue prepared by freeze substitution for electron microscopy. From ref. 25. Used by permission.

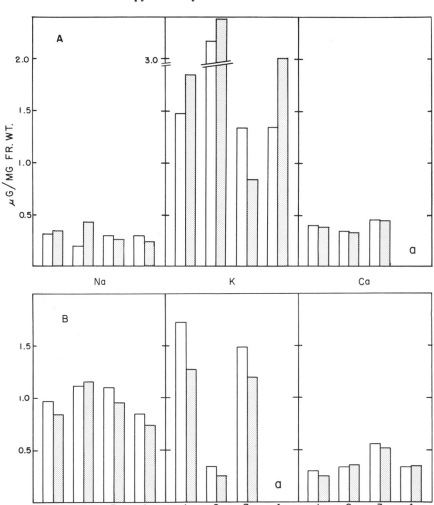

Figure 2.3. Sodium, potassium, and calcium contents of tomato leaf tissue unprocessed or processed by freeze substitution for ion microscopy. Sampling points: 1, 20% resin at 85°C; 2, 20% resin at 22°C; 3, 80% resin at 22°C; 4, 100% resin at 22°C. Analysis was by neutron activation of two samples of five specimens (ca. 1 mg fresh weight each) for each bar. From ref. 27. Used by permission.

follows. Five ml of acrolein:diethyl ether (1:4v/v) solution was frozen in a 20-ml vial under liquid nitrogen. The vial was filled with liquid nitrogen and submerged to the neck in liquid nitrogen contained in an ice bucket within a vacuum desiccator. The desiccator was closed and evacuated until rapid evaporation lowered the temperature sufficiently to create eutectic slush in the ice bucket and the vial. The desiccator was then opened and the specimen quenched. The specimen, usually 1 × 7 mm² strips cut from leaves of 8-week-old tomato plants, was quenched immediately by plunging rapidly into the eutectic slush.

The lowered temperature contributed to rapid freezing; more importantly, however, lowering the temperature below the boiling point of N_2 delays formation of a gas envelope around the specimen during quenching. Monitoring with a thermocouple showed approximately a 5-min period of acceptably low temperature. The loosely capped vials were transferred to a freezer at $-85°C$. As the vials warmed the specimens sank into the substitution fluid without disturbance. Infiltration of Spurr's low viscosity epoxy resin (Polysciences, Inc.) was begun by addition of cold 50% resin in diethyl ether to the cold vials to make 5% solution. At 24-hr intervals more cold resin solution was added to make 10 and 20% concentrations. After 24 hr in 20% resin at $-85°C$ the vials were allowed to warm to room temperature and the infiltration series continued conventionally in increments to 40, 80, and 100% with two changes of 100% before embedding. Specimens were sectioned dry with a diamond knife to 250-nm thickness for ion microscopy. Neutron activation analysis was used to monitor chemical retention because of its sensitivity and the capacity to analyze many samples simultaneously. Samples were activated at the Cornell University TRIGA reactor. Element retention averaged 90% or more (Fig. 2.3).

Anatomical and cytological preservation by freeze substitution in liquid N_2 slush was judged to be more than adequate for the 1 μm resolution of the ion microscope (27). Phase contrast light microscopy at 400× magnification (Fig. 2.4) was used to select sections for further study and to evaluate preservation through the thickness of the leaves. The upper and lower epidermis with guard cells, palisade, spongy mesophyll, and minor veins were present in good order. Transmission EM of post-stained sections showed good retention of cellular

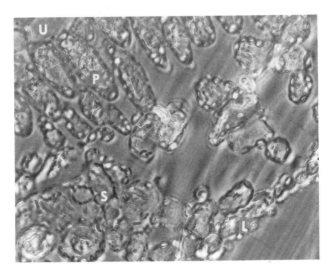

Figure 2.4.   Phase contrast micrograph of freeze-substituted tomato leaf cross sections illustrating overall anatomy of material used for ion microscopy.

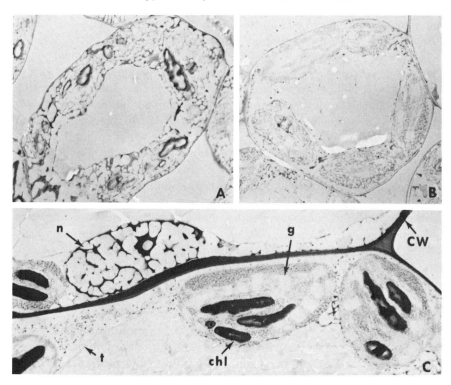

Figure 2.5. Transmission electron micrograph of specimen prepared by freeze substitution for ion microscopy. Typical spongy mesophyll cells (A,B), and intracellular detail (C) showing nucleus (n), chloroplast (chl) with intact grana (g), tonoplast (t), and cell wall (cw). Bar = 10 μm.

structure (Fig. 2.5), although ice damage was visible. Organelle boundaries were intact but structures smaller than the chloroplasts were distorted by ice crystals. The tonoplast, chloroplast envelope, and nuclear envelope were intact. The plasmalemma was closely appressed to the cell wall. Cell walls were differentiated such that middle lamellae were apparent. Within the chloroplasts the stroma, grana, and starch grains were well preserved. The reticulated appearance of much of the cell was due to retention of ice crystal boundaries through processing. The presence of the ice crystal boundaries is evidence that resin infiltration and embedding were completed without melting and relocation of free water.

2. MICROSCOPIC TECHNIQUES

2.1. Light

Light microscopy cannot directly detect soluble salt ions in cells. Thus, some form of chemical indicator must be used. There is a large body of elegant, rigorous work, which has been done by classical microscopists. For example, in

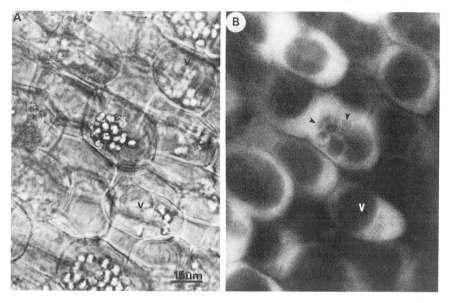

Figure 2.6. Visualization of membrane bound cytoplasmic calcium in *Zea* root cap cells by fluorescence of chlorotetracycline; accompanied by phase contrast image of same tissue. From ref. 29. Used by permission.

1904 potassium fluxes were demonstrated in active guard cells by Macallum (28) by precipitant staining. Precipitant treated material is now usually examined by electron microscopy. Fluorescence techniques also are useful. An example is the detection of membrane-bound calcium in the cytoplasm of corn root cap cells using chlorotetracycline (29) (Fig. 2.6).

2.2.   Electron Beam

Technical advances in electron microscopy (EM) have made this instrument the method of choice of chemical microscopy. So wide ranging are the developments that we can review here only a few aspects that are, or could be, important in plant salinity research.

Fixation and precipitation have been in use for years with all conceivable specimen types. Quantitation is problematic but a reasonable estimate may be made by counting precipitate grains per unit area. Sodium distributions in salt-treated halophytes have been studied with precipitation showing localization of sodium within the interior of mitochondria (30) (Fig. 2.7).

The passage of the electron microscope beam through a specimen also results in production of X-ray emissions that are unique and diagnostic of the atoms that were bombarded. Electron microscopes with X-ray analyzers function as electron microprobe units and have been the mainstay of chemical microscopy. Freeze-substitution sample preparation and electron microprobe analyses of plant tissues have formed a productive combination (31, 32). Microprobe

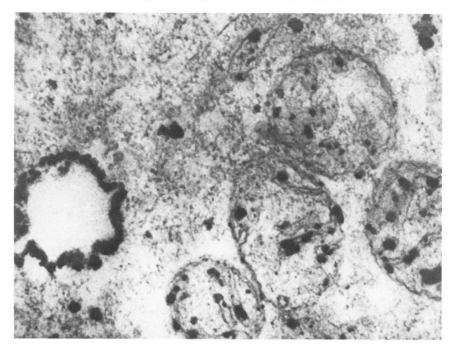

Figure 2.7.   Sodium precipitate within mitochondria of salt treated *Zea.* From ref. 30. Used by permission.

analyses can be of spots, line scans across the specimen, or surface scanning over larger areas.

Analyses of single spots within cells have been used to determine elemental contents of intracellular regions (33). Line scans can be performed on restricted regions (34) (Fig. 2.8a) to give analyses of individual cells (Fig. 2.8b) or on larger regions to investigate tissues in an organ (34) (Fig. 2.9). Scanning EM (SEM) and scanning transmission EM (STEM) can be coupled with X-ray analysis to perform chemical imaging of entire cross sections.

In electron microscopy, images are produced as a result of absorption by electron dense areas of the specimen. The specimen atoms also interact with the beam electrons in other ways. Both elastic and inelastic energy losses from the electrons occur. Inelastic interactions reduce the energy of the electrons by amounts which are specific to the atomic species. These properties form the basis for electron energy loss spectroscopy (EELS) which may become an extremely powerful tool. A mass spectrometer stage filters the image electrons from a conventional electron microscope. The filtering is by electron momentum and gives the equivalent of a monochromatic image of the locations of a specific element. Momentum filtering also reduces image aberration, thus increasing resolving power in addition to providing chemical information. The potential of EELS seems great. With computer image enhancement, it can resolve details to the level of single atoms and give direct views of molecular structure (35) (Fig.

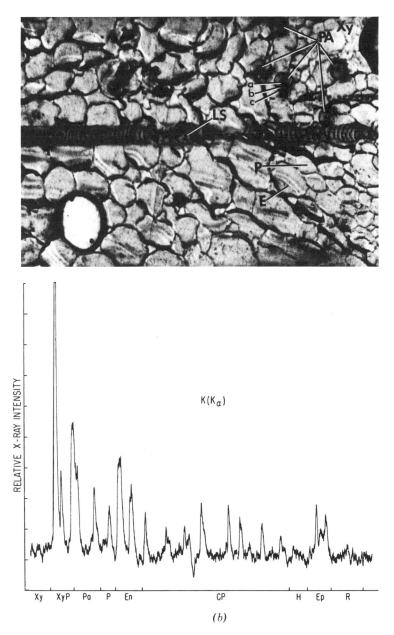

Figure 2.8. Freeze substituted *Zea* root tissue scanned by electron microprobe (*a*) for analysis of K (*b*). Adapted from ref. 34. Used by permission.

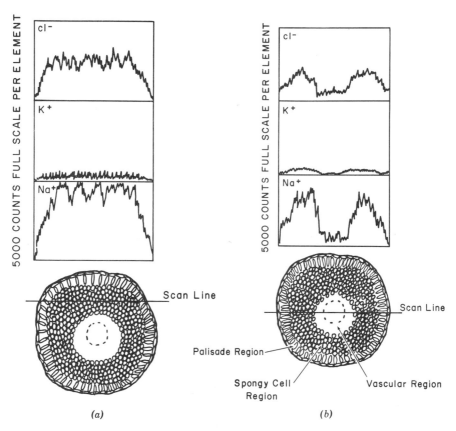

Figure 2.9.   Freeze substituted stem tissue of *Salicornia* analyzed for presence of Cl, K, and Na. From ref. 34. Used by permission.

2.10). With the exception of the EELS attachment to a dark field electron microscope, no unique techniques are required.

2.3.   Ion Beam

Secondary ion mass spectrometry (SIMS, Fig. 2.11) is used as an analytical probe and as a microscopical technique. Ion beam probes have properties similar to electron probes but with finer chemical resolution. A beam of accelerated ions is impinged on the specimen and the elements of the specimen surface are secondarily ionized and ejected. A mass spectrometer then separates and quantifies the secondary ions (36, 37). Ion microscopy uses the ion beam mass spectrometry principle with electrostatic lenses for image focusing and magnification (38). Continuous tone micrographs are produced with chemical resolution of less than one atomic mass unit. Magnification is now limited to $400\times$ magnification with 1-μm spatial resolution. Ion microscopy is presently

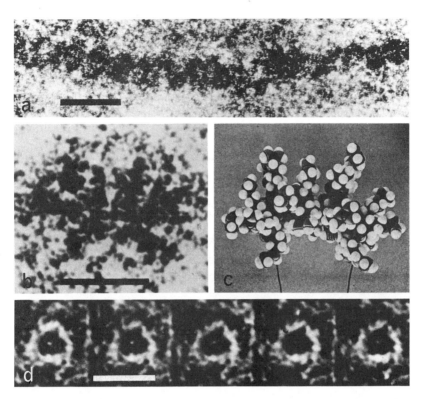

Figure 2.10. Computer enhanced images obtained by electron energy loss spectroscopy (*a*) phosphorus atoms in DNA helix, (*b*) molecular image and theoretical structural model, (*c*) valinomycin molecules containing single potassium atoms. From ref. 35. Copyright 1982 by the American Association for the Advancement of Science. Reprinted with permission.

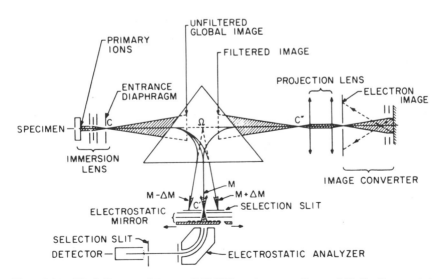

Figure 2.11. Block diagram of Cameca IMS 300 ion microscope. From ref. 22. Used by permission.

used for research and quality control analyses in many high technology materials applications but has had relatively little biological application. Problems inherent in using biological materials are chemical losses during processing and the requirement of a highly planar surface with minimal dielectric properties for imaging (22). Freeze sectioning and lyophilization have been used for sample preparation with good chemical retention but suffered severe physical distortion (22). Applications of ion microscopy in plant science have been those in which large ion losses during preparation have been recognized but tolerated because of other benefits (12), or studies of divalent ions, which are less prone to leaching (39).

Ion micrographs of normal and salt affected tomato leaves (27) (Fig. 2.12) suggest that salt treatment affects some, but not all, aspects of ion distributions

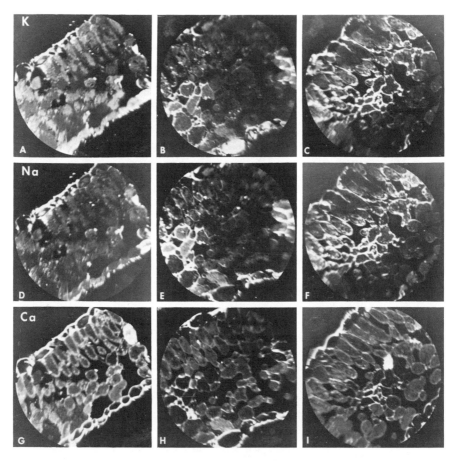

Figure 2.12. Ion micrographs of tomato leaf cross sections prepared by freeze substitution [method of Ross et al. (26)]. Treatments were control (C), half-strength Hoagland's solution and saline (S), half-strength Hoagland's solution + 0.1 M NaCl. Elements imaged were K (a, b, c), Na (d, e, f) and Ca (g, h, i). Bar = 10 μm. From ref. 27. Used by permission.

within tomato leaf cells. Potassium was located primarily in vacuoles and chloroplasts with highest concentrations in the lower epidermis (Fig. 2.12a). In salinated plants, potassium was also seen in the vacuoles and chloroplasts (Fig. 2.12b,c). The large epidermal cell in Fig. 2.12c exhibits a strong potassium accumulation. In this cell the partitioning of potassium to the vacuole is particularly apparent. Sodium was found to be primarily in vacuoles and chloroplasts in parallel with potassium location (Fig. 2.12d) but with greatly increased concentration in salinated plants (Fig. 2.12e,f). The epidermal cells (Fig. 2.12f), in contrast with the mesophyll cells, remained lower in sodium than the adjoining tissue. Calcium was contained in the cytoplasm and especially within the chloroplasts. Calcium content of the vacuoles was much less than that of the cytoplasm (Fig. 2.12g). A high concentration of calcium was seen in cutinized areas of epidermal cell walls. In cells of salinated plants the calcium distribution was very diffuse, with only vestiges of the compartmentation seen in the control plant cells (Fig. 2.12h,i). Calcium was found in the cell walls, especially in the pectate rich cell junctures, in control and salinated plants. The calcium concentration in the cutinized areas was still evident in salinated plants. Epidermal cells which were high in monovalent ions were notably low in calcium with either salt level.

3. SOME CONCLUSIONS

The greatest quantity of sodium was found in the vacuoles. This is similar to the reported distribution of sodium in leaves (34, 40, 41) and in roots (33, 42, 43). Sodium was also strongly partitioned into the chloroplasts, in agreement with other research (40, 41). The distribution of potassium was very similar to that of sodium, again consistent with other research (40, 41). Vacuoles of lower epidermal cells were especially high in potassium in both control and salinated plants and, whereas the mesophyll sodium image was substantially more intense in the salinated plants, the epidermal sodium appeared not to be increased relative to the control.

In contrast to the distributions in leaves, root cells have been found to have the majority of their potassium in the cytoplasm rather than in the vacuoles (33, 42, 43). A disparity between roots and leaves would be expected due to differences in functions and anatomical locations of the tissues. Root cells are under conditions of potassium and sodium flux rather than accumulation. Root cells of glycophytes are implicated as points of selectivity between sodium and potassium. Therefore, root cells are expected to pass potassium into the moving xylem transpiration stream and to accumulate sodium and partition it into the vacuoles, thereby retarding upward sodium transport. Leaf mesophyll cells have little or no opportunity for selective transport and thus are expected to partition high concentrations of either sodium or potassium into the vacuoles. Potassium movement within (41) and between (44) leaf cells is light dependent and can be expected to alter the distribution of potassium as it is accumulated within the leaf cells.

These studies provide direct evidence for intracellular ion interactions analogous to those proposed by Epstein (45) and Collander (46) for whole plants. Epstein concluded that sodium and calcium were competitive with regard to uptake and plant growth whereas sodium and potassium were not similarly competitive. Ion microscopical examination allowed comparisons of intracellular changes in the contents of these minerals in response to sodium chloride treatment. Calcium distributions were changed qualitatively as well as quantitatively in salinated plants. Although there were quantitative changes in potassium level, there existed the same compartmentation pattern in both salinated and control plants. Thus, calcium was displaced from its normal sites within the cells whereas potassium was not. Such results would be expected if calcium were loosely held, such as by ion exchange sites or by passive uptake into organelles (28). Potassium and sodium were distributed as if they were actively partitioned into the vacuoles and chloroplasts or were moving as counterions with mineral anions or organic acids (28, 41, 47, 48). Chloroplast H^+/cation exchange is predominantly H^+/K^+ but is relatively nonselective between monovalent cations taken in, and the envelope is relatively impermeable to them (49–51). It is to be expected then that either K^+ or Na^+ would accumulate in the chloroplasts in exchange for H^+ if either one were available. The images presented here show ion distributions consistent with the research cited above.

Mass action displacement of calcium by sodium would account for the reduced calcium compartmentation in salinated plants. Calcium dispersal could not be attributed to breakdown in compartmental integrity, as evidenced by the definite potassium and sodium gradients across the tonoplasts and chloroplast membranes within the same cells. Mass action displacement of potassium by sodium would also be expected and may be manifest in the reduced overall potassium content of the tissue. However, potassium distribution between cellular compartments was maintained in salinated and control plants. The chloroplasts and vacuoles thus appear to be selectively holding monovalent ions but allowing diffusion of calcium to equilibrate between compartments and cytosol.

The Na^+/Ca^{2+} interaction may be indicative of interferences of sodium with the metabolic actions of calcium and other ions such as magnesium. For example, magnesium, potassium, and proton fluxes that occur during photosynthesis may be vulnerable to disruption by sodium and, if altered, could be expected to have serious effects on regulation and function of the chloroplasts (49–51). Another example might be magnesium binding to thylakoids which is governed by the $(K^+ + Na^+)/Mg^{2+}$ ratio and which could be expected to change if the $(K^+ + Na^+)/Mg^{2+}$ ratio is substantially increased.

The techniques in chemical microscopy reviewed here are powerful tools for elucidating the cellular distributions of molecules of various kinds. For water soluble ions, however, all are still limited in applicability, especially where very high resolution is required. The real challenge now is to develop procedures for fixation which will hold the ions and cellular structure in position and allow examination at high magnification in high vacuum. These microscopy techniques are continually being refined. In the future, large improvements in the usefulness of chemical microscopy to salinity research can be expected.

REFERENCES

1. A. Hollaender. *The Biosaline Concept.* Plenum, New York, 1979, 400 pp.

2. A. San Pietro. *Biosaline Research.* Plenum, New York, 1982, 578 pp.

3. C. M. Rick. Tomato germplasm resources. In R. Cowell, ed., *Proceedings of the 1st International Symposium on Tropical Tomato.* Asian Veg. Res. and Dev. Center, Publication 78-59, Shanhua, Taiwan, 1979, pp. 214-224.

4. C. M. Rick. The tomato. *Sci . Am.* **239**, 76-89 (1978).

5. D. Pasternak, M. Twersky, and Y. De Malach. (1979) Salt resistance in agricultural crops. In *Stress Physiology in Crop Plants.* Wiley, New York, 1979, pp. 127-142.

6. D. W. Rush and E. Epstein. Genotypic responses to salinity: Differences between salt sensitive and salt tolerant genotypes of tomato. *Plant Physiol.* **57,** 162-166 (1976).

7. M. Tal. Salt tolerance in the wild relatives of the cultivated tomato: Responses of *Lycopersicon esculentum, L. peruvianum* and *L. esculentum minor* to sodium chloride solution. *Aust. J. Agric. Res.* **22,** 631-637 (1971).

8. R. F. Sacher, R. C. Staples, and R. W. Robinson. Saline tolerance in hybrids of *Lycopersicon esculentum* $\times$ *Solanum pennellii* and selected breeding lines. In A. San Pietro, ed., *Biosaline Research.* Plenum, New York, 1982, pp. 325-336.

9. R. F. Sacher, R. C. Staples, and R. W. Robinson. Ion regulation and response of tomato to sodium chloride: A homeostatic system. *J. Am. Soc. Hortic. Sci.* **108:**566-569 (1983).

10. J. Mielentz, K. Anderson, R. Tait, and R. Valentine. Potential for genetic engineering for salt tolerance. In A. Hollaender, ed., *The Biosaline Concept,* Plenum, New York, 1979, pp. 361-370.

11. D. W. Rains, L. Csonka, D. LeRudelier, T. P. Croughton, S. S. Yang, S. J. Stavarek, and R. C. Valentine. Osmoregulation of organisms exposed to saline stress: Physiological mechanisms and genetic manipulation. In A. San Pietro, ed., *Biosaline Research.* Plenum, New York, 1982, pp. 283-302.

12. N. A. Campbell, K. M. Stika, and G. H. Morrison. Calcium and potassium in the motor organ of the sensitive plant: Localization by ion microscopy. *Science* **204,** 185-187 (1979).

13. A. H. Caswell. Methods of measuring intracellular calcium. *Int. Rev. Cytol.* **56,** 145-181 (1979).

14. A. H. Caswell and J. D. Hutchinson. Selectivity of cation chelation to tetracyclines: Evidence for special conformation of calcium chelate. *Biochem. Biophys. Res. Commun.* **43,** 625-630 (1971).

15. J. E. Sass, *Botanical Microtechnique.* Iowa State University Press, Ames, Iowa, 1958, 228 pp.

16. D. A. Johansen. *Plant Microtechnique.* McGraw-Hill, New York, 1968, 523 pp.

17. C. J. Dawes. *Biological Techniques for Transmission and Scanning Electron Microscopy*. Ladd Research Industries, Burlington, Vermont, 1979, pp. 254–258.

18. R. L. Klein, S. Yen, and A. Thureson-Klein. Critique of the K-pyroantimonate method for semiquantitative estimation of cations in conjunction with electron microscopy. *J. Histochem. Cytochem.* **20**, 65–78 (1972).

19. L. Williams and S. Hodson. Quench cooling and ice crystal formation in biological tissues. *Cryobiology* **15**, 323–332 (1978).

20. L. Sevéus and T. Barnard. Preparation of biological material for X-ray microanalysis of diffusible elements. *J. Microsc. (Oxford)* **112**, 281–291 (1978).

21. A. R. Yeo, D. Kramer, A. Läuchli, and J. Gullasch. Ion distribution in salt-stressed mature *Zea mays* roots in relation to ultrastructure and retention of sodium. *J. Exp. Bot.* **28**, 17–29. (1977).

22. K. M. Stika, K. L. Bielat, and G. H. Morrison. Diffusible ion localization by ion microscopy: A comparison of chemically prepared and fastfrozen, freeze-dried, unfixed liver sections. *J. Microsc. (Oxford)* **118**, 409–420 (1980).

23. R. C. Staples and H. C. Hoch. A sensing mechanism in rust uredospore germlings responsive to host morphology starts the cell cycle. In J. Aist and D. W. Roberts, eds., *Infection Processes of Fungi*. Working Papers, Rockfeller Foundation, New York, 1984, in press.

24. H. C. Hoch and R. C. Staples. Ultrastructural organization of the non-differentiated uredospore germling of *Uromyces phaseoli* var. typica. *Mycologia*, in press (1983).

25. D. M. R. Harvey, J. L. Hall, and T. J. Flowers. The use of freeze substitution in the preparation of plant tissue for ion localization studies. *J. Microsc. (Oxford)* **107**, 189–198 (1976).

26. G. D. Ross, G. H. Morrison, R. F. Sacher, and R. C. Staples. Freeze substitution sample preparation for ion microscopy of plant tissue. *J. Microsc. (Oxford)* **129**, 221–228 (1983).

27. R. F. Sacher, G. D. Ross, and G. H. Morrison. Direct imaging if Na, K and Ca in normal and salt affected tomato leaves. *Plant Physiol.* in press (1983).

28. U. Lüttge and N. Higinbotham. *Transport in Plants*. Springer-Verlag, New York, 1979, 468 pp.

29. F. D. Sack and A. C. Leopold. (1982) Characteristics of statoliths from rootcaps and coleoptiles. *The Physiologist* **25**, Suppl. S-97–S-98 (1982).

30. A. Poljakoff-Mayber, Biochemical and physiological responses of higher plants to salinity stress. In A. San Pietro, ed., *Biosaline Research*. Plenum, New York, 1982, pp. 245–269.

31. A. Läuchli, A. R. Spurr, and R. W. Wittkipp. Electron probe analysis of freeze-substituted, epoxy resin embedded tissue for ion transport studies in plants. *Planta* **95**, 341–350 (1970).

32. J. Van Zyl, Q. G. Forrest, C. Hocking, and C. K. Pallaghy. Freeze-substitution of plant and animal tissue for the localization of water-soluble compounds by electron probe microanalysis. *Micron* **7**, 213–224 (1976).

33. M. G. Pitman, A. Läuchli, and R. Stelzer. Ion distribution in roots of barley seedlings measured by electron probe X-ray microanalysis. *Plant Physiol.* **68**, 673–679 (1981).

34. D. J. Weber, H. P. Rasmussen, and W. M. Hess. Electron microprobe analyses of salt distribution in the halophyte *Salicornia pacifica* var. *utahensis. Can. J. Bot.* **55,** 1516–1523 (1977).

35. F. P. Ottensmeyer. Scattered electrons in microscopy and micro-analysis. *Science* **215,** 461–466 (1982).

36. S. J. B. Reed. Trace element analysis with the ion probe. *Scanning* **3,** 119–127 (1982).

37. B. J. Garrison and N. Winograd. Ion bean spectroscopy of solids and surfaces. *Science* **216,** 805–811 (1982).

38. G. H. Morrison, and G. Slodzian. Ion microscopy. *Anal. Chem.* **47,** 932A–943A (1975).

39. S. Chandra, J. F. Chabot, G. H. Morrison, and A. C. Leopold. Localization of calcium in amyloplasts of root-cap cells using ion microscopy. *Science* **216,** 1221–1223 (1982).

40. D. M. R. Harvey, J. H. Hall, and T. J. Flowers. Ion localization in freeze substituted halophyte leaf tissue. In R. M. Spanswick, W. J. Lucas, and J. Dainty, eds., *Plant Membrane Transport: Current Conceptual Issues.* Elsevier/North-Holland, Amsterdam, 1980, pp. 493–494.

41. C. K. Pallaghy. Electron probe microanalysis of potassium and chlorine in freeze substituted leaf sections of *Zea mays. Aust. J. Biol. Sci.* **26,** 1015–1033 (1973).

42. W. D. Jeschke and W. Stelter. Measurement of longitudinal ion profiles in single roots of *Hordeum* and *Atriplex* by use of flameless atomic absorption spectroscopy. *Planta* **128,** 107–112 (1976).

43. M. G. Pitman. The determination of the salt relations of the cytoplasmic phase in cells of beetroot tissue. *Aust. J. Biol. Sci.* **16,** 615–617 (1963).

44. B. L. Sawhney and I. Zelitch. Direct determination of potassium ion accumulation in guard cells in relation to stomatal opening in light. *Plant Physiol.* **44,** 1340–1354 (1969).

45. E. Epstein and C. E. Hagen. A kinetic study of the absorption of alkali cations by barley roots. *Plant Physiol.* **26,** 457–474 (1951).

46. R. Collander. Selective absorption of cations by higher plants. *Plant Physiol.* **16,** 691–720 (1941).

47. B. Jacoby and G. C. Laties. Bicarbonate fixation and malate compartmentation in relation to salt-induced stoichiometric synthesis of organic acid. *Plant Physiol.* **47,** 525–531 (1971).

48. W. D. Jeschke. Involvement of proton fluxes in the K^+-Na^+ selectivity at the plasmalemma; K^+-dependent net extrusion of sodium in barley roots and the effect of anions and pH on sodium fluxes. *Z. Pflanzenphysiol.* **98,** 155–175 (1980).

49. J. Barber. Ionic regulation in chloroplasts and its control of photosynthesis. In R. M. Spanswick, W. J. Lucas, and J. Dainty, eds., *Plant Membrane Transport: Current Conceptual Issues* Elsevier/North-Holland, Amsterdam, 1980, pp. 83–96.

50. B. Halliwell. *Chloroplast Metabolism.* Oxford University Press, Oxford, 1981, 246 pp.

51. C. S. Huber. Photosynthetic carbon metabolism in chloroplasts. *Recent Adv. Phytochem.* **16,** 151–184 (1982).

3

K^+–Na^+ EXCHANGE AT CELLULAR MEMBRANES, INTRACELLULAR COMPARTMENTATION OF CATIONS, AND SALT TOLERANCE

W. Dieter Jeschke

Lehrstuhl Botanik
Universität Würzburg
Würzburg, Federal Republic of Germany

Salt tolerance is the ability of a plant to grow and complete its life cycle on saline substrates that contain high concentrations of salt, mostly NaCl, but sometimes also other salts including calcium salts and sulphates. In this habitat a plant has to meet two requirements: osmotic adaption and the acquisition of the mineral elements needed for growth and functional metabolism. In the case of NaCl salinity, this pertains particularly to uptake of potassium.

The relation between K/Na selectivity and salt tolerance has been reviewed by Rains (1), Flowers et al. (2), Greenway and Munns (3), Wyn Jones et al. (4), Jeschke (5), and Wyn Jones (6). During entry of ions into the root and the plant as a whole selectivity can be achieved at three membranes, the plasmalemma of the cortical root cells, the tonoplast of root (and shoot) cells and the plasmalemma of the xylem parenchyma cells. Five different, but possibly related mechanisms of K/Na selectivity, can be discerned at these sites:

1. Preference for K^+ during influx (influx selectivity).
2. K^+–Na^+ exchange at the plasmalemma (root cortex).

3. Selective Na$^+$ accumulation in vacuoles and Na$^+$–K$^+$ exchange across the tonoplast.
4. Selectivity during release of K$^+$ and Na$^+$ to the xylem vessels.
5. Selective reabsorption of Na$^+$ from the xylem sap.

In the whole plant, phloem transport particularly of K$^+$ decisively contributes to the overall selectivity. Selectivity here presumably is achieved during phloem loading, and it leads to a retranslocation of K$^+$ from leaves to growing tissues in the shoot, to inflorescences and to the root (7–10).

These processes have different consequences for the ability of a plant to withstand salinity stress. In this chapter, I want to give a short review of the processes and mechanisms of univalent cation selectivity and their implications for salt tolerance. Particular attention shall be given to the interrelationships between these processes in their importance for salt tolerance of the plant as a whole.

1. SELECTIVITY AT THE PLASMALEMMA OF CORTICAL ROOT CELLS

1.1. Selective Influx

The plasmalemma is the primary site that effects K$^+$–Na$^+$ selectivity in the root and the plant as a whole. Discrimination can be achieved here during influx [see the review by Rains (1)] or by K$^+$–Na$^+$ exchange. As shown by Epstein et al. (11) two systems are responsible for univalent cation uptake. System 1 has high affinity for K$^+$ and mediates uptake from low concentrations of K$^+$. Within its range (below 1 mM) K$^+$ uptake by barley roots is highly specific and not inhibited by Na (12, 13, 15) while K$^+$ almost totally inhibits Na$^+$ influx. At higher concentrations (system 2) K$^+$–Na$^+$ selectivity is less pronounced and both ions inhibit their uptake mutually.

The specificity for K$^+$ is widespread but the degree to which K$^+$ is favored at low concentrations or to which it is inhibited by Na$^+$ varies considerably even among glycophytes as has been reviewed recently (14). Moreover, the extent of specificity differs between salt-sensitive and -tolerant species as was shown for *Agropyrum* (15). Among halophytes, *Suaeda maritima* showed K$^+$ specificity during uptake (16), whereas in *Triglochin maritima,* K$^+$ influx from low concentrations was strongly stimulated by high concentrations of sodium (17). Quite differently *Atriplex nummularia* showed little specificity for K$^+$ also in the low concentration range (18).

No doubt these differences relate to the strategies of salt tolerance. Thus, the low selectivity in *Atriplex* possibly is related to the special form of salt removal by bladder hairs. On the other hand, influx selectivity often has been studied with low-salt roots, but the preference for K$^+$ changes with the salt contents of the roots (19). Similarly, the kinetic data of influx depend on the salt status and the apparent affinities for K^+ (20) and for Na^+ (14) are lower in high-salt roots. The significance of influx selectivity for salt tolerance will be considered in Section 1.4.

1.2. K^+–Na Exchange and K^+-Dependent Net Sodium Extrusion

Electrochemical data (17, 21, 22) predict that sodium must be actively extruded across the plasmalemma, although for *Helianthus* and *Allium* indications for active influx of sodium at low concentrations have been obtained (24, 25). At any rate, at high Na^+ concentrations under NaCl salinity and considering the substantial permeability of the plasmalemma for Na^+, active sodium extrusion is very likely and has been deduced from electrochemical data for the halophyte *Triglochin maritima* (17).

Net extrusion of sodium ions from sodium-loaded barley roots after an addition of K^+ ions has been demonstrated (26–28); an example is given in Figure 3.1. This Na^+ extrusion occurs in exchange for K^+ (26), it is transient and removes Na^+ from the symplast of the cortical root cells (the hatched area in Fig. 3.1 corresponds to the quantity of extruded sodium). The decrease in cytoplasmic Na^+ in turn induces a strong decrease in xylem transport of Na^+ (Fig. 3.1) and this is one basis of preferred K^+ transport to the shoot. In the continuous presence of K^+, sustained Na^+ extrusion maintains a low cytoplasmic Na^+ concentration in the roots.

The properties of the K^+–Na^+ exchange system have been described and

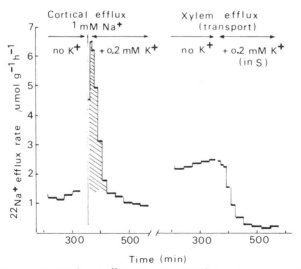

Figure 3.1.   Effect of 0.2 mM K^+ on the ^{22}Na efflux (1 mM Na^+) from the cortex (left) and the xylem (right) of sodium-loaded excised barley roots. Redrawn from ref. 5. The hatched area represents the amount of extruded sodium (2.5 μmol Na/g fresh wt), which is about 80% of the cytoplasmic Na^+ content.

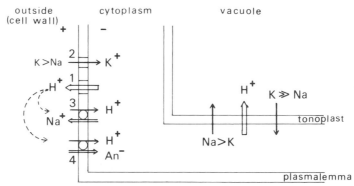

Figure 3.2.   Model of the proton-mediated K^+/Na^+ exchange system at the plasmalemma and of the fluxes that lead to vacuolar occlusion of Na^+ and facilitate of Na^+/K^+ exchange at the tonoplast. 1 = Proton pump; 2 = K^+ uniport = system 1 of K^+ influx; 3 = H^+–Na^+ antiport; 4 = H^+–anion symport. At the tonoplast the relative affinities of K^+ and Na^+ efflux are indicated. Redrawn from refs. 31 and 63.

discussed in detail (5, 29), and here I will not reconsider all the data which gave rise to the model of K^+–Na^+ exchange across the plasmalemma shown in Figure 3.2. The components of the model are (i) a proton pump, presumably powered by ATP, that generates an electrical potential difference and a protonmotive force across the plasmalemma; (ii) the electrical charge of H^+ is compensated by an influx of K^+ at a specific site or channel at which Rb^+ and Cs^+ and, with much lower affinity, Na^+ can compete; and (iii) the proton gradient provides the energy for extrusion of sodium from the cytoplasm by H^+–Na^+ antiport. This site has very low affinity for K^+; so far as I am aware K^+ extrusion has not been observed.

As has been discussed in detail (see Table III in ref. 5), common properties suggest a connection between the K^+–Na^+ exchange system and the cation-stimulated ATPases found in the plasmalemma fraction of plant roots (30). Moreover, it was shown (5) that the site mediating K^+ influx (Fig. 3.2) is identical to the high-affinity system of K^+ uptake (11) which is responsible for the influx selectivity (Section 1.1). Among other similarities a crucial one is the effect of Ca^{2+}; in its absence both the specificity of K^+ uptake is lost (13, 15) and K^+–Na^+ exchange at the plasmalemma is restrained (31).

1.3.   Specific Differences

Since active Na^+ efflux is considered a universal property of plant roots (23) K^+-dependent Na^+ extrusion should be found in other species, and particularly in salt-tolerant species. Table 3.1 contains data for those species that have been tested so far. Surprisingly, only for the three cereal genera *Hordeum*, *Triticum*, and *Secale* has a sizable K^+-dependent sodium extrusion been observed (Table 3.1). In some of the other species sodium extrusion might be restricted by a low

Table 3.1.   Plasmalemma Sodium Efflux ϕ_{co}, Cytoplasmic Sodium Content Q_c and K$^+$-Dependent Net Sodium Extrusion ϕ_{co} (K$^+$-dep) in Different Species [a]

Species	Steady-State Conditions 1 mM Na$^+$		Na$^+$ Extrusion 1 mM Na, 0.2 mM K		Reference
	ϕ_{co}	Q_c	ϕ_{co} (K$^+$-dep)	ϕ_{co}(K$^+$-dep)/Q_c	
Allium cepa	0.21	0.34	0.03	0.08	34
Helianthus annuus	1.1	1.08	0.3	0.23	34
Triticum aestivum	2.0	2.2	7.0	3.2	34
Hordeum distichon	2.1	3.6	8.2	2.3	96
Secale cereale	1.2	1.3	1.9	1.5	
Atriplex hortensis	2.5	4.5	0.9	0.2	47
Fagopyrum esculentum	0.07	0.3	N.D.	—	

[a]Fluxes and cytoplasmic contents were obtained by compartmental analysis (26); net sodium extrusion was induced by addition of 0.2 mM K$^+$. Fluxes: μmol g^{-1} fresh wt. hr^{-1}; content: μmoles g^{-1} fr. wt.; N.D. = not detectable.

cytoplasmic sodium content Q_c (Table 3.1) However, when relating the rate of Na$^+$ extrusion to Q_c, this ratio is high only for the three cereals (Table 3.1). Rye and sunflower, having a similarly low cytoplasmic sodium content, differ by a factor of five in the relative sodium extrusion. Thus, Table 3.1 shows efficient K$^+$–Na$^+$ exchange only for barley, wheat, and rye and discourages quick generalizations.

By analogy to its effects in animal cells ouabain has been used as a criterion for Na$^+$ extrusion in plants (22, 28) and on this basis a sodium efflux pump was proposed for maize (22). However, a sizable K$^+$-dependent Na$^+$ efflux so far could not be detected in maize roots (22; Behl, unpublished). Ouabain showed little effect on K$^+$–Na$^+$ exchange in barley (5, 32) and on the plasmalemma ATPase (30). Moreover, the K$^+$–Na$^+$ exchange system was shown to be quite different from the Na/K-ATPase in animal tissues (5, 30, 33). Nevertheless, critical measurements with ouabain appear to be needed.

One reason for the failure of some species to show efficient K$^+$–Na$^+$ exchange could be the use of excised roots and their low energy reserves could be responsible. However, a similarly low efficiency of K$^+$-dependent Na$^+$ extrusion was found in excised and in intact sunflower roots (34, 35), showing that a low energy status was not the basis of inefficient exchange.

The variation in Na$^+$ extrusion, therefore, suggests at least quantitative differences in membrane properties amongst different plant species. The high-affinity system mediating K$^+$ influx and the proton pump appear to be generally present in the plasmalemma of cortical root cells (36, 37). The graded responses of Na$^+$ efflux to an addition of K$^+$, therefore, suggest pronounced quantitative differences between species in the number or the efficiency of the sites mediating H$^+$–Na$^+$ antiport (Fig. 3.2).

Furthermore, depending on the species, an addition of K$^+$ to sodium-loaded roots might lead to other responses than sodium extrusion. Thus, uptake of K$^+$ could induce a transfer of Na$^+$ from the cytoplasm to the vacuole (see Section 2.2). Alternatively K$^+$ could lead to an increase in xylem transport of Na$^+$ as found in *Suaeda* (16, 38) (see also *Atriplex* in Section 3).

1.4. Importance of K$^+$–Na$^+$ Exchange and Influx Selectivity for Salt Tolerance

Remarkably, two of the cereals, barley and wheat, having an efficient K$^+$–Na$^+$ exchange system are among the most salt-tolerant or the moderately salt-tolerant crop species (39). On the other hand, the salt-tolerant *Atriplex hortensis* showed comparatively low K$^+$–Na$^+$ exchange and was similar in this respect to the salt-sensitive species onion (39), sunflower, and buckwheat. Thus, K$^+$–Na$^+$ exchange appears not to be related directly to salt tolerance. Indeed, Flowers et al. (2), Wyn Jones et al. (4), and Jeschke (5) suggested that the selective properties of the tonoplast were more important.

However, salt tolerance may be achieved by different strategies and in particular salt-tolerant members in the Poaceae, Juncaceae, and Cyperaceae achieve tolerance by restricting access of Na$^+$ to their shoots and by osmotic adaptation on the basis of sugar synthesis (40). The basis of this salt tolerance with exclusion of Na$^+$ from the shoot has been extensively studied for *Puccinellia peisonis* by Stelzer and Läuchli (41–43) and Stelzer (44). Also for barley varieties a higher degree of salt tolerance was suggested to be related to the ability to restrict the access of Na$^+$ and Cl$^-$ to the shoot (45, 46). In these species, an efficient K$^+$–Na$^+$ exchange at the plasmalemma, therefore, may be an important link in their ability to survive on saline soils. For judging an improvement of salt tolerance among cereals and pasture grasses belonging to the Pooideae a knowledge of K$^+$–Na$^+$ exchange at the plasmalemma thus appears to be of high importance. On the other hand, halophytic *Atriplex* species include large quantities of Na$^+$ in their shoot (18) and this is true also for *Atriplex hortensis,* the growth of which is strongly favored in the presence of Na$^+$ (47). In these species, K$^+$–Na$^+$ exchange could be less significant for salt tolerance, since excess Na$^+$ can be excreted by vesicular bladder hairs. As will be shown below, selectivity at the tonoplast and during xylem release is more important here. However, in species that rely on inclusion of salt for salt tolerance and osmotic adaptation, the intake of Na$^+$ (and Cl$^-$) must be under tight control (48). This is shown by the ability of *Suaeda* or *Atriplex* species to maintain a constant and sufficient K$^+$ uptake at high salinities (18, 16) and by the K$^+$ specificity of ion uptake in *Suaeda* (16).

A strong argument for the need and presence of a functional system of plasmalemma selectivity in halophytes and for salt tolerance in nonhalophytes is the interrelationship between tolerance and a high Ca/Na ratio (3) and the observation that salt tolerance, for instance, in beans, is highly increased at high levels of Ca^{2+} (49). Both functional $K^+–Na^+$ exchange and selective K^+ uptake depend on the presence of Ca^{2+} (13, 15, 31).

2. SELECTIVITY AT THE TONOPLAST

2.1. Intracellular Compartmentation of K^+ and Na^+

In recent years the hypothesis has been developed that vacuoles play a central role in salt tolerance (50), and that Na^+ occlusion in vacuoles and preferential retention of K^+ in the cytoplasm are crucial factors for osmotic adaptation and for providing a cytosolar milieu suitable for a functional metabolism. For glycophytes, this type of K^+ and Na^+ distribution has first been shown by compartmental analysis (21, 25, 51, 52). Flowers (38), Osmond (53), and Flowers et al. (2) reviewed the evidence that salt sensitivity of enzymes requires low cytoplasmic Na^+ and moderate K^+ concentrations. Wyn Jones et al. (4) showed that it is protein synthesis that needs a cytoplasmic K^+ concentration around 100 mM and low Na^+ and Cl^- levels. This scheme was strongly supported by the finding of relatively stable cytoplasmic K^+ concentrations between 90 and 110 mM and of sequestration of Na^+ in vacuoles in barley and *Atriplex hortensis* roots, as reviewed recently (5).

However, data about K^+ and Na^+ distribution in halophytes under saline conditions are scarce and are available only for *Triglochin* and *Suaeda* (Table 3.2), for *Plantago maritima* (Fig. 3.3) and recently for *Puccinellia* (64). The ob-

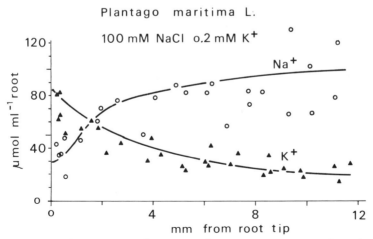

Figure 3.3. Longitudinal profile of K^+ (▲) and Na^+ (○) content (μl per ml of root in root of 12-day-old *Plantago maritima* seedlings. These were salinized for 1 day with 100 mM NaCl; further components of medium: 0.2 mM K^+ phosphate buffer, pH 5.8; 3 mM Ca(NO₃)₂; and 0.5 mM MgSO₄. K^+ and Na^+ were analyzed in 1-mm long sections. For method, see ref. 54.

Table 3.2. Intracellular Compartment of K$^+$ and Na$^+$ in Some Glycophytes and Halophytes: Comparison of Different Methods

Species	External Solution (mM)		Intracellular Concentrations (mM)				Methods	Reference
			K$^+$		Na$^+$			
	K$^+$	Na$^+$	Cyt[a]	Vac	Cyt[a]	Vac		
Glycophytes								
Hordeum vulgare								
roots	0	0	110	20	1.5	4	Longitudinal profile	63[b]
			95				Electron microprobe	56[c]
	0	1	56	4.2	22	76	Longitudinal profile	63[b]
			—	—	66	89	Compartment analysis	97[b]
	0.2	1	83	68	2.1	46	Longitudinal profile	63[b]
			230	71	11.8	56	Compartment analysis	85[b]
Atriplex hortensis								
roots	0	0	90	23	2	5	Longitudinal profile	63
	0	1	46	7	15	99	Longitudinal profile	47
			—	—	90	109	Compartment analysis	47
	1	10	81	33	4	44	Longitudinal profile	47

						Method	
Zea mays	1	126	122	1.8	2.5	Longitudinal profile	[e]
		108	67	15[d]	3.2	Compartment analysis	22
	1	118	122	5.8	16.4	Longitudinal profile	[e]
	30	—	—	51[d]	22.3	Compartment analysis	22
Halophytes							
Triglochin maritima roots	8 500	71	11.6	148	142	Compartment analysis	17
Suaeda maritima leaves	7 340	16 (22:chloropl.)	24	109 (104:chloropl.)	565	Electron microprobe	55
		34 (68:chloropl.)	34	165 (435:chloropl.)	600	Compartment analysis	86
						nonaqueously isolated	62

[a] Cytoplasmic concentrations (from compartment analyses calculated on the basis of 5% cytoplasm in the cell).
[b] *Hordeum vulgare*, convar. *distichon*.
[c] *Hordeum vulgare*, convar. *vulgare*.
[d] estimated on the basis of 3.5% cytoplasm.
[e] Behl (unpublished).

vious reasons for the lack of data are the difficulty to obtain these and the inherent limitations of the available methods. In Table 3.2, the K^+ and Na^+ distributions for a number of species are summarized.

Although the general pattern for high cytoplasmic K^+ around 100 mM (except *Suaeda,* see below) and preferred vacuolar accumulation of Na^+ are evident, deviations between comparable data obtained with different methods require a critical evaluation. Although *compartmental analysis* resulted in data of vacuolar content well in agreement with other methods, it invariably yielded high estimates of cytoplasmic concentrations (Table 3.2). This appears to be due to the basic assumption that tracer exchange from the cytoplasm is rate limited by the bordering membranes only and not in addition by intracytoplasmic organelles, and other membrane systems particularly the ER (cf. Chapter 1). In this latter case lower rate constants of cytoplasmic tracer exchange would be obtained and the cytoplasmic ion content Q_c* would be overestimated. The high cytoplasmic K^+ and Na^+ concentrations (Table 3.2.) strongly suggest that intracytoplasmic membranes retard tracer exchange. Nevertheless, as an *in vivo* method, compartmental analysis still has considerable advantages since it allows one to study effects of inhibitors, of phytohormones or of environmental conditions on ion fluxes and contents.

Longitudinal ion profiles within a root have been used to estimate cytoplasmic and vacuolar ion concentrations on the basis of low vacuolation in meristematic and high vacuolation in differentiated tissues (54). When applied to roots in saline media (Fig. 3.3) the presence of small vacuoles in meristematic tissues should be accounted for, since vacuolation can increase under these conditions as indicated by measurements with *Porphyra* at increased salinity (57). The method can be applied also to meristematic and differentiated tissues within shoots (58).

Using *nonaqueously isolated chloroplasts* (59) their ion contents have been estimated. For halophytes extremely high Na^+ and Cl^- contents were found in chloroplasts of *Limonium* (60), *Halimione* (61), and *Suaeda* (62). In this method corrections for cytoplasmic but not for vacuolar contaminations can be applied (59). The high ion contents could, therefore, be due to contaminations by electrostatic adsorption of salt crystals to chloroplasts during homogenization in nonaqueous media and the data could severely overestimate the true ionic contents as is suggested by the results for *Suaeda* (55, 62) (see Table 3.2).

Improved resolution of the *electron microprobe technique* recently allowed the study of ion distributions between cytoplasm and vacuole (55, 56). Although separate analyses of cytoplasm and vacuoles or cell walls still are at the limit of resolution (56), and estimates of cytoplasmic concentrations depend on several assumptions (56), there was good agreement between data obtained with this method and from longitudinal profiles (Table 3.2). As a great advantage, however, the electron microprobe analyzes single cells and reveals gradients in

---

*Q_c is calculated from an equation $Q_c = \phi_{co} + \phi_{cv} + \phi_{cx})/k_c$ that relates k_c to the fluxes at the membranes delimiting the symplasm.

ionic composition between individual cells and tissues (42, 55, 56). In this respect it is the most powerful method for studying intracellular ion distributions.

For many purposes, however, the less expensive technique measuring longitudinal profiles in roots still provides a versatile tool, and changes in ion distribution and vacuolar Na^+-K^+ exchange have been successfully studied in this way (63).

Even when allowing for shortcomings of some of the methods, the data of Table 3.2 reflect the general pattern of a preferential cytoplasmic localization of K^+ and a vacuolar occlusion of sodium and agree with the model of compartmentation suggested by Wyn Jones et al. (4). For potassium this distribution is maintained even when K^+ is missing in the external solution, as seen from the data for barley and *Atriplex*. In this case the vacuolar K^+ concentration decreased and potassium was shifted toward the cytoplasm. The functioning of Na^+ as a "cheap" vacuolar osmoticum clearly follows from the data for *Suaeda*, which shows very high vacuolar Na^+ concentrations (Table 3.2).

In two cases there were exceptions to this pattern. Firstly, Davis and Jaworski (22) found high $[Na]_{cyt}$ indicating a preferred cytoplasmic localization of Na^+ in maize roots. The data, however, were not confirmed by longitudinal profiles (Table 3.2). The high Na^+ concentrations (22) probably were due to the limitations of compartment analysis (see above), and to the incubation of 1-cm root segments in solutions containing up to 30 mM Na^+ (22) during which Na^+ could enter the stele without being "filtered" by the cortex and endodermis.

Second, K^+ was remarkably low in the cytoplasm of *Suaeda* leaf cells (55). From the tissue K^+ content (34 mM) (Table 3.2), and the vacuolar concentration of 24 mM, higher cytoplasmic K^+ concentrations would be estimated. This was not found by the transmission electron microprobe measurements (55). Harvey et al. (55) suggested that Na^+ might replace K^+ in some of its functions in this halophyte. However, it appears that more data on K^+ in the cytoplasm of halophytes are needed, before its functions or its necessity in the living cytoplasm of halophytes (2) can be judged. Moreover, the method applied by Harvey et al. (55), namely, quantitative X-ray microanalysis on freeze-substituted thin-sectioned specimens, is one of the most promising developments in quantitative elementary microanalysis. However, technical problems are not convincingly excluded in this paper so that it appears too early to draw such important conclusions.

Figure 3.3 shows the longitudinal profile in roots of *Plantago maritima* under moderate salinity (100 mM NaCl) but low K^+ (0.2 mM). Clearly K^+ was accumulated in the apical tissues. If it is assumed that small vacuoles in meristematic tissues (see above) contribute 10% to the intracellular volume, then cytoplasmic and vacuolar concentrations of 105 and 19 mM for K^+ and 25 and 117 mM for Na^+ can be estimated. Hence, *Plantago maritima* quite efficiently discriminates between K^+ and Na^+ in the root cytoplasm, whereas it resembles *Suaeda* to the extent that substantial amounts of Na^+ are translocated to the shoot (40).

Recently, Stelzer (64) found very low Na^+ and substantial K^+ concentrations in the meristematic tissues of the halophyte *Puccinellia* in the presence of NaCl by electron microprobe measurements, thus giving another example in halophytes for the strategy of excluding Na^+ from the cytoplasm while including K^+.

2.2.   Vacuolar K⁺–Na⁺ Exchange

Roots

Since the distribution of K^+ and Na^+ between vacuole and cytoplasm appears to be a crucial factor for salt tolerance (4, 5) and since vacuolar K^+ represents a potential reservoir that can be recovered by exchange for Na^+, the allocation of these ions no doubt must be regulated and two factors appear to be decisive; (i) as shown for barley roots the degree of vacuolar occlusion of Na or the ratio Na_v/Na_c depended on the internal K/Na ratio (Fig. 3.4) and (ii) this latter ratio, as would be expected, depended on and apparently was regulated with respect to the external K^+ and Na^+ concentrations (5). No matter, whether the tissue K/Na ratio was varied by changes in the external K^+ or Na^+ concentrations a higher K/Na ratio resulted in increasing occlusion of Na^+ in vacuoles (Fig. 3.4).

However, changes in the intracellular distribution of K^+ and Na^+ contain an irreversible component: while K^+ is passively distributed across the tonoplast (25, 52) and net movements in both directions are possible, Na^+ is actively transported into the vacuole (52, 65) and it is occluded there irreversibly (66).

Due to this situation net Na^+ uptake leading to replacement of vacuolar K^+ can occur under two conditions; (i) when a tissue contains a higher K/Na ratio, that is, a higher vacuolar K^+ content than corresponds to the external medium (5), and (ii) when a tissue containing a high K/Na level is exposed to solutions with higher Na levels. As will be shown below, both situations occur, the first during development of the root and the second during leaf development.

Net Na/K exchange across the tonoplast was directly observed in expanding tissues of excised barley roots in a solution containing K^+ and Na^+ (Table 3.3). The basis of this exchange was a rapid initial intake of K^+ that produced a tissue with a high K/Na ratio above that corresponding to the external solution [see

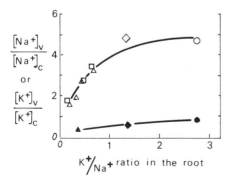

Figure 3.4.   Relation between the distribution of Na^+ (open symbols) and K^+ (closed symbols) between vacuole and cytoplasm and the K/Na ratio in the root. The K/Na ratio was varied by changes in the external K^+ ($\square$) or in the external Na^+ ($\triangle$) concentration (excised roots). $\diamondsuit,\blacklozenge$, roots of whole seedlings (1 mM Na, 0.2 mM K), data from ref. 82. $\bigcirc,\bullet$, data from ref. 21. Redrawn from ref. 5.

Table 3.3.   Net Vacuolar K^+/Na^+ Exchange in Expanding or Freshly Expanded Root Cells of Barley[a]

Roots	External Medium (mM)		Tissues Analyzed (mm from tip)	Vacuolar Loss (−) or Uptake (+) between 5 and 24 hr (μmol/ml root)		Reference
	K^+	Na^+		K^+	Na^+	
Excised	0.2	1	2.5	−43	+44.5	63
Excised	0.2	1	7	−13.1	+14.5	63
From whole seedlings without karyopsis	0.2	1	4–5	−28.4	+26.5	31
From whole seedlings with karyopsis	0.2	1	4–5	+20.1	+ 4.5	31

[a]Tissue content in 0.5-mm segments was determined in longitudinal profiles by means of flameless atomic absorption spectrometry (54). For details see ref. 63.

condition (i) above] and led to continued Na^+ uptake. Similarly vacuolar Na/K exchange was demonstrated in expanding tissues of excised *Atriplex* roots (47). Although these observations show the *possibility* for vacuolar exchange, it is uncertain whether and where such an exchange occurs in intact, growing roots.

In growing root tips, ion intake occurs only in part from the external solution (10, 67). Ion uptake by apical and differentiated root parts is similar (10, 68) and apparently is insufficient for the massive ion demand of expanding tissues. For the cation demand there is a substantial import of K^+ by the phloem and by symplastic transport (10, 67). Consequently, freshly expanded tissues in bean roots had a high K^+ content (67). Intact maize roots grown in 30 mM Na^+ and 1 mM K^+ contained 160 mM K^+ and 15 mM Na^+ in the newly expanded tissues (i.e., more K^+ in vacuoles than in the cytoplasm; cf. Table 3.2), and there was no Na/K exchange in these tissues. Similarly, in whole barley seedlings the expanded tissues were high in K^+ and low in Na^+ (31) and again there was no vacuolar Na/K exchange, by contrast to excised roots and to seedlings with excised karyopses (Table 3.3). Evidently, sufficient phloem supply of K^+ or energy (carbohydrates) prevented the vacuolar Na/K exchange that occurred in the freshly expanded tissues in the whole seedlings.

It is suggested, therefore, that the newly formed K^+-rich tissues represent a reservoir for vacuolar Na/K exchange which becomes effective when these tissues are in that part of the root with fully differentiated xylem vessels in which salt uptake and transport to the shoot occur. Here, selective uptake is then achieved by K/Na exchange at the plasmalemma and by Na/K exchange at the tonoplast, both cooperating in maintaining the cytoplasm low in Na+. This hypothesis is schematically represented in Fig. 3.5.

Although vacuolar Na/K exchange occurs in mature tissues of roots, to my knowledge it has not been directly measured. Two observations give circumstantial evidence. Firstly, the K/Na ratio in the entire barley root was substantially lower than in the freshly expanded tissues, indicating that Na+ had been taken up and K+ had been recovered from vacuoles and transported to the shoot in the older tissues. Second, Pitman et al. (56) found a radial increase in the vacuolar K/Na ratio in barley roots, suggesting that the proposed vacuolar Na/K exchange at first proceeded in the outer cortical cells.

Similar events could occur in salt-tolerant species under saline conditions (see *Puccinellia* in Section 4.1), provided that phloem retranslocation of K+ is sufficient for the formation of high-K cells in growing roots. Clearly, here is a need for measurements of the longitudinal K+ and Na+ distribution in roots under saline conditions.

Leaves

Vacuolar Na/K exchange was suggested to occur also during leaf ageing (5), see also the review of Greenway and Munns (3). Leaves develop and expand close to the shoot apex and derive minerals mainly from the phloem (70), particularly since this differentiates prior to xylem elements (71). The ion composition of young leaves, therefore, reflects that of the phloem sap, which is rich in K+ and

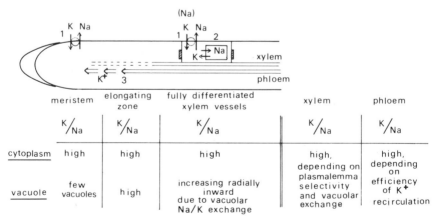

Figure 3.5. Idealized model of the root of a whole barley seedling indicating the location of the processes that result in selective transport of K+ to the shoot. 1, K+–Na+ exchange at the plasmalemma in meristem (5) and in differentiated tissues. 2, Vacuolar Na/K exchange. 3, acropetal transport of K+ in the phloem and symplasm. The tabulation summarizes the relative magnitude of the K/Na ratio in the cytoplasm and vacuole in the regions of the root from data of Pitman et al. (56), Jeschke and Stelter (54), and Jeschke (31).

low in Na^+ (72) even under saline conditions (73). With increasing leaf age, minerals are imported by the xylem with lower K^+ and possibly higher Na^+ concentrations, particularly at high transpiration rates (74) or under saline conditions. In response to the altered mineral supply the K/Na ratio decreases with leaf age (69), as shown also by the example of *Atriplex hortensis* in Fig. 3.6A. At this stage leaves become sources, from which assimilates and minerals are exported (7, 75). Especially, K^+ is retranslocated by phloem transport from older to younger leaves and to inflorescences in barley plants even under salt stress (7, 92). This occurs similarly in the halophyte *Suaeda maritima* (76).

Vacuolar Na/K exchange could contribute to the retranslocation of K^+ (5). The high K/Na ratio in young leaves can be expected to be in excess of the equilibrium ratio corresponding to the composition of the xylem sap [see above section on Roots, condition (ii)], and this would allow sodium uptake and vacuolar Na/K exchange and thus provide K^+ for retranslocation.

However, there are some objections: (i) The decrease in K^+ content (Fig. 3.6A) by itself does not yet prove that vacuolar K^+ is reexported. It can be due also to a "dilution" of the K^+ content by growth. (ii) Mature barley leaves showed high rates of K^+ retranslocation; net import of K^+ was low but not negative (7, 77). Only the latter case would indicate export of (vacuolar) K^+ from the leaves. (iii) Vacuolar Na/K exchange delivers K^+ to the symplast and not necessarily to the apoplast from which phloem loading appears to occur (8).

Nevertheless, some contribution of vacuolar Na/K exchange can be seen from the changes in the molar amounts of ions in a leaf with time (Fig. 3.6B).

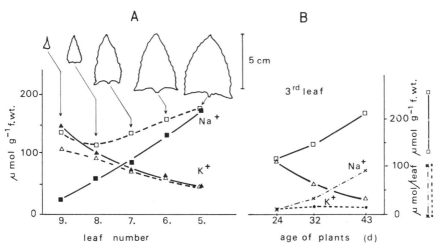

Figure 3.6. (A) Change in K^+ ($\triangle$, $\blacktriangle$) and Na^+ ($\square$, $\blacksquare$) content in leaves of *Atriplex hortensis* with leaf age and growth. Open symbols, leaves with bladder hairs; closed symbols, leaf blades, bladder hairs removed. Growth is indicated by the outline of leaves. Plant age, 43 days; external $[K^+]$ and $[Na^+]$, 1 mM each. From ref. 47. (B) Change in the K^+ ($\triangle$) and Na^+ ($\square$) content and the moles of K ($\bullet$) and Na (x) per leaf with leaf age. Leaves with bladder hairs were analyzed. External $[K^+]$ and $[Na^+]$, 1 mM each. From ref. 47.

During 19 days the *Atriplex* leaf increased in weight from 0.09 to 0.43 g, K^+ remained almost constant, Na^+ increased from 9.3 to 91 μmol/leaf, and the ratio K/Na (without bladder hairs) dropped from ca. 6 to 0.15. The total increase in weight and size* proceeded thus almost without net uptake of K^+. The actual extent of vacuolar Na/K exchange is not known, but it appears highly probable that a proportion of the initial 100 μmol K^+/g fr. wt. was in vacuoles and a considerable amount of the final 30μmol K^+/g fr. wt. (Fig. 3.6B) was in the cytoplasm. In this case leaf growth was supported by massive uptake of Na^+ combined with some vacuolar Na/K exchange.

No matter to which degree Na/K exchange was involved, the massive Na^+ uptake allowed the retranslocation† of nearly all of the K^+ imported with the xylem stream. This suggests that K^+ retranslocation occurred mainly by direct xylem/phloem exchange (8) and that this was facilitated by the capacity of the leaf to withdraw Na^+ from the xylem sap or the apoplast.

The significance of young leaves with a high K/Na ratio and with the potential to withdraw Na^+ from the xylem and possibly to exchange vacuolar K^+ and Na^+ is impressively underlined by the decisive contribution of bladder hairs to the high K/Na ratio in *Atriplex hortensis* leaves (Fig. 3.6A). Bladder hairs contained most of the Na^+ in young leaves and it may be inferred that removal of Na^+ from *young* leaf blades may be a crucial function of these hairs.

In summary, it is suggested that maturing leaves in salt-tolerant species can function as sinks for Na^+ due to growth and vacuolar Na/K exchange. Net export of K^+ from the leaf as a result of vacuolar Na/K exchange could occur in senescing leaves (77).

2.3. Factors Affecting the Compartmentation of K^+ and Na^+

Compatible solutes are an important factor in the osmotic balance of the cytoplasm under salt stress (4), and sodium salts are suggested to play a complementary osmotic role in the vacuole (2, 4, 5). Compatible solutes or cytosolutes such as glycinebetaine, therefore, might regulate the intracellular Na^+ distribution under salt stress and could induce increased vacuolar Na^+ accumulation. Evidence for this was obtained by Ahmad et al. (78) showing that externally applied glycinebetaine increased the vacuolar Na^+ concentration in barley roots.

An impressive amount of evidence shows abscisic acid (ABA) to interfere with and regulate ion fluxes (79). In beetroot slices ABA shifted the selectivity in favor of Na^+ by increasing Na^+ and inhibiting K^+ uptake (80). As judged by the

---

*Part of the weight increase was due to an increase in succulence with leaf age.
†Corresponding to a rough estimate K^+ was retranslocated from the leaf at a rate of 3 μmol/day, similar to that quoted for barley (76).

Table 3.4.   Intracellular Distribution of K^+ and Na^+ in Excised Barley Roots in the Absence and Presence of Abscisic Acid[a]

	External Medium (mM)		Control	5 μM ABA	Reference
	K^+	Na^+			
$[K^+]_v/[K^+]_c$	1	0	0.76	1.95	73
$[Na^+]_v/[Na^+]_c$	0	1	2.7	17.8	73

[a]The vacuolar and cytoplasmic contents of K^+ and Na^+ were determined by compartmental analysis. For details see ref. 73.

duration of the experiments vacuolar Na^+ accumulation was affected. Similarly excised bean roots showed an increase in Na^+ but not in K^+ uptake, whereas in intact bean seedlings Na^+ uptake was increased and that of K^+ was inhibited (81, 82).

Also, in barley roots ABA had differential effects on K^+ and Na^+ (83). Apart from inhibiting xylem transport of both ions, ABA increased the accumulation of K^+ without affecting its uptake rate. By contrast, ABA stimulated Na^+ uptake and greatly increased its (vacuolar) accumulation. Interestingly, Na^+ uptake was more sensitive to ABA than xylem transport, indicating that ABA affected two sites independently and one of these was the tonoplast. For K^+ and Na^+, the ratio of vacuolar to cytoplasmic concentration was increased in the presence of ABA (84) (see Table 3.4). These data were obtained with solutions containing either K^+ or Na^+, and further experiments are needed to show whether ABA can shift Na^+ into vacuoles and thereby lead to a recovery of K^+. Since ABA levels can increase under salinity stress (86, 87), and excess of Na^+ is one of the problems a salt-stressed plant has to cope with, ABA-induced mobilization of K^+ could be of vital importance under saline conditions.

3.   SELECTIVITY DURING XYLEM RELEASE

3.1.   Conditions of Root Pressure Exudation

During entry of salt into the shoot selectivity can also be affected at the plasmalemma of the xylem parenchyma cells in two ways. First, K^+ or Na^+ could be favored during xylem release. Second, Na^+ can be reabsorbed from the xylem sap. This latter process will be discussed in Chapter 9.

In glycophytic species, which have high K/Na ratios in their shoots, a preference for K^+ during release to the xylem vessels could improve the overall selectivity. Indications for this are limited, however (25) (see Table 3.4). On the other hand, sodium contents in leaves of dicotyledonous halophytes are high (2, 40, 48) and salt-tolerant barley varieties (45) or tomato species (88) accumulate Na$^+$ in their leaves. In these plants Na$^+$ release to the xylem could be preferred. For barley, little if any selectivity during xylem secretion was found (21, 69), and high Na$^+$ contents could be due to effects of transpiration, see next Section 3.2.

For *Suaeda maritima,* Yeo (16) obtained indirect evidence for a preference for Na$^+$ in xylem transport. Here sodium transport (at 5 mM NaCl) was increased by K$^+$ particularly at high external K$^+$ concentrations.

K/Na selectivity during xylem release can be unequivocally described only when cytoplasmic concentrations [k]$_c$ and [Na]$_c$ and the rates of xylem transport for both ions ϕ_{cx}(k) and ϕ_{cx}(Na) are known. As a measure of selectivity the ratio

$$S(\text{transport}) = \frac{\phi_{cx}(\text{K})[\text{Na}]_c}{\phi_{cx}(\text{Na})[\text{K}]_c}$$

is chosen and the available data are shown in Table 3.5. A high S(transport) indicates preference for K$^+$ and a ratio far below 1 suggests that Na$^+$ is favored during xylem release.

For *Allium* roots S(transport) was 2.8 and this could suggest some preference

Table 3.5.   K/Na Selectivity of Xylem Transport, S(transport),[a] Rates of Xylem Transport, and Cytoplasmic Concentrations of K$^+$ and Na$^+$

Species	External Solution (mM)		Cytoplasm (mM)		Xylem Transport ϕ_{cx} μmol g^{-1} fr. wt. hr^{-1}		S(transport)	Reference
	K	Na	K	Na	K	Na		
Allium cepa	1	1	170[b]	15.2[b]	21.6	0.65	2.8	25
Hordeum distichon	1	0.2	230[b]	11.8[b]	6.2	0.14	2.3	98
Atriplex hortensis	0	0	90[c]	2[c]	0.8[d]	0.9[d]	0.02	47
Atriplex hortensis	0	1	46[c]	15[c]	3.7[d]	8.7	0.14	47
Atriplex hortensis	1	0	91[c]	0.5[c]	8.3	0.7[d]	0.06	47
Atriplex hortensis	1	1	81[c]	2.5[c]	5.8	5.4	0.03	47

[a]S(transport) $= \dfrac{\phi_{cx}(\text{K})}{\phi_{cx}(\text{Na})} \times \dfrac{[\text{Na}^+]_c}{[\text{K}^+]_c}$

[b]Data of compartment analysis; cytoplasmic concentrations based on 5% cytoplasm in the roots.

[c]Cytoplasmic concentration from longitudinal profiles (54).

[d]Rate of endogenous ion transport from the root.

for K^+. For barley roots we found a ratio of 2.3, but considering the possible errors in calculating the cytoplasmic content by means of compartmental analysis I am inclined to take this value to be close enough to one as to suggest that there is little discrimination in K^+ and Na^+ release to the xylem sap in barley as proposed earlier (21, 69). Selectivity here is set up at the plasmalemma and the tonoplast (see Sections 1 and 2).

Quite differently, S(transport) was clearly below one for the salt-tolerant *Atriplex hortensis* under all conditions tested (Table 3.4), indicating preferential Na^+ transport. This agrees with the situation suggested for *Suaeda* (16) and may have important consequences. *Atriplex hortensis* shows little if any K^+/Na^+ exchange at the plasmalemma (Table 3.1); nevertheless, the cytoplasmic Na^+ content in its roots was similarly low as in barley (Table 3.2). Preferential xylem transport *and* vacuolar occlusion of Na^+ (Table 3.2) could be a means by which this species removes Na^+ from the root cytoplasm. In addition Na^+ (and Cl^-) transport in the xylem aids in the osmotic adaptation of the shoot under saline conditions and, as a special adaptation, *Atriplex hortensis* similar to other *Atriplex* species can excrete excess quantities of Na^+ into bladder hairs (see Fig. 3.6).

3.2. Conditions of Transpiration and Xylem Transport

Under field conditions xylem transport usually is driven by transpiration and then the selectivity of transport is altered (74, 89). As has been discussed by Munns et al. (48), in halophytes the entry of Na^+ and Cl^- to the roots or their release to the xylem sap must be tightly regulated under conditions of salinity and high transpiration in order to avoid excess ion supply to the shoots. This pertains also to passive ion entry. In relation to this morphological changes such as an increased width of the casparian strip (90) or the formation of a double endodermis (42) have been reported. In seedlings of *Hordeum* and *Sinapis,* high transpiration increased the transport of Na^+ but not of K^+, thereby shifting the selectivity toward Na^+ (74, 89). As discussed by Pitman (77) transpiration favors passive components of ion uptake and this pertains particularly to Na^+ (and Ca^{2+}). The way in which water flow affects ion movements depends also on its pathway across the root. Recent measurements of the hydraulic conductivity L_P of the root cells and of the root as a whole (91) suggest that the majority of water movement in the root occurs through transcellular and symplasmatic pathways. In this case water flow could affect also the transport of ions that are actively taken up and transported symplastically through the root. In order to test this, xylem transport and intracellular fluxes of K^+ and Na^+ under varied conditions of transpiration were measured. Interestingly transpiration increased the xylem transport of both Na^+ and K^+, but the fluxes of these ions at the cellular membranes were altered in different ways.

4. MECHANISMS OF SALT TOLERANCE IN TWO SELECTED SPECIES

4.1. Puccinellia peisonis

Puccinellia shows a high degree of salt and flooding tolerance (41), and as a member of the Poaceae it is characterized by slow growth under saline conditions and by a high K/Na ratio in its leaves (40). Osmotic adaptation in the leaves is aided by accumulation of sugars (40). Stelzer and Läuchli (41–43) and Stelzer (44) studied the basis of its salt tolerance and termed *Puccinellia* a highly salt-tolerant sodium excluder.

As an adaptation to regulated salt uptake, *Puccinellia* showed a pronounced and fast development of an endodermis with suberin lamellae except in the passage cells. In addition, the structure of the inner cortical cells was comparable to that of the passage cells and was rich in cytoplasm and organelles. In the tertiary state the inner cortical cells developed into a second endodermis.

Data of electron microprobe analyses (43) revealed a remarkable gradient in the K/Na ratios from the outer cortex to the stele (Fig. 3.7) showing that most of the sodium was retained in the outer cortical cells and little entered the endodermis and the xylem parenchyma cells (43). According to this ion distribution Stelzer and Läuchli (43) attributed a special role to the endodermis and the inner cortical layer and concluded "both cells appear to be barriers to Na⁺ transport to the stele and hence to the shoot, while K⁺ and Cl⁻ probably move more easily across the endodermis."

These interesting data, however, can be interpreted also on the basis of an efficient compartmentation of K⁺ and Na⁺. As is seen by a comparison of the electron micrographs (42) and the electron microprobe data (43) the gradient in the K/Na ratio (Fig. 3.7) is paralleled by a gradient in vacuolation from the

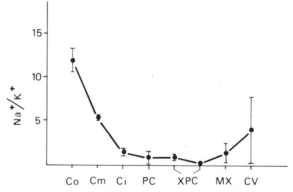

Figure 3.7. Na/K ratios measured by X-ray microanalysis in individual root cells of *Puccinellia peisonis*, Na/K in growth medium = 33; Co, outer cortex; Cm, middle cortex; Ci, inner cortex; PC, passage cell; XPC, xylem parenchyma; MX, metaxylem vessel; CV, central vessel. Redrawn from ref. 43.

highly vacuolated cortical aerenchyma cells to the endodermis and the xylem parenchyma cells which both are rich in cytoplasm.

Assuming that (i) xylem parenchyma cells are almost nonvacuolated and their K/Na ratio (Fig. 3.7) reflects that in the cytoplasm, and (ii) the Na/K ratio is 20 in the vacuole of the outer cortical cells, the following Na/K ratios can be estimated: inner cortical cell (30% cytoplasm):Na/K = 0.43 (found 0.77), outer cortical cells (5% cytoplasm):Na/K = 0.099 (found 0.084).

I would suggest, therefore, that *Puccinellia* achieves exclusion of Na^+ from the cytoplasm and thereby from the xylem sap in a similar way as barley, that is, by efficient K/Na selectivity at the plasmalemma and powerful occlusion of Na^+ in vacuoles. Due to the serial arrangement of the cortical aerenchyma cells (42), ions have to pass each cell consecutively, and in this way Na^+ ions can be repetitively "filtered" from the symplasm and occluded in vacuoles.

This scheme does not assign a special ion barrier function to the endodermis apart from forcing all solutes to pass the symplast, a function that apparently is improved by the double endodermis in *Puccinellia*. Moreover, this scheme is strengthened by recent results of Stelzer (64) showing low Na^+ and high K^+ concentrations in the cytoplasm of *Puccinellia* in the presence of NaCl.

Furthermore, the early development of a double, tertiary endodermis (42) can be accommodated into this model. Due to this tight barrier ion uptake is restricted to those younger parts of the root in which sodium accumulation in cortical cells and vacuolar Na/K exchange still can occur.

Depending on the efficiency of the plasmalemma selectivity, this model demands that the plant provides new K^+-rich cells capable of Na^+ accumulation in exchange for K in order that sufficient root growth be achieved. Interestingly the root to shoot ratio strongly increased with salinity above 100 mM NaCl, the concentration of optimal growth (41).

Selectivity in the shoot. Sodium exclusion by no means is perfect in *Puccinellia,* and at higher salinities, Na^+ concentrations in the leaf rise considerably (44). In this case Na^+ ions were retained in the voluminous bundle sheath cells. In these and to a lesser degree in mesophyll cells the increase in Na^+ content at higher salinities was accompanied by a decrease in K^+ content. Assuming occlusion of Na^+ in vacuoles, K^+ could be recovered by Na/K exchange, or as discussed in Section 2.3, Na^+ withdrawal from the xylem sap could improve K^+ recirculation in the phloem. Stelzer (44) suggested recirculation to the root and to the xylem paranchyma cells. In line with the present interpretation, K^+ could be retranslocated to shoot and root meristems and in the latter be available for root growth and for the formation of new cell material for the maintenance of K/Na selectivity during uptake.

In conclusion, *Puccinellia* appears to be a suitable plant for studying K/Na selectivity in the Poaceae. Several questions warrant further experimentation. Does *Puccinellia* show efficient K/Na exchange at the plasmalemma and Na/K exchange at the tonoplast? What are the rates of K^+ retranslocation in this species?

4.2. Hordeum

Barley was mentioned in all sections of this chapter and these data shall not be reiterated here, but I want to discuss the crucial factors that establish and limit salt tolerance in this species. As follows from the review of Greenway and Munns (3), there is not a unique answer, although many facets of tolerance in barley have been studied.

Even in the most tolerant varieties growth was retarded at low NaCl, and a 50% reduction occurred at about 100 mM NaCl in tolerant varieties (45). Grain formation was subject to salinity damage substantially more than vegetative growth (45).

Although barley takes up considerable amounts of Na^+ and Cl^-, and thereby achieves at least the vacuolar component of osmotic adaptation (45, 46), Storey and Wyn Jones (46) concluded from the higher Na^+ and Cl^- contents in leaves of sensitive compared to tolerant varieties (45, 46) that Na^+ and Cl^- exclusion rather than uptake was the strategy of salt tolerance in barley. Here, comparative measurements of K^+–Na^+ exchange at the plasmalemma in varieties differing in tolerance appear to be needed. On the other hand, low Na^+ transport to the shoot can be achieved also by efficient occlusion of Na^+ within root cells (see Section 2.2) and measurements with the sensitive (45) *H. distichon* cv. Kocherperle and the tolerant *H. vulgare* cv. Mariout (Bleifuss, unpublished) revealed that vacuolar Na^+ accumulation in roots of Kocherperle was less efficient than in Mariout, and this could contribute to the higher Na^+ contents in shoots of the more sensitive varieties (45, 46).

Another and possibly crucial facet of salt tolerance is the ability to retranslocate K^+ from mature leaves to younger ones, to shoot and root apices and into inflorescences as indicated by the early data of Greenway (45). As shown by higher K^+ and lower Na^+ contents in inflorescences of tolerant compared to sensitive varieties (92), K^+ retranslocation appears to be more efficient in the tolerant cultivars and could be decisive for their performance (see below).

Greenway and Munns (3) discussed the possible causes for salt-induced growth reduction and they condensed the various possible adverse effects to the alternative "ion excess or water deficit." Delane et al. (73) and Munns et al. (48) proposed that ultimately water deficit in expanding tissues was responsible for reduced growth.

This conclusion was based on measurements of elongating rates and solute contents in expanding and expanded leaves of the tolerant variety Beecher and in brief on the following results. Ion contents were high in expanding tissues, but while Na^+ and Cl^- increased, K^+ decreased at higher salinity (180 mM NaCl). The high ion (and similarly sugar) contents, however, were due to reduced growth rather than to high ion (or sugar) intake: rates of net ($Na^+ + K^+$) uptake into elongating cells decreased at high salinity, but net uptake into the shoots was less reduced. This showed that low ion supply to the expanding tissues restricted their ion content and henceforth osmotic water uptake and cell elongation (73). Based on the observation that expansion growth was resumed within 1 hr after a stepdown from high to lower salinity, Munns et al. (48) concluded that water deficit due to insufficient ion content was the ultimate limitation of growth.

Delane et al. (73) considered either hindered ion transport in the phloem or saturated uptake in the expanding cells to be the limiting step in ion supply but could not decide between these alternatives. They based their discussion exclusively on sufficient transport of Na^+ and Cl^- in the phloem although they quoted a high K^+ (70–120 mM) concentration in the phloem sap.

As an alternative, I would suggest that relative K^+ deficiency and insufficient retranslocation of K^+ was a crucial factor. Although most analyses (48, 73) gave the sum of $(K^+ + Na^+)$ only, the data showed a shift from K^+ to Na^+ or a salinity-dependent decrease in K^+ transport to the growing tissues. Since growing tissues derive their mineral supply from the phloem (see Section 2.2), which contains K^+ as the major cation (72) even under the saline conditions of the present experiments (73), the decreased K^+ supply points to decreased phloem transport.[*] Furthermore, because photosynthesis is unimpaired or at least not limiting growth under salinity (3, 73), the basis of reduced phloem transport must be restricted phloem loading. This could be due to low K^+ or high Na^+ levels in the apoplast of the mature leaves or to an inhibition by ABA (93) which can increase at high salinities (86, 87).

According to present knowledge, phloem loading occurs by sugar-proton symport powered by a proton pump (e.g., 94) that is electrically neutralized by K^+ influx to the sieve tubes. This process will be retarded if the K^+ levels become limiting (95).[†] Alternatively, if the Na^+ levels in the apoplast increase due to increased xylem import[†] and/or to insufficient withdrawal of Na^+ by the leaf cells (Section 2.2), high Na^+ may inhibit K^+ influx, as it does at the plasmalemma of root cells (96), and thereby indirectly phloem loading. Furthermore, phloem loading with Na^+ is less efficient than with K^+ (94).

This proposal is strengthened by the findings that grain formation, which needs assimilate transport, is more salt sensitive than growth (45) and that net K^+ uptake to the shoot was severely decreased at high salinities even in a salt-tolerant barley variety.[2†] Moreover, leaf growth occurs by cell division and expansion, and thus is dependent on protein synthesis which was shown to be highly dependent on sufficient K^+ levels (4). Finally, Na^+ could adversely affect expansion growth as was observed in barley roots (54). Also in the experiments of Delane et al. (73) elongation was retarded despite high Na^+ and Cl^- levels,

---

[*] In the expanding tissues the carbohydrate levels were high, but as stated by Delane et al. (73), this could be attributed to reduced growth and this does not contradict reduced phloem transport.

[†] As calculated from the data of Storey and Wyn Jones (46) the net flux of K^+ to the shoot of barley cv. Mariout at 200 mM NaCl was severely decreased to 10% of the control (0 NaCl) and was about one-half on the net flux of Na^+ to the shoot.

indicating that K^+ was relatively deficient for the generation of turgor (97) for extension growth.

In conclusion, I would suggest that the common basis of salt-induced reduction in leaf growth and reproduction could derive from the inhibition of transport of assimilate and K^+ in the phloem* due to high apoplastic Na^+ or low K^+ concentrations in barley leaves. The more efficiently a variety of barley can withdraw Na^+ from the apoplast (Section 2.2) or prevent its access to the shoot (46), the more it is suited to maintain (slow) growth at higher salinities and to produce grains even though at lower number and weight (45).

This proposal is not at variance with the conclusions of Munns et al. (58), it only specifies that growing tissues in barley suffer from water deficit due to insufficient K^+ in addition to a low salt supply.

5. CONCLUDING REMARKS

In the last decade, particular processes of $K^+–Na^+$ selectivity during influx and by K/Na exchange have been studied. These occur by vacuolar Na^+ accumulation and Na/K exchange, or during xylem release and by reabsorption of Na^+ from the xylem sap. In addition a considerable body of knowledge about preferential K^+ retranslocation has been obtained. The importance of the intracellular compartmentation of K^+ and Na^+ appears to be well founded.

For the selectivity within the plant and for its performance under saline conditions, however, a cooperation of these processes appears to be of crucial importance.

The selectivity of a root, for example, of barley or *Puccinellia,* appears to be achieved by the combination of a continued formation of K^+-rich cells in the root tip due to sufficient K^+ import by the phloem, and to the ability of these cells at a later stage to maintain a symplasm low in Na due to K^+/Na^+ exchange at the plasmalemma and vacuolar Na/K exchange.

Similarly, in the shoot the first step appears to be the generation of young leaves having sufficiently high cytoplasmic *and* vacuolar K^+ content, by virtue of which the maturing or mature leaves can withdraw excess Na^+ from the xylem sap by growth and vacuolar Na/K exchange and thereby allow K^+ recirculation for renewed growth. In *Atriplex* the generation of K^+-rich cells appears to be aided by early formation of bladder hairs that withdraw Na^+ (and Cl^-) particularly from the juvenile leaf (Fig. 3.6A). As long as the circle of Na^+ withdrawal by maturing or mature leaves can be maintained and a K^+-rich

---

*I have centered my interpretation on retranslocation of K^+ although the phloem can contain appreciable amounts of Na^+ under saline conditions (48, 73), and the possible importance of Na recirculation has been discussed and suggested as a means of desalinination of leaves (48) possibly by transport to the roots. However, the estimated Na^+ concentrations in the phloem always were below the substrate concentrations so that excretion of this sodium into the medium at least would not be possible in a passive way.

phloem be sustained, slow growth and eventually reproduction can continue under saline conditions. A crucial basis of salt tolerance, therefore, appears to be sufficient phloem transport, which not only retranslocates K^+ to shoot and root meristems and to inflorescences but provides the assimilates and energy for maintaining the selective circle.

One major question for future research appears to be the efficiency of K^+ recirculation, and the composition of the phloem sap under saline conditions. Another question is the efficiency of Na^+ and K^+ compartmentation in halophytes.

ACKNOWLEDGMENT

The experimental results reported in this paper were supported by funds of the Deutsche Forschungsgemeinschaft. Thanks are expressed to Miss Andrea Schniering for technical assistance and to Dr. R. Behl for critical reading of the manuscript.

REFERENCES

1. D. W. Rains. Salt transport by plants in relation to salinity. *Annu. Rev. Plant Physiol.* 23, **367** (1972).

2. T. J. Flowers, P. F. Troke, and A. R. Yeo. The mechanism of salt tolerance in halophytes, *Annu. Rev. Plant Physiol.* **28,** 89 (1977).

3. H. Greenway and R. Munns. Mechanisms of salt tolerance in nonhalophytes. *Annu. Rev. Plant Physiol.* **31,** 149 (1980).

4. R. G. Wyn Jones, C. J. Brady, and J. Speirs. Ionic and osmotic relations in plant cells. In D. L. Laidman and R. G. Wyn Jones, eds., *Recent Advances in the Biochemistry of Cereals.* Academic Press, New York, 1979, p. 63.

5. W. D. Jeschke. Univalent cation selectivity and compartmentation in cereals. In D. L. Laidman and R. G. Wyn Jones, eds., *Recent Advances in the Biochemistry of Cereals.* Academic Press, New York, 1979, p. 37.

6. R. G. Wyn Jones. Salt tolerance. In C. B. Johnson, ed., *Physiological Processes Limiting Plant Productivity.* Butterworths, London, 1980, p. 271.

7. H. Greenway and M. G. Pitman. Potassium retranslocation in seedlings of Hordeum vulgare. *Aust. J. Biol. Sci.* **18,** 235 (1965).

8. J. S. Pate. Exchange of solutes between phloem and xylem and circulation in the whole plant. In M. H. Zimmerman and J. A. Milburn, eds., *Encyclopedia of Plant Physiology, New Series,* Vol. 1. Springer-Verlag, Berlin, 1975, p. 451.

9. H. Greenway, A. Gunn, M. G. Pitman, and D. A. Thomas. Plant responses to saline substrates VI. Chloride, sodium, and potassium uptake and redistribution within the plant during ontogenesis of Hordeum vulgare. *Aust. J. Biol. Sci.* **18,** 525 (1965).

10. C. Richter and H. Marschner. Umtausch von Kalium in verschiedenen Wurzelzonen von Maiskeimpflanzen. *Z. Pflanzenphysiol.* **70,** 211 (1973).

11. E. Epstein, D. W. Rains, and O. E. Elzam. Resolution of dual mechanisms of potassium absorption by barley roots. *Proc. Natl. Acad. Sci. USA.* **49,** 684 (1963).

12. D. W. Rains and E. Epstein. Sodium absorption by barley roots: role of the dual mechanisms of alkali cation tranport. *Plant Physiol.* **42,** 314 (1967).

13. D. W. Rains and E. Epstein. Sodium absorption by barley roots: Its mediation by mechanism 2 of alkali cation transport. *Plant Physiol.* **42,** 319 (1967).

14. W. D. Jeschke. Cation fluxes in excised and intact roots in relation to specific and varietal differences. In M. R. Saric, ed., *Genetic specificity of Mineral Nutrition of Plants.* Serbian Academy of Sciences and Arts, Belgrade, 1982, p. 57.

15. O. E. Elzam and E. Epstein. Salt relations of two grass species differing in salt tolerance I. Growth and salt content at different salt concentrations. II. Kinetics of the absorption of K, Na, and Cl by their excised roots. *Agrochimica* **13,** 190 and 196 (1969).

16. A. R. Yeo. Salt tolerance in Suaeda maritima. Ph.D. thesis, University of Sussex, England (1974).

17. R. L. Jefferies, The ionic relations of seedlings of the halophyte Triglochin maritima L. In W. P. Anderson, ed., *Ion Transport in Plants.* Academic Press, New York, 1973, p. 297.

18. C. B. Osmond, O. Björkmann, and D. J. Anderson. Absorption of ions and nutrients. In *Physiological Processes in Plant Ecology Towards a Synthesis with Atriplex,* Chap. 7, Springer-Verlag, Heidelberg, 1979, p. 191.

19. M. G. Pitman, A. C. Courtice, and B. Lee. Comparison of potassium and sodium uptake by barley roots at high and low salt status. *Aust. J. Biol. Sci.* **21,** 871 (1968).

20. A. D. M. Glass. Regulation of potassium absorption in barley roots. An allosteric model. *Plant Physiol.* **58,** 33 (1976).

21. M. G. Pitman and H. D. W. Saddler. Active sodium and potassium transport in cells of barley roots. *Proc. Nat. Acad. Sci. USA* **57,** 44 (1967).

22. R. F. Davis and A. Z. Jaworski. Effects of ouabain and low temperature on the sodium efflux pump in excised corn roots. *Plant Physiol.* **63,** 940 (1979).

23. U. Lüttge and N. Higinbotham. *Transport in Plants.* Springer-Verlag, New York, 1979, p. 155.

24. D. J. F. Bowling and A. Q. Ansari. Evidence for a sodium efflux pump in sunflower roots. *Planta* **98,** 323 (1971).

25. A. E. S. Macklon. Cortical cell fluxes and transport to the stele in excised root segments of Allium cepa L. I. Potassium, sodium and chloride. *Planta* **122,** 199 (1975).

26. W. D. Jeschke. K⁺-stimulated Na⁺ efflux and selective transport in barley roots. In W. P. Anderson, ed., *Ion Transport in Plants.* Academic Press, New York, 1973, p. 285.

27. W. D. Jeschke and W. Stelter. K⁺-dependent net Na⁺ efflux in roots of barley plants. *Planta* **114,** 251 (1973).

28. H. Nassery and D. A. Baker. Extrusion of sodium ions by barley roots. I. Characteristics of the extrusion mechanism. *Ann. Bot.* **36,** 881 (1972).

29. W. D. Jeschke. Involvement of proton fluxes in K⁺-Na⁺ selectivity at the plasmalemma; K⁺-dependent net extrusion of sodium in barley roots and the effect of anions and pH on sodium fluxes. *Z. Pflanzenphysiol.* **98,** 155 (1980).

30. T. K. Hodges. ATPases associated with membranes of plant cells. In U. Lüttge and M. G. Pitman. eds., *Encyclopedia of Plant Physiology, New Series,* Vol. 2A. Springer-Verlag, Berlin, 1976, p. 260.

31. W. D. Jeschke. Roots, cation selectivity and compartmentation, involvement of protons and regulation. In R. M. Spanswick, W. J. Lucas, and J. Dainty, eds., *Plant Membrane Transport: Current Conceptual Issues.* Elsevier, Amsterdam, 1979, p. 17.

32. W. D. Jeschke. Evidence for a K^+-stimulated Na^+ efflux at the plasmalemma of barley root cells. *Planta* **94,** 240 (1970).

33. A. Ratner and B. Jacoby. Effect of K^+, its counteranion and pH on sodium efflux from barley root tips. *J. Exp. Bot.* **27,** 843 (1976).

34. W. D. Jeschke and H. Nassery. K^+-Na^+ selectivity in roots of Triticum, Helianthus and Allium. *Physiol. Plant.* **52,** 217 (1981).

35. W. D. Jeschke and W. Jambor. Determination of unidirectional sodium fluxes in roots of intact sunflower seedlings. *J. Exp. Bot.* **32,** 1257 (1981).

36. E. Epstein. *Mineral Nutrition of Plants: Principles and Perspectives.* Wiley, New York, 1972.

37. F. A. Smith and J. A. Raven. H^+ transport and regulation of cell pH. In U. Lüttge and M. G. Pitmen, eds., *Encyclopedia of Plants Physiology, New Series,* Vol. 2A, Springer-Verlag, Berlin, 1976, p. 317.

38. T. J. Flowers. Halophytes. In D. A. Baker and J. L. Hall, eds., *Ion Transport in Plant Cells and Tissues.* North-Holland, Amsterdam, 1975, p. 309.

39. E. V. Maas and G. I. Hoffmann. Crop salt tolerance current assessment. *J. Irrig. Drain. Div.* **103,** 115 (1977).

40. R. Albert and M. Popp. Chemical composition of halophytes from the Neusiedler Lake region in Austria. *Oekologia* **27,** 157 (1977).

41. R. Stelzer and A Läuchli. Salz- und Überflutungstoleranz von Puccinellia peisonis. I. Der Einfluβ von NaCl- und KCl-Salinität auf das Wachstum bei variierter Sauerstoffversorgung der Wurzel. *Z. Pflanzenphysiol.* **83,** 35 (1977).

42. R. Stelzer and A. Läuchli, Salz- und Überflutungstoleranz von Puccinellia peisonis. II. Strukturelle Differenzierung der Wurzel in Beziehung zur Funktion. *Z. Pflanzenphysiol.* **84,** 95 (1977).

43. R. Stelzer and A. Läuchli. Salt- and flooding tolerance of Puccinellia peisonis. III. Distribution and localization of ions in the plant. *Z. Pflanzenphysiol.* **88,** 437 (1978).

44. R. Stelzer. Ion localization in the leaves of Puccinellia peisonis. *Z. Pflanzenphysiol.* **103,** 27 (1981).

45. H. Greenway. Plant responses to saline substrates. I. Growth and ion uptake of several varieties of Hordeum during and after sodium chloride treatment. *Aust. J. Biol. Sci.* **15,** 16 (1965).

46. R. Storey and R. G. Wyn Jones. Salt stress and comparative physiology in the Gramineae. I. Ion relationships of two salt- and water-stressed barley cultivars, California Mariout and Arimar. *Aust. J. Plant Physiol.* **5,** 801 (1978).

47. W. Stelter. Untersuchungen zur Kalium-Natrium-Selektivität von Atriplex hortensis L. Dissertation, University of Würzburg, 1979.

48. R. Munns, H. Greenway, and G. O. Kirst. Halophytes: Algae and higher plants. In

O. L. Lange and B. Osmond, eds., *Encyclopedia of Plant Physiology, New Series,* Vol. 12C. Springer-Verlag, Berlin, 1983.

49. P. A. Lahaye and E. Epstein. Calcium and salt tolerance by bean plants. *Physiol. Plant.* **25,** 213 (1971).

50. D. H. Jennings. Halophytes, succulence and sodium in plants—A unified theory. *New Phytol.* **67,** 899 (1968).

51. C. K. Pallaghy and B. I. Scott. The electrochemical state of cells of broad bean roots. II. Potassium kinetics in excised root tissues. *Aust. J. Biol. Sci.* **22,** 585, (1969).

52. A. E. S. Macklon and N. Higinbotham. Active and passive transport of potassium in cells of excised pea epicotyls. *Plant Physiol.* **45,** 133 (1970).

53. C. B. Osmond. Ion absorption and carbon metabolism in cells of higher plants. In U. Lüttge and M. G. Pitman, eds., *Encyclopedia of Plant Physiology, New Series,* Vol. 2A. Springer-Verlag, Berlin, 1976, p. 347.

54. W. D. Jeschke and W. Stelter. Measurement of longitudinal ion profiles in single roots of Hordeum and Atriplex by use of flameless atomic absorption spectroscopy. *Planta* **128,** 107 (1976).

55. D. M. R. Harvey, J. L. Hall, T. J. Flowers, and B. Kent. Quantitative ion localization within Suaeda maritima leaf mesophyll cells. *Planta* **151,** 555 (1981).

56. M. G. Pitman, A. Läuchli, and R. Stelzer. Ion distribution in roots of barley seedlings measured by electron probe X-ray microanalysis. *Plant Physiol.* **68,** 673 (1981).

57. C. Wiencke and A. Läuchli. Growth, cell volume, and fine structure of Porphyra umbilicalis in relation to osmotic tolerance. *Planta* **150,** 303 (1980).

58. R. Munns, H. Greenway, R. Delane, and J. Gibbs. Ion concentration and carbohydrate status of the elongating leaf tissues of Hordeum vulgare growing at high NaCl. *J. Exp. Bot.* **33,** 574 (1982).

59. U. Heber. Vergleichende Untersuchungen an Chloroplasten, die durch Isolierungsoperationen in nicht-wässrigem und wässrigem Milieu erhalten wurden. II. Kritik der Reinheit und Enzymlokalisation in Chloroplasten. *Z. Naturforsch.* **15b** 100 (1960).

60. A. W. D. Larkum and A. E. Hill. Ion and water transport in Limonium. V. The ionic status of chloroplasts in the leaf of Limonium vulgare in relation to the activity of the salt glands. *Biochim. Biophys. Acta* **203,** 133 (1970).

61. M. Maier and L. Kappen. Cellular compartmentalization of salt ions and protective agents with respect to freezing tolerance of leaves. *Oecologia* **38,** 303 (1979).

62. D. M. R. Harvey and T. J. Flowers. Determination of the sodium. potassium and chloride ion concentrations in the chloroplasts of the halophyte Suaeda maritima by non-aqueous cell fractionation. *Protoplasma* **97,** 337 (1978).

63. W. D. Jeschke. K⁺-Na⁺ selectivity in roots, localization of selective fluxes and their regulation. In E. Marre and O. Cifferi, eds., *Regulation of Cell Membrane Activities in Plants.* Elsevier, Amsterdam, 1977, p. 63.

64. R. Stelzer. Röntgenanalysen der Zellkompartimente von Wurzel- und Blattzellen von mit NaCl kultivierten Poaceen, Kurzfass. d. Poster, 152, Versamml. Dtsch. Bot. Ges., Freiburg, (1982).

65. W. S. Pierce and N. Higinbotham. Compartments and fluxes of K^+, Na^+ and Cl^- in Avena coleoptile cells. *Plant Physiol.* **46**, 666 (1970).

66. L. J. A. Neirinckx and G. G. J. Bange. Irreversal equilibration of barley roots with Na^+ ions at different external Na^+ concentrations. *Acta Bot. Neerl.* **20**, 481 (1971).

67. B. I. H. Scott, H. Galline, and C. K. Pallaghy. the electrochemcial state of cells of broad bean roots. I. Investigations of elongating roots of young seedlings. *Aust. J. Biol. Sci.* **21**, 185 (1968).

68. A. Läuchli, D. Kramer. M. G. Pitman, and U. Lüttge. Ultrastructure of xylem parenchyma cells of barley roots in relation to ion transport in the xylem. *Planta* **119**, 85 (1974).

69. M. G. Pitman. Sodium and potassium uptake by seedlings of Hordeum vulgare. *Aust. J. Biol. Sci.* **18**, 10 (1965).

70. J. A. Webb and P. R. Gorham. Translocation of photosynthetically assimilated C^{14} in straight-necked squash. *Plant Physiol.* **39**, 663 (1964).

71. K. Esau. *Pflanzenanatomie.* G. Fischer Verlag. Stuttgart, 1969.

72. E. A. C. MacRobbie. Phloem translocation. Facts and mechanisms. A comparative survey. *Biol. Rev.* **46**, 429 (1971).

73. R. Delane, H. Greenway, R. Munns, and J. Gibbs. Ion concentration and carbohydrate status of the elongating leaf tissues of Hordeum vulgare growing at high NaCl. *J. Exp. Bot.* **33**, 557 (1982).

74. M. G. Pitman. Transpiration and the selective uptake of potassium by barley seedlings (Hordeum vulgare c.v. Bolivia). *Aust. J. Biol. Sci.* **18**, 987 (1965).

75. T. G. Mason and E. J. Maskell. Further studies on transport in the cotton plant. I. Preliminary observations on the transport of phosphorus, potassium and calcium. *Ann. Bot.* **45**, 125 (1931).

76. A. R. Yeo. Salt tolerance in the halophyte Suaeda maritima L. Dnm.: Intracellular compartmentation of ions. *J. Exp. Bot.* **32**, 487 (1981).

77. M. G. Pitman. Whole plants. In D. A. Baker and J. L. Hall, eds., *Ion Transport in Plant Cells and Tissues.* North Holland, Amsterdam, 1975, p. 267.

78. N. Ahmad, R. G. Wyn Jones, and W. D. Jeschke. Effect of exogenous glycine betain and related compounds on salt tolerance and Na^+ transport in cereals. II. Excised roots in press. *Plant Physiol.*

79. R. F. M. Van Steveninck. Effect of hormones and related substances on ion transport. In U. Lüttge and M. G. Pitman, eds., *Encyclopedia of Plant Physiology, New Series,* Vol. 2b, Springer-Verlag, Berlin, 1976, p. 307.

80. R. F. M. Van Steveninck. Abscisic acid and stimulation of ion transport and alteration in K^+/Na^+ selectivity. *Z. Pflanzenphysiol.* **67**, 282 (1972).

81. J. L. Karmoker and R. F. M. Van Steveninck. Stimulation of volume flow and ion flux by abscisic acid in excised root systems of Phaseolus vulgaris L. c. v. Redland Pioneer. *Planta* **141**, 37 (1978).

82. J. L. Karmoker and R. F. M. Van Steveninck. The effect of abscisic acid on the uptake and distribution of ions in intact seedlings of Phaseolus vulgaris c. v. Redland Pioneer. *Physiol. Plant* **45**, 453 (1979).

83. R. Behl and W. D. Jeschke. On the action of abscisic acid on transport,

accumulation and uptake of K$^+$ and Na$^+$ in excised barley roots; effects of the accompanying anions. *A. Pflanzenphysiol.* **95**, 335 (1979).

84. R. Behl and W. D. Jeschke. Influence of abscisic acid on unidirectional fluxes and intracellular compartmentation of K$^+$ and Na$^+$ in excised barley root segments. *Physiol. Plant.* **53**, 95 (1981).

85. W. D. Jeschke. Shoot-dependent regulation of sodium and potassium fluxes in roots of whole barley seedlings. *J. Exp. Bot.* **33**, 601 (1982).

86. Y. Mizrahi, A. Blumenfeld, A. E. Richmond. Abscisic acid and transpiration in leaves in relation to osmotic root stress. *Plant. Physiol.* **46**, 169 (1970).

87. G. W. Roeb, J. Wieneke, and F. Führ. Auswirkungen hoher NaCl-Konzentrationen im Nährmedium auf die Transpiration, den Abscisinsäure-, Cytokinin- und Prolingehalt zweier Sojabohnensorten. *Z. Pflanzenernähr. Bodenk.* **145**, 103 (1982).

88. D. W. Rush and E. Epstein. Genotypic responses to salinity: Differences between salt-sensitive and salt-tolerant genotypes of the tomato. *Plant Physiol.* **57**, 162 (1976).

89. M. G. Pitman. Uptake of potassium and sodium by seedlings of Sinapis alba. *Aust. J. Biol. Sci.* **19**, 257 (1966).

90. A. Poljakoff-Mayber. Morphological and anatomical changes in plants as a response to salinity stress. In A. Poljakoff-Mayber and J. Gale, eds. *Plants in Saline Environments.* Springer-Verlag, Berlin, 1975, p. 97.

91. E. Steudle and W. D. Jeschke. Water transport in barley roots. Measurement of root pressure and hydraulic conductivity of roots in parallel with turgor and hydraulic conductivity of root cells. *Planta* **158**, 237 (1983).

92. H. Greenway. Plant responses to saline substrates. VII. Growth and ion uptake throughout plant development in two varieties of Hordeum vulgare. *Aust. J. Biol. Sci.* **18**, 763 (1965).

93. K. Dörffling. Growth. *Fortschr. 11 Botanik* **42**, 111 (1960).

94. F. Malek and D. A. Baker. Proton co-transport of sugars in phloem loading. *Planta* **135**, 297 (1977).

95. K. Mengel. Effect of potassium on the assimilate conduction to storage tissues. *Ber. Deutsch. Bot. Ges.* **93**, 353 (1980).

96. W. D. Jeschke. K$^+$-Na$^+$ exchange and selectivity in barley root cells: Effect of Na$^+$ on the Na$^+$ fluxes. *J. Exp. Bot.* **28**, 1289 (1977).

97. A. Läuchli and R. Pflüger. Potassium transport through plant cell membranes and metabolic role of potassium in plants. In *Potassium Research—Review and Trends,* Proc. 11th Congr. Int. Potash Inst. Bern, 1978, p. 111.

4

EVIDENCE FOR TOXICITY
EFFECTS OF SALT
ON MEMBRANES

A. C. Leopold and R. P. Willing

Boyce Thompson Institute
Cornell University
Ithaca, New York

There are at least three components of salt stress to plants, including osmotic effects, nutritional effects, and a more vague component of toxic effects. Alterations of the osmotic relations between the plant tissues and the soil solution are a part of salt stress, as well as alterations in the osmotic pools inside the plant. Nutritional alterations due to salt stress may represent changes in the ability of the plant to process needed ions in the presence of large amounts of soluble salts. As noted by Levitt (1), the primary toxic lesion produced by salt is not known.

Our experiments on the induction of leakiness of membranes by salt were undertaken in an effort to seek a toxic effect that may be involved in salt stress. Here we consider the possibility that salt-induced lesions are the result of the interaction of salt ions with the biological macromolecules associated with membranes.

1. SALT-INDUCED LEAKINESS

Experiments on solute leakage were carried out using 7-mm discs of soybean cotyledonary leaves. The leakage of organic solutes from the tissue was followed as a time course, using a peristaltic pump that continuously recycled the ambient solution past a UV monitor that recorded the solute concentration as absorbance at 254 μm. Leakage rates were determined from the tracing of the time course using a computer-generated equation of best fit for the slope of the leakage curve between 5 and 20 min.

When soybean leaf discs are placed in a solution of NaCl at concentrations ranging from 20 to 200 mM, it is observed that the leakage rate from the tissue is increased in a manner generally proportional to the concentration of the salt (Fig. 4.1). We interpret these results as suggesting that membranes of the leaf tissue become increasingly leaky as the concentration of NaCl is increased.

In previous studies of stress effects on membrane lesions, we have found that the extent of damage to membranes is often exaggerated under conditions in which the membranes are forced to expand. For example, drought stress effects on membrane leakiness were greatly increased when the stressed tissue was made to expand (2). Similarly, low temperature stress effects on membrane leakiness were greatly amplified when the tissue was forced to expand (3). As a probe of the possible effects of NaCl on the ability of membranes to withstand osmotic expansion, tissues were placed in 200 mM NaCl, and solute leakage from the tissues was monitored as they recovered osmotic equilibrium in water. The time course for leakage during deplasmolysis is shown in Figure 4.2 along with the time course of leakage while immersed in the NaCl. It is evident that the leakage rate was about fourfold greater during the deplasmolysis, even though the salt was no longer present. We interpret this result as indicating that salt causes a condition of greater membrane leakiness, and that this condition was greatly exacerbated under conditions of osmotic expansion.

A difficulty with many studies of tissue lesions induced by salt is the possible confusion of salt toxicity with osmotic effects (1). In order to compare the leakage effects of NaCl with equivalent levels of another osmoticum, we have used sorbitol at 400 mM, which has approximately the same osmotic value as

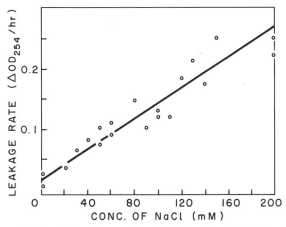

Figure 4.1.   Rates of solute leakage from soybean leaf discs placed in NaCl solutions of various concentrations. Ten leaf discs in 10 ml water; leakage expressed as tangent of the time course at 5 min.

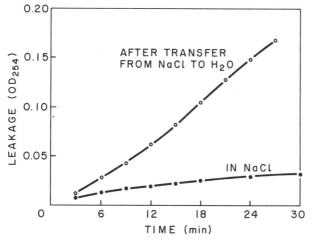

Figure 4.2. A comparison of leakage time course curves for leaf discs either in 200 mM NaCl or after transfer from 200 mM NaCl into water. Ten leaf discs in 10 ml water.

NaCl at 200 mM. In Figure 4.3, we report the time course of leakage from leaf discs upon transfer to water from 200 mM NaCl as compared with transfer from 400 mM sorbitol. One can see that the prior exposure of the tissue to NaCl powerfully potentiated the induction of leakage during deplasmolysis. Even 50 mM NaCl substantially increased leakage when the residual osmotic value had been supplied as sorbitol (300 mM). We conclude that the susceptibility of membranes to damage was specifically expanded by the salt, and was not due solely to osmotic effects.

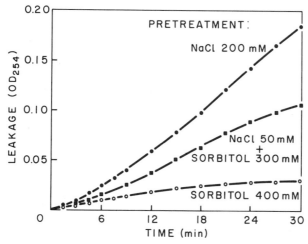

Figure 4.3. A comparison of leakage from leaf discs transferred into water from comparable osmotic pretreatments including NaCl, sorbitol, or both.

2.   COMPARATIVE EFFECTS OF DIFFERENT SALTS

As a first comparison of leakage induced by NaCl with other salts, NaCl, CaCl$_2$, and sorbitol were utilized at equivalent osmolarity. As shown in Figure 4.4, deplasmolysis after NaCl exposure was much more damaging to the soybean tissue discs than after sorbitol, and approximately an equiosmolar concentration of CaCl$_2$ (130 mM) resulted in no increase in leakage rates. It is evident, then, that there is a real salt specificity to the membrane damage done by the osmotic shock treatment.

A wider comparison of salts was therefore attempted. We exposed leaf discs to equivalent osmotic concentrations of eight salts, then returned them to water and compared the amount of solute leakage after 30 min (Fig. 4.5). Although the differences between monovalent cations are small, the data suggest that the greatest leakage was obtained following exposure to NH$_4$Cl, with smaller effects for Na, K, and Li chlorides. Among the divalent cations tested (using 133 mM each), MgCl$_2$ caused a small amount of leakage, and CaCl$_2$ caused almost none. A comparison of the Cl$^-$ with the SO$_4^{2-}$ anion can be seen in these data, with Na$_2$SO$_4$ causing slightly less leakage than NaCl, and MgSO$_4$ causing less leakage than MgCl$_2$.

From these preliminary experiments, we suggest that several monovalent ions can damage membranes. The data also show that ammonium ions were most effective in inducing lesions whereas calcium and magnesium ions were the least effective among the salts tested here. The data also indicate the anions can contribute to membrane damage, with Cl$^-$ being more damaging than SO$_4^{2-}$.

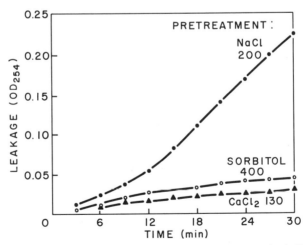

Figure 4.4. A comparison of leakage from leaf discs after pretreatment in similar osmotic concentrations of NaCl, sorbitol, or CaCl$_2$.

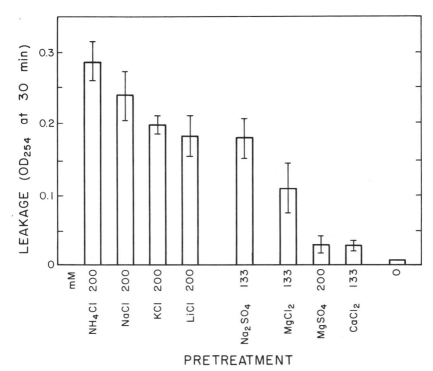

Figure 4.5.   Leakage rates from leaf discs after pretreatment in similar osmotic concentrations of eight different salts. Leakage into water was measured after 30 min.

3.   PROTECTIVE EFFECTS OF CALCIUM

A distinctive characteristic of ionic effects in the lyotropic series is that ions from one end of the series can relieve the lyotropic effects of ions from the other end of the series (4). For example, the leakage of solutes from plant tissues induced by $(NH_4)_2SO_4$ can be relieved by the addition of $CaCl_2$ (5). If the membrane damage being examined here is related to lyotropic effects of the salts, one might expect the NaCl damage to be relieved in part by the inclusion of calcium in the challenging solution. In order to examine this possibility, leaf discs were given 200 mM NaCl, with additional amounts of $CaCl_2$ ranging from 1 to 100 mM. The relative leakage of solutes from these various treatments after transfer to water is shown in Figure 4.6; in two different experiments it is evident that $CaCl_2$ between 1 and 50 mM served partially to protect the tissue from the NaCl damage. There was a tendency toward increased damage again when the $CaCl_2$ concentration was elevated to 100 mM.

A further experiment was carried out in which the osmotic value of the stress

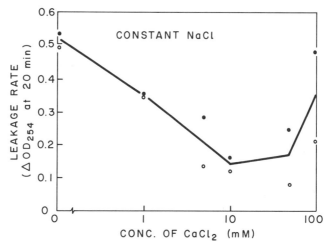

Figure 4.6.   Leakage rates from leaf discs pretreated with 200 mM NaCl plus varying amounts of CaCl$_2$. Leakage into water was measured after 20 min.

treatment was kept constant, and CaCl$_2$ was substituted for various amounts of NaCl. The data in Figure 4.7 indicate that at equal osmolarities, the substitution of Ca for Na resulted in an impressive protection effect, with about 50% protection being obtained with 3 mM CaCl$_2$, and almost complete protection with 33 mM CaCl$_2$.

From these experiments it can be concluded that CaCl$_2$ can protect leaf membranes against the leakiness induced by NaCl.

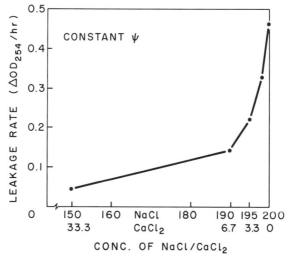

Figure 4.7.   Leakage rates from leaf discs pretreated with equal osmotic concentrations of salts, with various levels of CaCl$_2$ replacing NaCl so as to maintain a constant water potential in the pretreatment.

4. SALT INTERFERENCE WITH PHOTOSYNTHESIS

In view of the evidences that salt can cause lesions in membranes, one might expect salt to have detrimental effects on a membrane-organized function such as photosynthesis. Previous reports have indicated that photosynthetic mechanisms can be damaged by NaCl (6).

Effects of salt on photosynthesis were examined in four different modes. Following the methods of Heber and Santarius (7) and Alscher-Herman and Strick (8), protoplasts were isolated from pea seedlings and the effects of 250 mM NaCl on CO_2-dependent oxygen production were measured. Three additional modes were utilized with chloroplast preparations made from the protoplasts, including the CO_2-dependent O_2 production, uncoupled electron transport (O_2 production in the presence of NH_4Cl uncoupler plus ferricyanide), and coupled electron transport (in the presence of ferricyanide).

The results of some preliminary experiments on these aspects of photosynthesis are presented in Table 4.1. The data suggest that NaCl at 250 mM is not particularly inhibitory of photosynthesis in whole protoplasts or in uncoupled electron transport, but is severely inhibitory of CO_2-dependent O_2 production and of coupled electron transport in isolated pea chloroplasts. These data suggest that NaCl has particularly damaging effects on the photophosphorylation component of photosynthesis, which is structurally located on the thylakoid membranes. The lesser effect of NaCl on uncoupled electron transport may indicate that the photosystem II may be relatively less susceptible to NaCl damage.

Although these results are only preliminary, they do suggest that NaCl may produce structural damage to membrane-bound components of the photosynthetic system, which is consistent with the concept of a salt toxic reaction involving lesions to membranes.

5. DISCUSSION

The experiments presented here indicate that NaCl at concentrations commonly encountered in plant tissues under salt stress, causes a pronounced leakage of solutes from soybean leaf tissue. Similar increases in leakage of solutes from roots have been observed by Sacher (personal communication) and by Wainwright (9). The data also indicate that the inferred alterations of membranes by the salt are markedly amplified under conditions of cellular expansion; a similar amplification of membrane damage has been reported for both drought-induced membrane lesions and chilling-induced membrane lesions (2, 3). We suggest the interpretation that NaCl causes a membrane damage, which may be considered a basic toxic component of salt injury. Our experiments indicate that there are small but significant differences between leakage rates produced by monovalent cation salts, and marked differences

Table 4.1.   Effects of NaCl on Photosynthetic Effectiveness of Pea Protoplasts and Chloroplasts

Experimental Material	Photosynthetic Rate (μM O_2/mg chlorophyll/hr)	
	Basal Medium	+ 250 mM NaCl
Protoplasts	52.9	47.8
Chloroplasts		
CO_2-dependent O_2 production	35.6	6.6
Uncoupled electron transport	265.6	240.3
Coupled electron transport	392.0	68.3

between monovalent and divalent cations, as well as between anions, and that the damaging effects of NaCl to membranes can be relieved by calcium.

Preliminary results indicate that NaCl may interfere with photosynthetic effectiveness of chloroplasts, presumably by interfering with photophosphorylation. Effects of high concentrations of NaCl inhibiting photophosphorylation have been previously reported by Muller and Santarius (10). Such photosynthetic inhibitions are likewise consistent with the concept of a general damaging or toxic effect of NaCl on membranes.

Field experiments on salt stress have indicated a common relief of NaCl damage by calcium salts (1, 11). Our studies suggest that such a nutrient interaction may have at least a partial basis in the interactions of NaCl and CaCl$_2$ at the cellular level.

Among possible interpretations of the damage to membranes by salt would be a role of salt in altering lyotropic effects. Consistent with this possibility is the evidence that ammonium salts may be more damaging than sodium, and that the effects of the monovalent sodium salt could be relieved by calcium, since sodium is near the middle of the lyotropic series and calcium is a destabilizing salt (4). A possible involvement of lyotropic effects in salt damage has been suggested by Pollard and Wyn Jones (12). Another possibility is that the monovalent sodium may serve to weaken membrane structure by displacing one or more of the divalent bridges provided by calcium or another divalent cation. Evidence has been published to indicate that calcium may serve to bind phospholipids together and thus limit membrane permeability (13), and the ability of calcium to modify the structural organization of proteins in solution is well documented (4). If sodium were able to displace calcium ions from such double-valenced bridge positions, it would be expected to weaken the extent to which the bridge units were held together, and in the case of membrane components, to result in greater permeability.

The suggestion that salt toxicity may be a consequence of membrane damage is consistent with some symptoms of salt damage. For example, damage to the photosynthetic system and the uncoupling of cyclic phosphorylation (14) may well reflect damage to chloroplast membranes. Salt damage to respiratory systems (12, 15, 16) may reflect salt damage to mitochondrial membranes; the increase in alternative respiration with salt stress (17) would be consistent with other types of stress in which membrane damage is reportedly associated with larger proportions of alternative respiratory activity (18, 19).

In conclusion, we suggest from these experiments that a basic toxicity reaction to salt may be the production of lesions in membranes generally. Lesions in the plasmalemma would result in general leakage of solutes from the cells. A disturbance of the membranes of chloroplasts might account for observed damage to photosynthetic effectiveness with salt stress; similar disturbance of mitochondria might account for observed alteration of respiratory effectiveness with salt stress. We feel that the toxic component of salt stress may involve salt-induced damage to membranes.

REFERENCES

1. J. Levitt. *Responses of Plants to Environmental Stresses.* Academic Press, New York, 1972, 697 pp.

2. A. C. Leopold, M. E. Musgrave, and K. M. Williams. Solute leakage resulting from leaf desiccation. *Plant Physiol.* **68**, 1222–1225 (1981).

3. R. P. Willing and A. C. Leopold. Cellular expansion at low temperature as a cause of membrane lesions. *Plant Physiol.* **71**, 118–121 (1982).

4. P. H. von Hippel and T. Schleich. The effects of neutral salts on the structure and conformational stability of macromolecules in solution. In S. N. Timashef and G. D. Fasman, eds., *Structure and Stability of Biological Macromolecules.* Dekker, New York, 1969, pp. 417–573.

5. B. W. Poovaiah and A. C. Leopold. Effects of inorganic salts on tissue permeability. *Plant Physiol.* **58**, 182–185 (1976).

6. J. S. Boyer. Effects of osmotic water stress on metabolic rates of cotton plants. *Plant Physiol.* **40**, 229–234 (1965).

7. U. Heber and K. A. Santarius. Direct and indirect transfer of ATP and ADP across the chloroplast envelope. *Z. Naturforsch.* **B25**, 718–728 (1970).

8. R. Alscher and C. Strick. Diphenyl-ether-chloroplast interactions. *Plant Physiol.* submitted (1983).

9. S. J. Wainwright. Plant in relation to salinity. *Adv. Bot. Res.* **8**, 221–261 (1980).

10. M. Muller and K. A. Santarius. Changes in chloroplast membrane lipids during adaptation of barley to salinity. *Plant Physiol.* **62**, 326–329 (1978).

11. S. Z. Hyder and H. Greenway. Effects of Ca^{++} on plant sensitivity to high NaCl concentrations. *Plant Soil* **23**, 258–260 (1965).

12. A. Pollard and R. G. Wyn Jones. Enzyme activities in concentrated solutions of glycinebetaine and other solutes. *Planta* **144**, 291–298 (1979).

13. C. M. Gary-Bobo. Effect of Ca^{++} on the water and non-electrolyte permeability of phospholipid membranes. *Nature* **228**, 1101–1102 (1970).

14. D. W. Krogmann, A. J. Jagendorf, and A. Avron. Uncouplers of spinach chloroplast photosynthetic phosphorylation. *Plant Physiol.* **54,** 272–277 (1959).

15. T. J. Flowers. Salt tolerance in Suaeda maritima, a comparison of mitochondria from Suaeda and Pisum. *J. Exp. Bot.* **101,** 101–110 (1974).

16. A. Livne and N. Levin. Tissue respiration and mitochondrial phosphorylation of NaCl treated pea. *Plant Physiol.* **42,** 407–414 (1967).

17. I. M. Moller. Balance between cyanide-sensitive and insensitive respiration influenced by salt concentration. *Physiol. Plant.* **42,** 157–162 (1978).

18. H. Lambers and G. Smakman. Respiration of the roots of flood-tolerant and intolerant Senecio species: Affinity for oxygen and resistance to cyanide. *Physiol. Plant.* **42,** 163–166 (1978).

19. A. C. Leopold and M. E. Musgrave. Respiratory changes with chilling injury in soybeans. *Plant Physiol.* **64,** 702–705 (1979).

5

FUNCTIONING OF PLANT CELL MEMBRANES UNDER SALINE CONDITIONS: MEMBRANE LIPID COMPOSITION AND ATPases

Pieter J. C. Kuiper

Department of Plant Physiology
Biological Centre
University of Groningen
The Netherlands

1. WHY PERFORM A STUDY OF PLANT CELL MEMBRANES IN PLANTS UNDER SALINE STRESS?

Salinity exerts several effects on plant metabolism. Large quantities of NaCl in the root environment will interfere with the nutritional status of the plant. Uptake and translocation of nutritional ions such as K^+ and Ca^{2+} are greatly reduced by salinity stress and the salt tolerance of the plants will be reduced under conditions of low nutrient availability (1). Addition of extra Ca^{2+} to the root environment stimulates growth of maize (2) and barley (3) in combination with improved K^+ uptake and translocation. On the other hand, translocation of a nutritional ion such as phosphate from its pool does not limit plant growth under salt stress (4).

Plant cells often react upon a salt stress by increased vesiculation of the plasmamembrane (5, 6) and by increased amounts of rough endoplasmic reticulum (7). All these observations point to the importance of plant cell membranes in regulation of internal ion fluxes under conditions of salt stress in order to maintain a favorable ionic composition in plant cells and plant tissues.

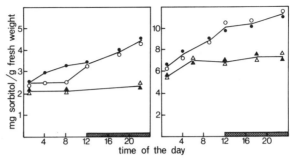

Figure 5.1. Sorbitol content of roots (left) and shoots (right) of salt-sensitive *Plantago media* and salt-tolerant *P. maritima* as affected by salinity. ▲, *P. media*, control; △, *P. maritima*, control; ●, *P. media*, 100 m*M* NaCl, added at time 0; ○, *P. maritima*, 100 m*M* NaCl.

A second consequence of salinity stress is the effect of the high ion concentration. Plants respond to high salt concentrations by synthesis of organic molecules such as proline, betaine, and polyalcohols in order to maintain an osmotic balance between the plant cell cytoplasm and its environment (8).

Plants differ in their capacity to retain the osmoregulator within the cell. When salt-sensitive *Saccharomyces cerevisiae* cells are exposed to NaCl, the yeast will lose considerable quantities of glycerol to the medium, in contrast to the drought-tolerant *S. rouxii* (9) or the salt-tolerant *Debaryomyces hansenii* (10).

A similar example can be given for higher plants. Both the halophyte *Plantago maritima* and the salt-sensitive but drought-resistant *Plantago media* react to salinity by synthesis of sorbitol as an osmoticum (Fig. 5.1) in roots and shoots. Nevertheless, in contrast to *P. maritima*, *P. media* plants will die within a short time due to plant cell membrane breakdown (1).

The present chapter will deal with two aspects of plant cell membranes in plants under salinity stress. First, composition of the lipid matrix and lipid metabolism will be discussed in relation to maintenance of ionic composition and concentration of organic osmoregulators within the plant cell cytoplasm. Second, the role and function of plant cell ATPases will be discussed as a controlling factor in salinity. For practical reasons attention will also be given to genetic differentiation and physiological adaptability of plants to salinity stress.

2. SALINITY AND COMPOSITION AND STRUCTURE OF THE LIPID MATRIX OF PLANT CELL MEMBRANES

In general, plant cell membranes are organized as a bimolecular lipid layer in which protein molecules are partially or more fully imbedded, a description also known as Singer's model. Structure and permeability characteristics of the lipid matrix are important factors in determining ionic composition and organic osmoregulator concentration of the cell cytoplasm under saline conditions. In addition, ions and organic solutes may permeate membranes because of the presence of proteins in the membrane which extend through the lipid matrix. In this connection the ion-stimulated membrane ATPases that are supposed to function as ion pumps (11) are of particular interest.

Structure and permeability properties of bimolecular lipid layers strongly depend on the lipid composition. For this reason properties of the various groups of lipid species will be discussed and relevant literature on the role of such lipid species in plant cell membranes will be taken into consideration.

2.1. Sterols and Sterol Metabolism

Free sterols are characteristic for plasma membrane and tonoplast. They are effective in regulation of membrane stability and ionic permeability (12–19). Sterol esters are ineffective in this respect (20, 21). They are found in mitochondria and the nucleus (22).

As an example, free cholesterol added to phosphatidylcholine bilayers will decrease the K^+-permeability relative to Na^+, thus contributing to a higher K^+/Na^+ ratio inside the liposomes (23). Also, the nonelectrolyte permeability of biomembranes is decreased by added sterol. De Kruyff et al. (24) observed that passive diffusion of erythritol into *Acholeplasma laidlawii* is decreased with increased content of 3-hydroxysterol.

Literature indicates that a high level of free sterols in plant tissues may contribute to salt tolerance. In grape varieties differing in salt resistance the level of free sterols was lowest in the most sensitive variety (25; Table 5.1). The same observation was made in a comparative study on bean, barley, and sugar beet (26). The sterol/phospholipid ratio is higher in the salt-resistant *Debaryomyces hansenii* than in the nontolerant *Saccharomyces cerevisiae* (10). This parameter should be handled carefully since sterol/phospholipid ratios higher than one, which may be observed in senescing tissue (27), are associated with a loss of regulation of ion permeability in membranes due to formation of sterol clusters in the lipid matrix.

Table 5.1. Lipid Composition of the Roots of Five Grape Rootstocks[a]

Variety	Cl^- Accumulation Ratio	Sterols	Phosphatidyl-choline	Monogalactosyl Diglyceride
SC	0.06	19.4	20.8	9.5
1613.3	0.11	15.6	14.6	22.4
DR	0.28	20.7	7.0	23.0
TS	0.39	22.4	3.7	33.9
Cardinal	0.86	8.7	6.0	39.0

[a]From ref. 25. Data are expressed as % of total lipids. The data on Cl^- accumulation ratio are expressed as (meq Cl^-/100 g dry wt. of leaves)/(meq Cl^-/liter in the root solution).

The level of free sterols and of sterol esters in plant tissue often is affected by salinity. In the above-mentioned yeast species it is increased upon exposure of the cells to salinity. The level of free sterols in the roots of the halophytic *Plantago maritima* and *P. coronopus* was maintained upon exposure to salinity whereas it was decreased in the salt-sensitive *P. media* (28). The level of sterol esters in the roots of the halophytic *Plantago* species was even increased upon exposure to salinity, an observation also made with sugar beet (29). It remains to be investigated whether the elevated sterol ester level has anything to do with the number of functioning of mitochondria in salt-tolerant plants under saline conditions. Plants of *P. maritima* and *P. major* spp. *major* which are exposed to a low nutritional ion level in the root environment show a depressed level of sterol esters in the roots together with an unchanged level of free sterols, in agreement with the above-mentioned function of free sterols for maintenance of membrane stability. The decrease in sterol ester levels of the roots under these conditions may point again to metabolic disturbance of the roots (30).

Sterol species differ in degree of control of permeability of plant cell membranes. Grunwald (18) showed that leakage of betacyanin from red beet slices, induced by methanol, is reduced by added sterols, with cholesterol and stigmasterol being most effective and sitosterol being least effective. In *P. maritima* roots the sitosterol level is reduced with increasing salinity, indicating that cholesterol may become more effective in regulation of ion permeability (28). In *Citrus* rootstocks, the sitosterol/stigmasterol ratio is reduced under saline conditions. In the studied varieties this salt-induced change correlated well with salt exclusion capacity (31). Moreover, in experiments with *Plantago* grown at different levels of mineral nutrition it was observed that in plants of a low nutrient status the roots were much higher in cholesterol in combination with a lowered level of sitosterol (30). Clearly, regulation of sterol metabolism is a factor in regulation of ionic permeability of plant roots, since changes in the level of ions in the root environment were matched with corresponding changes in the levels of cholesterol and sitosterol of the roots.

2.2.   Phospholipid Composition and Phospholipid Metabolism

Several phospholipids like phosphatidylcholine, saturated phosphatidylethanolamine, and phosphatidylglycerol form bilayer structures with a high degree of control of semipermeability because of a favorable balance between the size of the polar groups and the size of the hydrophobic parts of the molecules (32). Cardiolipin (in the presence of Ca^{2+}), unsaturated phosphatidylethanolamine, monoglucosyl diglyceride (32), and chloroplast monogalactose diglyceride (33) are organized as hexagonal packed cylinders, with the polar heads of the molecules directed inside. Lipid molecules with a head group dominating over the hydrophobic part of the molecule, for example, lysophospholipids, tend to be organized as globular micelles with the polar head groups facing outward.

Such structural differences greatly affect permeability characteristics of the lipid matrix. It is not surprising that a lipid-like monogalactose diglyceride because of the hexagonal (H II) structure allows a far greater transport of Cl⁻ across the lipid membrane than phosphatidylcholine and (saturated) phosphatidylethanolamine (34).

In addition, fatty acid composition of phospholipids affects the permeability barrier. Permeability of phospholipid membranes for polyols like glycerol decreased with decreasing unsaturation of phospholipids (19, 24, 35). Increasing unsaturation of phosphatidylcholine causes a significant increase of passive K^+ permeability together with a much smaller increase in Na^+ permeability (23).

Literature indicates that the above-mentioned phospholipid permeability parameters have physiological meaning in salinity responses of several plant species. As in Section 2.1, a distinction will be made between comparative studies on species, varieties, and genotypes (genetic differentiation) and plant phospholipid responses to salinity (salt-induced changes in phenotypic plasticity).

In a comparison of five grape rootstocks differing in Cl⁻ transport to the shoot, it has been found that the level of phosphatidylcholine (Table 5.1) and of saturated phosphatidylethanolamine in the roots of the rootstocks was inversely related to Cl⁻ transport to the shoot. In addition, it has been observed that considerable quantities of extra-long chain fatty acids like behenic acid and lignoceric acid (22 and 24 C atoms, respectively) are exclusively found in the phosphatidylcholine and phosphatidylethanolamine fractions (25). These observations point to the importance of a strict bilayer structure in prevention of excess Cl⁻ transport from the root to the shoot. In agreement with this conclusion is the observation that in sugar beet inbred lines (29) the phospholipid level of the shoot of the salt-sensitive genotype also was lower than that of the salt-resistant genotype (29, nonsaline conditions). However, in a comparative study of bean, barley, and sugar beet no consistent correlation between phospholipids and salt resistance is visible (26).

The level of phospholipids in plant tissue often is affected by salinity. The phospholipid level (and specifically phosphatidylcholine) on a total lipid basis is increased by NaCl in barley (36) and wheat (36, 37). On a dry weight basis the phospholipid level of the roots of halophytic *Plantago* species (28) and the salt-tolerant sugar beet line FIA (29) is increased upon exposure to salinity. In these species (genotypes) the phospholipid level of the shoot is maintained under saline conditions even though less phosphate is absorbed by the roots (38) and further transported to the shoot (39). At higher salinity levels and in salt-sensitive species (28, 40) a decrease in phospholipids in roots and shoots is observed. This decrease, however, is quantitatively similar to the NaCl induced decrease in total lipids (Fig. 5.2). Evidently the decrease of phospholipid content (per g dry matter) is not due to decreased synthesis or to breakdown of phospholipids, but may be caused by intracellular changes in the cell upon exposure to salt stress (e.g., production of more cell wall material or more

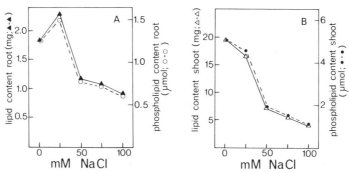

Figure 5.2.   Content of total lipids and phospholipids of the roots (A) and shoots (B) of *Plantago media* as affected by salinity. After ref. 38.

starch) and by accumulation of NaCl in the cell (40). Also, the phospholipid composition of roots and shoots of salt-tolerant as well as salt-sensitive *Plantago* species remains unchanged by long-term exposure to NaCl, indicating that salt stress does not lead to "adaptive" changes in phospholipid composition in *Plantago* species. This is in contrast to adaptive changes in phospholipid composition observed in plants upon exposure to low-temperature stress (41).

The contrasting data on *Plantago* and barley regarding the effect of NaCl on phospholipid level (per mg total lipid) and on phospholipid composition may relate to the possible role of phosphatides in glycine–betaine synthesis. *Plantago* produces sorbitol (42) as an osmoregulator under salinity stress, and only traces of betaine are observed. Barley produces betaine via the serine–ethanol-amine–choline–betaine aldehyde–methylation pathway (43) and part of the phosphatides (20%) act as intermediates (44, 45). Thus, elevated levels of phosphatidylcholine in barley plants under salinity stress may be caused by betaine synthesis via the phospholipid pathway. However, it should be noted that the phospholipid pathway is not the only way for betaine synthesis, for example, in spinach the involvement of phospholipids in betaine synthesis is very minor (46). Involvement of water-soluble phosphorylglycero intermediates in betaine synthesis seems more logical, considering the complications of NaCl induced phospholipid turnover.

In this connection it is of interest that choline application to the roots leads to phosphatidylcholine accumulation in leaves of wheat (47). At the same time the degree of unsaturation of phosphatidylcholine and other phospholipids is increased, thus compensating the fluidizing effect of the choline head group (48). Addition of ^{32}P-phosphoryl glycerocholine to the roots of bean results in elevated levels of ^{32}P-lipids in roots and leaves together with a reduced transport of Cl$^-$ to the shoot under saline conditions (49).

In salt-sensitive yeast cells the degree of unsaturation of phospholipids is decreased upon exposure to salinity, a factor that may decrease permeability of glycerol and ions such as K$^+$. A similar decrease in unsaturation of lipids as a response to NaCl stress is noted in sugar beet (29) and thus may contribute to a more favorable K^+/Na^+ balance in roots and shoots (80, 84). The opposite tendency is observed in the growing axes of groundnut seedlings, which react to salinity by an elevated level of linoleic acid. Possibly an increased permeability for K^+ may facilitate K^+ transport from the cotyledons to the young embryo under salinity stress (50).

In summary, phospholipid composition is not much affected by exposure of the plants to salinity. This conclusion is surprising, since phospholipid biosynthesis is greatly influenced by divalent cations (51). Ca^{2+}, for example, already at 1 mM concentration, inhibits synthesis of phosphatidylcholine and phosphatidylethanolamine by the phosphotransferases (52). The NaCl-induced decrease in transport of this ion to the shoot (1) as well as the growth stimulation of salt-stressed plants by this ion (2, 3) could indicate an effect via phospholipid biosynthesis. Because this is not the case, the above observations point to a high degree of regulation of phospholipid biosynthesis and breakdown in plants under salinity stress.

2.3. Glycolipid Composition and Glycolipid Metabolism

In the previous section it was mentioned that the hexagonal (H II) structure of glycolipids does not contribute to a high degree of regulation of ionic permeability. Numerous data in the literature indicate, indeed, that a high level of monogalactose diglyceride and digalactose diglyceride in plant tissue is correlated with sensitivity of the plants to saline conditions. In five grape rootstocks of different salt sensitivity the monogalactose diglyceride content of the roots is directly related with Cl^- transport of the rootstock to the shoot (Table 5.1). Addition of monogalactose diglyceride and digalactose diglyceride to the roots of bean plants leads to elevated levels of these lipids in roots and shoots together with an increased uptake and transport of Cl^- from the saline nutrient solution. As expected, the more surface-active digalactose diglyceride is more effective in this respect.

In contrast to phospholipids the level of glycolipids is often affected by salinity. In salt-sensitive species/genotypes and in salt-tolerant species/genotypes under more extreme saline conditions, the glycolipid level of roots and shoots decreased with salinity stress: barley (53), sugar beet (29), sunflower (54, 55), alfalfa (56), and *Plantago* (28). Synthesis of chloroplast galactolipids is reduced by inhibition of the envelope enzymes galactosyltransferase and -acylase (53). If one accepts the above statement regarding the hexagonal structure of monogalactose diglyceride, a reduced level of monogalactosyl diglyceride would indicate a higher degree of control of ionic permeability in the chloroplasts. In sunflower the level of linolenic acid, being the principal fatty acid of chloroplast galactolipids, is decreased by NaCl, indicating that linoleic acid desaturase is inhibited by salt stress. In sugar beet genetic differences in the effect of NaCl on leaf glycolipid level and degree of unsaturation of leaf fatty

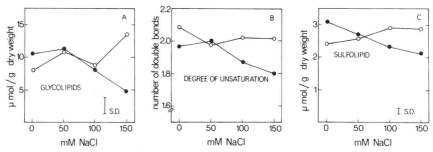

Figure 5.3. Content of glycolipids (shoots, A), degree of unsaturation of fatty acids (shoots, B), and content of sulfolipid (roots, C) of the salt-sensitive genotype of sugar beet, ADA (●), and the salt-tolerant genotype, FIA (○), as affected by salinity. After ref. 28.

acids are evident (Fig. 5.3). Possibly senescence of chloroplasts is accelerated by NaCl in salt-sensitive species/genotypes. It is also of interest in this connection that a rather salt-tolerant species such as sugar beet has more linoleic acid than linolenic acid in the leaves, even though the crop is also well adapted to low temperature.

Sulfolipids seem to have some connection with adaptation of plants to salinity. It is therefore very disappointing that so little is known about plant sulfolipids. Knowledge of biosynthesis of sulfoquinovosyl diglyceride, the chloroplast sulfolipid, is very limited (57) and its physical properties (bilayer structure or hexagonal structure) have not been studied. In roots a much lower sulfolipid level (in proplastids?) has been observed than in the shoots. It is unknown whether other sulfolipids such as sterol sulfate and phosphatidyl sulfocholine, which have been identified in the marine diatom *Nitschia alba* (58, 59), also occur in higher plants (halophytes). Halophytic *Plantago* species have a higher sulfolipid level in the roots than the other species (30). In sugar beet genetic differentiation of the effect of NaCl on the sulfolipid level of the roots is evident: sulfolipid level of the roots increases with NaCl in the salt-resistant genotype and it decreases in the salt-sensitive genotype (Fig. 5.3). Such effects warrant further research since $(Na^+ + K^+)$-stimulated ATPases in sugar beet roots (60, 61) and in the salt gland of the ducks (62) seem to require sulfolipid (sugar beet) or sulfatides (duck) for proper functioning.

2.4. Structure of Biomembranes and Salinity

Research on structure of plant cell membranes in plants under saline conditions is very limited. Membrane configuration may be involved in salt resistance of plants, since mycostatin, which reacts with sterols, induces K^+ efflux of bean leaves (salt sensitive), but not in sugar beet (salt tolerant; 63). Possibly the sterol molecules in sugar beet cell membranes are inaccessible for this compound.

In low temperature research on plant cell membranes much attention is given to results obtained by physical methods (electron spin resonance, nuclear magnetic resonance, differential scanning calorimetry, and Arrhenius plots). It is surprising that this type of data is completely absent in salinity research on plant cell membranes.

3. ION-STIMULATED ATPases IN HIGHER PLANTS UNDER SALINE CONDITIONS

Ion-stimulated ATPases seem to be involved in energy-dependent ion transport in plant cells, but in many cases a quantitative coupling between ATPase activity and ion flux remains to be demonstrated (64, 65). For salinity studies the ($Na^+ +$ K^+)-activated ATPase and its relation to transport of these ions are most important. In several halophytes as well as salt-sensitive species a more or less complex interaction between these ions and ATPase activity has been observed [sugar beet (66, 67), mangrove leaves (68), bean (69), English ryegrass (70), and soybean (71)]. It should not be surprising that ($Na^+ + K^+$)-activated ATPases occur in both salt-tolerant and salt-sensitive species, since the latter often possess a mechanism of exclusion of Na^+ from the plant tissue (72, 73). Moreover, activation of an ATPase by Na^+ does not tell anything about the direction of transport. In most cases the above-mentioned ATPases require Mg-ATP as a substrate (66, 68), in others Mg^{2+} inhibited stimulation of the ATPase by $Na^+ + K^+$ (69–71). In halophytic *Plantago* species Mg-ATP as well as Ca-ATP can serve as a substrate for ($Na^+ + K^+$)-stimulated hydrolytic activity (74). In the presence of Mg-ATP the ATPase activity is stimulated by Na^+ and by K^+ alone, but their combination reduces the enzyme activity considerably as is also observed in English ryegrass roots (70) and soybean (71). In sugar beet (66, 67) and *Avicennia* (68) the ATPases respond in a more complex way. In the presence of Ca-ATP the ATPase activity of the halophytic *Plantago* species is stimulated by the combined action of $Na^+ + K^+$, even though the additive effects are slightly lower than predicted (competition between Na^+ and K^+). Mg-ATP and Ca-ATP as substrates probably serve as allosteric affectors, inducing different types of enzyme responses to $Na^+ + K^+$. As shown in Figure 5.4 the translocators for Na^+ and K^+ may interact with the hydrolytic unit (ATPase) and with each other in a regulatory way when Mg-ATP is the substrate. In the presence of Ca-ATP the interaction between the Na^+ and K^+-translocators may be sterically hindered (see also 65). As presented here, the model may serve as an allosteric regulator of ion movements in and out of the plant cell (75). It should be noted, however, that there is strong evidence that a primary function of Ca-ATPases is in counteracting high external calcium (76).

Na^+ stimulation of the ATPase is suppressed when the plants are exposed to salinity (74, 77), indicating that long-term exposure of the plant to NaCl induces a conformational change of the Na^+ translocator of the ATPase complex. Also, the more general Ca^{2+}, Mg^{2+}-stimulated ATPase activity is depressed in

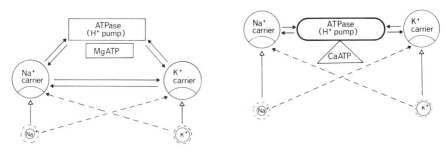

Figure 5.4. Schematic presentation of the $(Na^+ + K^+)$-stimulated transport ATPase system, demonstrating the response of Mg-ATP and Ca-ATP as allosteric affectors of enzyme activity. After ref. 74.

halophytic *Plantago* species when the plants are grown under extreme saline stress (28). It is interesting to observe that the ATPase activity of the roots (on a fresh weight basis) is stimulated at a minor salt stress due to an increased amount of microsomal fraction of the roots together with a small increase in phospholipid and sulfolipid level. The ATPase activity at higher salinity and the sulfolipid level of the roots decrease in a similar way at more extreme salinity, indicating a role for sulfolipid in regulation of ion permeability (see also Section 2.3).

The sugar beet inbred lines mentioned earlier (see Section 2) differ in K^+/Na^+ selectivity (78, 79). In microsomal preparation from the roots patterns of $(Na^+ + K^+)$-activated ATPases can be demonstrated, which reflect different transport rates for Na^+ and K^+ in the two lines (see also 80).

4. PLANT CELL MEMBRANES: POSSIBLE PARAMETERS FOR BREEDING FOR SALT TOLERANCE IN CROP PLANTS

The question arises whether there is a need for selection for salt tolerance in crop plants on the basis of one or more of the membrane parameters discussed in Sections 2 and 3. It could be much simpler just to grow seedlings of the crop under investigation, under various degrees of salinity stress, and to select the proper individual plants for further breeding.

Several aspects have to be taken into consideration before it can be decided which approach leads to the required goal. Salt tolerance may vary with age of the plant and with stage of development. Because of the slow turnover of plant lipids, a lipid parameter might be useful for selection for a more extended long-term salt resistance. Second, one might wish to choose between selection of genotypes adapted to a certain (optimal) salt stress and the possibility of selection of genotypes that are adaptable to a rather wide range of salinity stress because of a more flexible physiological adaptation (Fig. 5.5). The latter characteristic is called phenotypic plasticity. (Practically all plants show a certain degree of physiological plasticity to the factor temperature, since

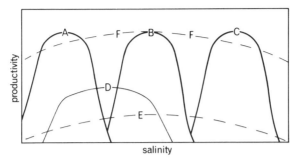

Figure 5.5. Schematic presentation of the effects of genetic differentiation and physiological plasticity on plant productivity as affected by salinity. A, B, and C, genotypes of similar plasticity, but differing in optimal NaCl concentration for productivity. D, plasticity greater than for A, B, and C, but productivity is reduced. E, extreme degree of plasticity, in combination with a very low productivity. F, combination of extreme plasticity with high productivity.

habitats of constant temperature are rare; plants have no strict temperature control.)

Sugar beet inbred lines constitute a good example of the above-mentioned possibilities for selection. In a growth experiment (76) the commercial variety Monohill shows maximal fresh weight at 50 mM NaCl (650 g), the salt-sensitive line ADA at 0 mM NaCl (450 g), and the salt-tolerant line FIA shows constant growth between 0 and 150 mM NaCl (170 to 180 g). It seems that in FIA the greatest phenotypic plasticity for varying NaCl concentrations has been realized, however, at the expense of a high productivity (curve D in Fig. 5.5). It is not a necessity that large phenotypic plasticity is coupled with low productivity. The halophyte *Spartina townsendii* (its origin located in England) in Western Europe shows vigorous growth in a range from the frequently flooded salt marsh (with *Salicornia*), up to infrequently flooded salt marshes, outcompeting numerous native halophytes over a wide range of frequency of flooding by sea water (curve F in Fig. 5.5).

Recent research on *Plantago major* reveals that the genotypes of this species differ in physiological (phenotypic) plasticity for the factor mineral nutrition (81–83) as measured by parameters of growth and respiration (81, 83) and ATPase activity of the roots (81, 82). The question whether a genotype of great physiological plasticity for mineral nutrition also exhibits a sizable degree of plasticity for salinity is currently under investigation. The greater plasticity of the sugar beet genotype FIA is due to its greater capacity to replace K^+ by Na^+; its growth is doubled by replacement of K^+ (5 mM) by K^+ (0.25 mM) + Na^+ (4.75 mM; 84).

Physiological plasticity of crop plants for NaCl may be of advantage under conditions where only occasionally the NaCl concentration of the irrigation water becomes a stress for the plant. In such a situation growth may continue without interruption. Selection for plasticity will be based on genetic differentiation in NaCl-induced responses in the plant (Table 5.2).

Whether the physiological adaptation to NaCl is gene regulated, regulating

Table 5.2.   Lipid Composition and Metabolism: Genetic Differentiation for Salt Stress and for NaCl-Induced Physiological Adaptation

Lipid Class	Genetic Differentiation for Salt Stress	Genetic Differentiation for NaCl-Induced Adaptation
Sterols	Yes (sugar beet, grape, yeast)	Yes (yeast, *Plantago*, *Citrus*)
Phospholipids	Yes, especially phosphatidylcholine and -ethanolamine (grape)	No (exception barley); apparent effects due to morphogenetic changes
Galactolipids	Yes (grape)	Yes (sugar beet)
Sulfolipid	Yes (*Plantago*)	Yes (sugar beet, *Plantago*)

the physiology of the plant toward the proper direction, remains to be investigated. The other possibility is that NaCl induces changes in the physiology of the plant via effects on mineral nutrition and (osmotic) water relations, leading to hormonally controlled adaptive changes in growth and development. At the moment no quantitative analysis of these two possibilities for an explanation of the nature of physiological plasticity for NaCl can be given; the separate effects of genetic differentiation for salt stress and genetic differentiation for salt-induced physiological adaptation are summarized for various lipid groups (Table 5.2).

REFERENCES

1. L. Erdei and P. J. C. Kuiper. *Physiol. Plant.* **47,** 95 (1979).
2. T. Kawasaki and M. Moritsugu. *Ber. Ohara Inst. Landwirtsch. Biol., Okayama Univ.* **17,** 57 (1978).
3. T. Kawasaki and M. Moritsugu. *Ber. Ohara Inst. Landwirtsch. Biol., Okayama Univ.* **17,** 73 (1978).
4. H. Nassery, G. Ogata, R. H. Nieman, and E. V. Maas. *Plant Physiol.* **62,** 229 (1978).
5. T. J. Flowers and J. L. Hall. In W. P. Anderson, ed., *Ion Transport in Plants.* Academic Press, New York, 1973, p. 357.
6. D. Kramer, A. Läuchli, and A. R. Yeo. *Ann. Bot.* **41,** 1031 (1977).
7. A. R. Yeo, D. Kramer, and A. Läuchli. *J. Exp. Bot.* **28,** 17 (1977).
8. R. G. Wyn Jones. In C. B. Johnson, ed., *Physiological Processes Limiting Plant Productivity.* Butterworths, London, 1981, p. 271.
9. A. D. Brown. In A. H. Rose and J. G. Morris, eds., *Advances in Microbial Physiology.* Academic Press, New York, 1978, p. 181.

10. L. Adler and C. Liljenberg. *Physiol. Plant.* **53**, 368 (1981).

11. T. K. Hodges. In U. Lüttge and M. G. Pitman, eds., Transport in plants II, Part A, Cells, *Encyclopedia of Plant Physiology*, Vol. 2. Springer-Verlag, Berlin, 1976, p. 260.

12. D. L. Hendrix. *Annu. Rev. Phytopath.* **8**, 111 (1970).

13. D. L. Hendrix. *Plant Physiol.* **52**, 93 (1973).

14. E. Heftmann. *Lipids* **6**, 128 (1971).

15. H. Brockerhoff. *Lipids* **9**, 645 (1974).

16. B. Aloni, A. Eitan, and A. Livne. *Biochim. Biophys. Acta* **465**, 46 (1977).

17. A. Darin-Bennett and J. G. White. *Cryobiology* **4**, 466 (1977).

18. C. Grunwald. *Plant Physiol.* **43**, 484 (1968).

19. B. Deuticke. *Rev. Physiol. Biochem. Pharmacol.* **78**, 1 (1977).

20. R. J. Kemp, I. J. Good, and E. J. Mercer. *Phytochemistry* **6**, 1609 (1967).

21. C. Grunwald. *Plant Physiol.* **48**, 653 (1971).

22. R. J. Kemp and E. J. Mercer. *Biochem. J.* **110**, 119 (1968).

23. A. Scarpa and J. de Gier. *Biochim. Biophys. Acta* **241**, 789 (1971).

24. B. de Kruijff, W. J. de Greef, R. V. W. van Eyck, R. A. Demel, and L. L. M. van Deenen. *Biochim. Biophys. Acta* **298**, 479 (1973).

25. P. J. C. Kuiper. *Plant Physiol.* **43**, 1367 (1968).

26. C. E. E. Stuiver, P. J. C. Kuiper, and H. Marschner. *Physiol. Plant.* **42**, 124 (1978).

27. G. L. Lees and J. E. Thompson. *Physiol. Plant.* **49**, 215 (1980).

28. L. Erdei, C. E. E. Stuiver, and P. J. C. Kuiper. *Physiol. Plant.* **49**, 315 (1980).

29. C. E. E. Stuiver, P. J. C. Kuiper, H. Marschner, and A. Kylin. *Physiol. Plant.* **52**, 77 (1981).

30. D. Kuiper and P. J. C. Kuiper. *Physiol. Plant.* **44**, 81 (1978).

31. T. J. Douglas and R. R. Walker. *Physiol. Plant.* **58**, 69 (1983).

32. J. de Gier, C. J. A. van Echteld, A. T. M. van der Steen, P. C. Noordam, A. J. Verkleij, and B. de Kruijff. In J. F. G. M. Wintermans and P. J. C. Kuiper, eds., *Biochemistry and Metabolism of Plant Lipids*. Elsevier, Amsterdam, 1982, p. 315.

33. P. R. Cullis and B. de Kruijff. *Biochim. Biophys. Acta* **559**, 399 (1979).

34. P. J. C. Kuiper. *Plant Physiol.* **43**, 1372 (1968).

35. J. de Gier, J. G. Mandersloot, and L. L. M. van Deenen. *Biochim. Biophys. Acta* **150**, 666 (1968).

36. W. S. Ferguson. *Can. J. Plant Sci.* **46**, 639 (1966).

37. S. Chetal, D. S. Wagle, and H. S. Nainawatee. *Phytochemistry* **19**, 1393 (1980).

38. E. V. Maas, G. Ogata, and M. H. Finkel. *Plant Physiol.* **64**, 139 (1979).

39. M. Twerskey and R. Felhendler. *Physiol. Plant.* **29**, 396 (1973).

40. C. E. E. Stuiver, L. J. de Kok, A. E. Hendriks, and P. J. C. Kuiper. In J. F. G. M. Wintermans and P. J. C. Kuiper, eds., *Biochemistry and Metabolism of Plant Lipids*. Elsevier, Amsterdam, 1982, p. 455.

41. S. Yoshida and A. Sakai. *Plant Physiol.* **53**, 509 (1974).

42. H. Lambers, T. Blacquière, and C. E. E. Stuiver. *Physiol. Plant.* **51**, 63 (1981).

43. C. C. Delwiche and H. M. Bregoff. *J. Biol. Chem.* **223**, 430 (1958).

44. A. D. Hanson and N. A. Scott. *Plant Physiol.* **66**, 342 (1980).

45. W. D. Hitz, D. Rhodes, and A. D. Hanson. *Plant Physiol.* **68**, 814 (1981).

46. S. J. Coughlan and R. G. Wyn-Jones. *Planta* **154**, 6 (1982).

47. I. Horváth, L. Vigh and T. Farkas. *Planta* **151**, 103 (1981).

48. I. Horváth, L. Vigh, T. Farkas, L. J. Horváth, and D. Dutits. *Planta* **153**, 476 (1981).

49. P. J. C. Kuiper. *Plant Physiol.* **44**, 968 (1969).

50. G. Rama Gopal and G. Rajeswara Rao. *Z. Pflanzenphysiol.* **106**, 1 (1982).

51. P. Mazliak, A. Jolliot, and C. Bonnerot. In J. F. G. M. Wintermans and P. J. C. Kuiper, eds., *Biochemistry and Metabolism of Plant Lipids.* Elsevier, Amsterdam, 1982, p. 89.

52. A. Oursel, A. Tremolières, and P. Mazliak. *Physiol. Vég.* **15**, 377 (1977).

53. M. Müller and K. A. Santarius. *Plant Physiol.* **62**, 326 (1978).

54. M. Gharsalli and A. Cherif. *Physiol. Vég.* **17**, 215 (1979).

55. M. Ellouze, M. Gharsalli, and A. Cherif. *Physiol. Vég.* **18**, 1 (1980).

56. F. Harzallah-Shkiri, T. Guillot-Salomon, and M. Signol. In J. F. G. M. Wintermans and P. J. C. Kuiper, eds., *Biochemistry and Metabolism of Plant Lipids.* Elsevier, Amsterdam, 1982, p. 423.

57. J. B. Mudd, R. Dezacks, and J. Smith. In P. Mazliak, P. Benveniste, C. Costes, and R. Douce, eds., *Biogenesis and Function of Plant Lipids.* Elsevier/North-Holland, Amsterdam, 1980, p. 57.

58. R. Anderson, B. P. Livermore, M. Kates, and B. E. Volcani. *Biochim. Biophys. Acta* **528**, 77 (1978).

59. R. Anderson, M. Kates, and B. E. Volcani. *Biochim. Biophys. Acta* **528**, 89 (1978).

60. A. Kylin, P. J. C. Kuiper, and G. Hansson. *Physiol. Plant.* **26**, 271 (1972).

61. G. Hansson, P. J. C. Kuiper, and A. Kylin. *Physiol. Plant.* **28**, 430 (1973).

62. K.-A. Karlsson, B. E. Samuelsson, and G. O. Steen. *J. Membr. Biol.* **5**, 169 (1971).

63. H. Marschner and G. Mix. *Z. Pflanzenernähr. Bodenkd.* **136**, 203 (1973).

64. J. Fisher and T. K. Hodges. *Plant Physiol.* **44**, 385 (1969).

65. R. T. Leonard and C. W. Hotchkiss. *Plant Physiol.* **58**, 331 (1976).

66. G. Hansson and A. Kylin. *Z. Pflanzenphysiol.* **60**, 270 (1969).

67. J. Karlsson and A. Kylin. *Physiol. Plant.* **32**, 136 (1974).

68. A. Kylin and R. Gee. *Plant Physiol.* **45**, 169 (1970).

69. Y. F. Lai and J. E. Thompson. *Biochim. Biophys. Acta* **233**, 84 (1971).

70. P. V. Nelson and P. J. C. Kuiper. *Physiol. Plant.* **35**, 263 (1975).

71. D. L. Hendrix and R. M. Kennedy. *Plant Physiol.* **59**, 264 (1977).

72. H. Marschner. In R. L. Bieleski, A. R. Ferguson and M. M. Creswell, eds., *Mechanisms of Regulation of Plant Growth*, Bulletin 12. Royal Society of New Zealand, 1974, p. 99.

73. O. Tanczos, L. Erdei, and J. Snijder. *Plant Soil* **63**, 25 (1981).

74. L. Erdei and P. J. C. Kuiper. *Physiol. Plant.* **49**, 71 (1980).

75. A. D. M. Glass. *Plant Physiol.* **58**, 33 (1976).

76. M. Monestiez, A. Lamant, and R. Heller. *Physiol. Plant.* **55**, 445 (1982).

77. C. T. Horowitz and Y. Waisel. *Experientia* **26,** 941 (1970).

78. A. Kylin and G. Hansson. *8th Coll. Int. Potash Instit., Bern, 1971*, p. 64.

79. G. Hansson, Patterns of Ionic Influences on Sugar Beet ATPases, Ph.D. thesis, University of Stockholm, Sweden, 1975.

80. H. Marschner, A. Kylin, and P. J. C. Kuiper. *Physiol. Plant.* **51,** 234 (1981).

81. D. Kuiper. In N. R. Sarić, ed., *Genetic Specificity of Mineral Nutrition of Plants*, Vol. 13. Serbian Academy of Science and Arts, 1982, p. 215.

82. D. Kuiper. *Physiol. Plant.* **56,** 436 (1982).

83. D. Kuiper. *Physiol. Plant.* **57,** 222 (1983).

84. H. Marschner, P. J. C. Kuiper, and A. Kylin. *Physiol. Plant.* **51,** 239 (1981).

6

TRANSPORT ACROSS THE ROOT AND SHOOT/ROOT INTERACTIONS

M. G. Pitman

School of Biological Sciences
The University of Sydney
Sydney, Australia

A typical experiment in salinity studies is to grow a plant species or variety on a range of solutions containing basic nutrients (e.g., Hoagland's solution) to which varied concentrations of NaCl are added.

In the majority of papers published with these data, the aim usually concerns the amounts or concentrations of solutes in the shoot or leaf (C_s), possibly in relation to reduction in growth or the balance of ions between cytoplasm and vacuole. Very few papers address the problem of the relationship between concentrations in the leaf, the amount of solutes in the shoot, and, in particular, the rate of supply of solutes from the root. The purpose of this chapter is to consider the overall process of solute transport across the root to the shoot, and its relation to growth.

Plant growth sets up a "demand" for solutes to maintain nutrient levels or osmotic concentrations in the leaves. In saline conditions, Na^+ and Cl^- can be particularly important as osmotic solutes in the leaves. It can be seen intuitively that the plant's requirements for the major ions (K^+, Na^+, and Cl^-) will be determined both by the rate of growth and by the level of these solutes in the leaves.

A major characteristic of solute transport by plants in saline conditions is the degree of selectivity between potassium and sodium. Another is the extent to which Cl^- is taken up by the plant and whether it is transported to the shoot. Plants vary in selectivity between potassium and sodium ranging from virtual exclusion of Na^+ to preferential accumulation of Na^+. There are also degrees of uptake or exclusion of Cl^-. For many halophytes, uptake of Na^+ and Cl^- serves a useful function by providing osmotic solutes. Many nonhalophytes, however, exclude Na^+ and Cl^-, and there seems to be an adaptive choice in plant evolution between use of ions as osmotic solutes with their potential toxic effects and the exclusion of solutes, solving the toxicity problem but requiring the plant to produce organic solutes to withstand water stress (1).

The cellular basis for ionic compartmentation and its role in K^+/Na^+ selectivity has already been covered in Chapters 3, 4, and 5 on cell membranes and on K^+/Na^+ selectivity in cells, but certain aspects will be emphasized in what follows. In this chapter it will only be possible to refer to a very few examples of the infinite variety of plant performance, but these are selected to show the relevance of general models for salt transport to problems of salinity.

1.   THE ROOT AS A TRANSPORT SYSTEM

The outstanding difference between the properties of plant roots and the properties of cells that make up the root is the behavior of the root as an organ. The root is supplied with nutrients from the shoot through the phloem and in turn it supplies the shoot with water, ions, and nutrients scavenged from the soil. The ability to handle this two-way traffic both longitudinally in the stele and laterally across the cortex depends on tissue organization as well as on the properties of the cells. This point is shown in Figure 6.1, which compares growth on varied salinities of whole plants and of cells cultured from the plants (2).

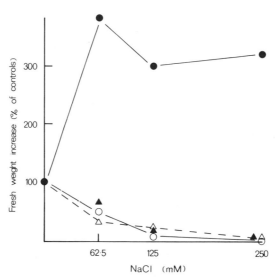

Figure 6.1. Comparison of growth of whole plants of *Atriplex undulata* (●) and *Phaseolus vulgaris* (▲) with tissue cultures of these plants (○,△). From ref. 2.

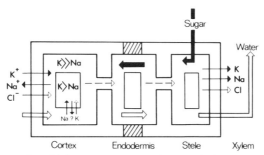

Figure 6.2.   Conceptual model for transport across the root. Active transport into the symplast leads to higher selectivity of K relative to Na than in the solution. Other processes maintain differences in selectivity and possibly concentration (see Chapter 3) between vacuoles and cytoplasm. The endodermis is a barrier to movement in the symplast. Within the stele release of solutes to the xylem vessels occurs where ions are carried to the shoot in transpiration, which leads to water flow across the root. Sugars are supplied to the root via the phloem and then outwards in the symplast.

Concepts used in discussing transport across the root are summarized in Figure 6.2. More details of structure and organization of plant roots are given in various books (3–6) and in reviews (7–12).

Transport of solutes into the root occurs at the plasmalemmas of epidermal cells and to some degree at cortical cells. Plasmodesmata between cells in the root connect the cytoplasms of adjacent cells, forming the symplast, which provides a high-selectivity pathway across the root for solutes. Entry to the symplasm, and to some extent the release to the xylem, is controlled by transport at cell membranes. The symplast is also part of the pathway for water flow across the root. The cell walls form another pathway in which water can flow, carrying with it solutes from the external solution; this nonselective pathway is referred to as the apoplast. Barriers to diffusion in the endodermis (casparian strips and suberization) restrict nonselective flows of solutes, which could otherwise bypass the symplast. Within the stele, release of solutes can take place from the symplast into the cell walls (apoplast) of the stele and from there to the xylem vessels, or else there may be release to the xylem directly from the xylem parenchyma, which can develop transfer-cell structures. Xylem parenchyma may also absorb solutes from the xylem sap, modifying its composition (Chapter 9), while transport in the phloem may supply K^+ preferentially to the root tip (Chapter 3).

The symplast also transports outward from the stele certain solutes involved in the metabolism of root cells (e.g., sugars). Supply of nutrients in this way can then affect the rates of ion transport into the root across the cell membranes. Blocking phloem transport to the root can lead to substantial inhibition of ion uptake (13).

Questions particularly relevant to salinity problems are concerned with the mechanisms for maintaining selectivity between K^+ and Na^+ and factors controlling transport of Cl^- or total $(K^+ + Na^+)$ to the shoot. In each case the interaction between solute flows and water flows across the root is important.

2. SELECTIVE PROCESSES FOR K$^+$ UPTAKE AND TRANSPORT IN A GLYCOPHYTE: BARLEY

It is doubtful whether high Na$^+$ levels in the soil depress K$^+$ uptake to such an extent that this reduces plant growth, but the selective mechanisms of K$^+$ and Na$^+$ uptake to the root also affect the extent to which the root can control uptake to the shoot in saline conditions.

The selective process has been studied extensively, using barley seedlings (see Chapter 3) though barley does not appear to possess other, important processes affecting selectivity, such as reabsorption of solutes in the xylem as commonly found in legumes (Chapter 9). Information has already been given about cellular regulation of K$^+$ and Na$^+$ selectivity (Chapter 3), and details are repeated here only to give a context for transport in the whole plant.

2.1. Transport of K$^+$ and Na$^+$ in Barley Roots: The Symplast

When barley seedlings were grown on a solution containing 2.5 mM K$^+$ and 7.5 mM Na$^+$ plus basic nutrients the average ratios of Na$^+$/K$^+$ in the root and shoot were lower than in the solution, as shown in Table 6.1. Kinetic analysis of K$^+$ and Na$^+$ efflux from the roots (e.g., Chapter 3) showed that the ratio of Na$^+$/K$^+$ in the cytoplasmic component was 0.09 while in the vacuoles it was 0.3. This is

Table 6.1. Potassium and Sodium Distribution in Barley Roots and Whole Plants

	Content			Ratio	
	K$^+$	Na$^+$	Units	Na$^+$/K$^+$	Reference
Whole plant (2 days old)					14
Shoot				0.12	
Root	62	22	μmol g$_{FW}^{-1}$	0.36	
Exudate from	30	4	mM	0.14	
excised root system					
Root					15
Kinetic analysis					
"Cytoplasm"	4.6	0.4	μmol g$_{FW}^{-1}$	0.09	
"Vacuole"	79	24		0.3	
Electron probe					16
Cytoplasm	—	—	X-ray counts	0.11	
Vacuole	—	—	X-ray counts	0.67	
Cortex (average)	—	—	X-ray counts	0.30	
Stele (average)	—	—	X-ray counts	0^a	
Central vessel	—	—	X-ray counts	0.09	

^{a}Na was very low and the ratio Na/K near zero.

necessarily an average for the root, but electron probe microanalysis has since confirmed this difference in selectivity between cytoplasm and vacuoles of cortical cells (Table 6.1) though it has also shown some general differences in K$^+$ and Na$^+$ distribution across the root (see also 17). First, the ratio of Na$^+$/K$^+$ in cells near the epidermis may be larger than in the inner cortex. Second, there can be a sharp discontinuity at the endodermis in Na$^+$/K$^+$ ratio, or in the ratio of Cl$^-$/(K$^+$ + Na$^+$) (see 18 for *Atriplex*). However, there is good support for a generalization of the type shown in Figure 6.2 (for barley) that locates high selectivity for K$^+$ in the symplast and within the stele, recognizing that communication between stele and cortex is largely restricted to the endodermis.

This distribution can be interpreted in terms of cell processes. Jeschke (Chapter 3) presented electrochemical and other evidence in favor of active Na$^+$ efflux and K$^+$ influx as the major processes establishing high selectivity for K$^+$ in the symplast. In addition to the transport processes at the plasmalemma, there is also a need for transport of Na$^+$ out of the cytoplasm into the vacuoles (Chapter 3). Jeschke (19) also presents evidence for involvement of H$^+$ transport in the selective K$^+$ and Na$^+$ transport which has consequences for the determination of plasmalemma potential differences (P.D.'s).

One separate consequence of the model in Figure 6.2 both for the root cells and the whole plant is that the low levels of Na$^+$ or its restricted transport across the root are not entirely due to low permeability of the membranes since K$^+$ can be replaced by Na$^+$ in the cell vacuoles and in transport to the shoot (Fig. 6.3).

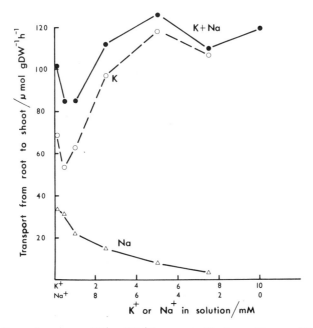

Figure 6.3. Rates of transport of K$^+$ and Na$^+$ from roots of barley seedlings aged 7–10 days grown on solutions containing 10 mM (K$^+$ + Na$^+$) in varied ratios. Data from ref. 14.

The efficiency of selective transport processes for K^+ and Na^+ also involves the diffusive permeability of the membranes. Put simply, the "leakier" the membrane to Na^+, the larger the efflux pump will have to be to maintain a low concentration of Na^+ in the cytoplasm. Too low a permeability, on the other hand, could restrict Na^+ transport into the symplast and hence into the plant as a whole, when external concentration was low.

A number of arguments put the ratio of diffusive permeability for Na^+ to K^+ (P_{Na}/P_K) at about 0.3 in barley roots. Calculation of K^+ efflux and Na^+ influx and observations of the effect of hyperpolarization induced by fusicoccin on fluxes of K^+ and Na^+ both support this view (20). The absolute values of P_K or P_{Na} depend on the area across which these fluxes occur; it has been generally assumed to be the cell surface area in the cortex, but under some conditions it may be more appropriately restricted to the outer surface of the epidermal cells. Other estimates of permeability coefficients have been given (3, 10).

It should be noted that the ratio of Na^+/K^+ measured in roots is largely due to the vacuoles and that the important ratio in the symplast may be quite different (Table 6.1). The vacuole of cortical cells is in some respects a dead end; continued selective transport across the root depends on selective transport at entrance to, and exit from, the symplast. The vacuole's role may be more in using Na^+ as an osmoticum instead of K^+, and in providing a source of stored K^+ under salination, rather than as part of the selective system of transport across the root.

2.2. Transport from Root to Shoot

Figure 6.3 gave estimates of rates of transport of K^+ and Na^+ from root to shoot of barley seedlings when the ratio of K^+/Na^+ in the external solution was varied but total (K^+/Na^+) was constant. Total $(K^+ + Na^+)$ transport was little affected, but the ratio of K^+/Na^+ transported varied with the ratio of K^+/Na^+ in the solution. Similar results have been found with a number of plant species (e.g., 21).

Effects of increasing the external salt concentration while maintaining a constant ratio of K^+/Na^+ in the solution are shown in Figure 6.4. The ratio of K^+/Na^+ in the root was little affected by the increase in $(K^+ + Na^+)$ from 10 to 100 mM but there was a reduction in selectivity of transport to the shoot as concentration increased.

It is not a straightforward matter to assess whether selectivity has been affected by increasing salt concentration when the common experimental procedure of adding NaCl to a constant K^+ concentration is followed. One approach is to calculate $S_{K,Na}$ (defined as the ratio of K^+/Na^+ in the plant part divided by the ratio of K^+/Na^+ in the solution). In the example of Figure 6.4 there clearly is a reduction in $S_{K,Na}$ with increasing total salt concentration.

However, salinity problems generally are concerned with the response of

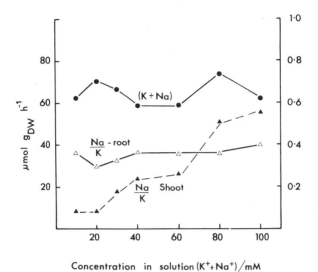

Figure 6.4. Rate of transport of (K$^+$ + Na$^+$) from roots of barley seedlings 7 to 10 days old grown on solution in which Na$^+$/K = 3 but (K$^+$ + Na$^+$) varied. Also shown is the ratio of Na$^+$/K in the roots (△) and in transport from root to shoot (▲). Data from ref. 22.

plants to high concentrations of NaCl. Barley is noted for its range of tolerance of salinity and many experiments have investigated the relation between growth and the content of root and shoot. Greenway (23) measured K$^+$, Na$^+$, and Cl$^-$ content of two varieties ('Chevron' and 'Pallidum') during their growth to grain set. From these data, rates of transport from root to shoot can be estimated showing the changes in rate of transport that occurred during growth. There was a lower rate of transport of (K$^+$ + Na$^+$) to the shoot of the more tolerant cv. 'Pallidum', and a much lower ratio of Na$^+$/K$^+$ transported to the shoot of this variety at later stages of development. The proportion of Cl$^-$/(K$^+$ + Na$^+$) was little affected (Table 6.2).

Table 6.2.   Comparison of Two Cultivars of Barley ('Chevron' and 'Pallidum') Grown on Solution Containing 125 mM NaCl[a]

Age (days)	'Chevron'				'Pallidum'			
	RGR[b]	$\frac{\Delta Na^+}{\Delta K^+}$	$\frac{\Delta Cl^-}{\Delta (K^+ + Na^+)}$	(K$^+$ + Na$^+$)[c]	RGR	$\frac{\Delta Na^+}{\Delta K^+}$	$\frac{\Delta Cl^-}{\Delta (K^+ + Na^+)}$	(K$^+$ + Na$^+$)
43–50	70	3.3	0.51	67	91	3.0	0.41	47
50–64	70	6.3	0.71	30	91	2.2	0.55	39
64–93	96	7.1	0.77	44	103	0.98	0.79	18

[a]From Ref. 23.
[b]RGR, relative growth rate as % of control.
[c]Rate of transport from root to shoot; μmol g$_{DW}$$^{-1}$hr^{-1}.

Table 6.3.   Comparison of Ratio of Na^+/K^+ in Shoot and Root, and $S_{K,Na}$ for the Shoot (25) of Three Barley Varieties Grown on Solutions with Varied NaCl Concentrations

Barley Variety	NaCl in Solutions (mM)					
	0	50	100	150	200	250
'California Mariout'						
Na^+/K^+						
Shoot	0.06	0.78	1.60	1.95	2.15	2.40
Root	0.23	1.23	2.50	6.3	9.1	14.3
$S_{K,Na}$	0.6	10.6	10.5	12.7	15.6	17.5
'Arimar'						
Na^+/K^+						
Shoot	0.02	1.05	1.60	1.45	1.85	2.20
Root	0.04	1.0	2.9	4.8	6.7	14.3
$S_{K,Na}$	2.2	8.1	10.5	17.0	15.0	18.7
'Chevron'						
Na^+/K^+						
Shoot	0.07	2.05	4.6	5.3	5.6	5.0
Root	0.27	1.10	1.8	3.3	4.4	6.7
$S_{K,Na}$	0.5	4.1	3.6	4.8	6.0	8.3

Storey and Wyn-Jones (24, 25) compare the response of two other varieties of barley ('California Mariout' and 'Arimar'), though only single harvests are given and rates of transport cannot be calculated. Their results cover a wider range of NaCl concentrations; in general the content of $(K^+ + Na^+)$ in the shoot on a dry weight basis did not increase with increased NaCl but there was an almost linear increase in content of Cl^-, as found subsequently for cv. 'Beecher' by Delane et al. (26). Changes in fresh weight/dry weight ratio led to some increase in $(K^+ + Na^+)$ concentration relative to tissue water with external NaCl concentration. Since relative growth rate decreased with increased salinity, it seems likely that there also was a decreased rate of transport of $(K^+ + Na^+)$ from the root to the shoot. Storey and Wyn-Jones also reported estimates of $S_{K,Na}$ for the shoot and ratios of Na^+/K^+ in the root and shoot (Table 6.3) showing maintenance of high selectivity for K^+ for all varieties, even in very high concentration of NaCl. 'Chevron' showed the least selectivity and the ratio of Na^+/K^+ was in this case higher in the shoot than in the root. Transport to the shoot appeared to be less able to exclude Na^+ (and possibly Cl^-) than in the other varieties, yet the root cells of 'Chevron' were highly selective.

2.3.   Retranslocation and Distribution in the Shoot

The results of Section 2.2 emphasize the export process from the root and ignore the different types of sinks in the shoot. Mature leaves of barley have less requirement for growth than cells in the growing regions at the base of the leaves and shoot apex. Delane et al. (26) and Munns et al. (27) have investigated the uptake of K^+, Na^+, and Cl^- to mature and expanding parts of barley leaves and the extent to which these ions contribute to osmotic adjustment. They showed that the elongating tissue was high in K^+, Na^+, and Cl^- (about 55% of osmotic solutes) and suggested that growth was limited by water deficit in the tissue related to the ionic content, rather than to toxic effects of the ions on growth. The position of the developing tissue in monocotyledonous leaves could allow ions to be supplied either from the phloem or the xylem. Phloem transport of K^+ from old to young leaves is well established, whereas retranslocation of Na^+ and Cl^- has been measured at lower but substantial rates in barley (28) and other species (e.g., 29).

The redistribution of ions in this way is an important aspect of ionic adjustment in different parts of the shoot, but need not affect conclusions about the average transport from root to shoot unless there is export from the shoot back to the root. Retranslocation of K^+ from the seed is a major source of K^+ in developing seedlings but does not appear to contribute major flows of K^+ from shoot to root of more mature barley plants (30).

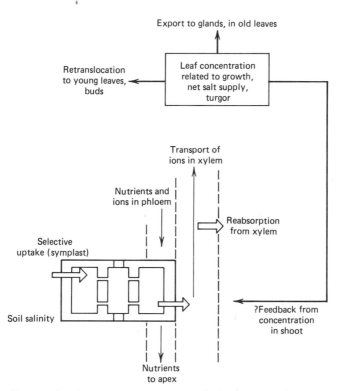

Figure 6.5. Processes involved in overall uptake and distribution of salt in plants and interactions in determination of concentrations in the plant. Reabsorption from the xylem and translocation of K to the root seems low in barley.

2.4.   Explaining Selective Uptake to the Shoot

Studies with barley roots and intact barley plants lead to the generalization about transport summarized in Figure 6.5. The symplast appears to contain K^+ and Na^+ at a much higher selectivity (for K^+) than in the root vacuoles. It is assumed that the endodermis acts as a barrier to flow in the apoplast but not in the symplast so that the ratio of Na^+/K^+ in the symplast can affect and perhaps determine the ratio of Na^+/K^+ in the stele and in the xylem. Barley happens to be a plant in which there seems to be little differential reabsorption from the xylem to the xylem parenchyma either of K^+ or Na^+, but in other glycophytes reabsorption may be particularly important in reducing total salt content or eliminating Na^+ (Chapter 9).

Reabsorption has been shown very clearly using *Phaseolus* seedlings, where absorption of Na^+, but not Cl^-, occurs from the xylem into the upper root and to the stem. Jacoby (31), for example, measured absorption of Na^+ and Cl^- at different distances from the base of the stems of *Phaseolus vulgaris* in varied concentrations of NaCl. Absorption characteristically showed progressive saturation of sites along the pathway due to specific absorption of Na^+. Kramer et al. (32) studied distribution of Na^+ in roots of *P. coccineus* and showed that the xylem parenchyma accumulated Na^+ and had higher ratios of Na^+/K^+ than the rest of the root.

As will be seen below, reabsorption or selective efflux from the symplast may be a more important factor determining Na^+, K^+ and Cl^- transport in halophytes.

3.   HALOPHYTES

Halophytes show extreme behavior with respect to K^+ and Na^+, since they are able to tolerate high salinity and to take up relatively large amounts of Na^+ and Cl^-. In recent years the cellular basis for selectivity in these plants has been studied in some detail.

3.1.   Potassium and Sodium Selectivity in Cells of Halophytes

Species of *Atriplex* have been used in many studies of response to salinity. Characteristically, growth increases if moderate amounts of NaCl (about 100 mM) are added to culture solutions (Hoagland's) and then becomes reduced at higher concentrations of NaCl. Uptake to root cells shows higher selectivity for K^+ relative to Na^+ than transport to the shoot (Fig. 6.6) as has been found for other halophytes [*Triglochin maritima* (34), *Suaeda maritima* (35), *Aster tripolium* (36)].

Electron microprobe analysis has been used to investigate K^+ and Na^+

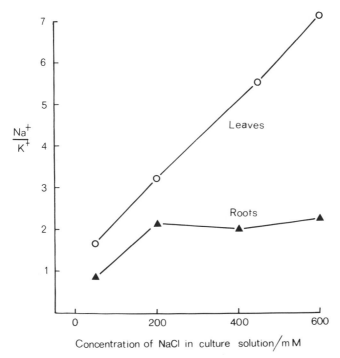

Figure 6.6. Ratio of Na^+/K^+ in leaves and roots of *Atriplex spongiosa* plants grown on culture solution containing varied concentrations of NaCl. Data from ref. 33.

distribution in roots and leaves of *Atriplex spongiosa* (33, 37). In the leaves and in meristemmatic regions of shoot and root, the cytoplasmic concentration of $(K^+ + Na^+)$ was estimated to be in the range 75–150 mM compared with concentrations in the vacuole that rose to about 700 mM when grown on solution containing 600 mM NaCl. The ratio of Na^+/K^+ measured in the cytoplasm in various parts of the plant (Table 6.4) was consistently lower than in adjacent vacuoles (bundle sheath) or vacuoles of mature cells of the cortex (root). The values estimated for cytoplasmic $(K^+ + Na^+)$ concentration agreed well with estimates made using *Suaeda maritima* by Harvey et al. (35). Measurements of Cl^- distribution using electron microprobe analysis of *Atriplex spongiosa* showed an increasing contribution of Cl^- to charge balance in vacuoles of roots and leaves as well as an increasing concentration in the cytoplasm as the external concentration of NaCl was increased.

There was also an evident gradient of ratio of Na^+/K^+ across the root from the epidermis, to the cortex, and into the stele. Similar results were reported for *Atriplex hastata* (18), and there was a clear change in ratio of $Cl^-/(K^+ + Na^+)$ at the endodermis, as found in barley roots (16).

Electrophysiological measurements on roots of *Atriplex hastata* grown on varied saline solutions made by Anderson et al. (38) yielded the results shown in

**Table 6.4.   Ratio of Na^+/K^+ and $Cl^-/(K^+ + Na^+)$ in Plants of
Atriplex spongiosa Grown on Solution Containing 400 mM NaCl
Concentration and Estimated with Electron Probe Microanalysis[a]**

	Na^+/K^+	$Cl^-/(K^+ + Na^+)$
Root		
Meristemmatic cytoplasm	0.35	0.30
Vacuoles		
Epidermis	2.05	0.69
Cortex	0.97	0.38
Stele	0.12	0.12
Whole root average	2.84	0.45
Leaves		
Bundle sheath cytoplasm	0.43	0.25
Bundle sheath vacuole	8.0	0.19
Bladder vacuole	14.0	1.33

[a]From refs. 33 and 37.

Table 6.5. Features of these results of interest are the large metabolically
determined component (about $-$ 60 mV) and the insensitivity of the inhibited
potential to changes in external Na^+ concentration. Anderson et al. (38) pointed
out that the ratio of P_{Na}/P_K must be very low and estimated it as 0.03 based on
the vacuolar concentrations.

Estimates of cytoplasmic and vacuolar content in *Triglochin* are used to
calculate ratios of Na^+/K^+ in Table 6.6. There is less difference between
cytoplasm and vacuole than shown in Table 6.4 and found for *Suaeda* (35). This
may be a reflection of the difficulties of using flux analysis with halophyte roots,
but it is evident that there is clear selectivity for K^+ in the cytoplasm. In addition,
Na^+ is not completely excluded from the cytoplasm. The resting potentials and
their response to variation in Na^+ external concentration are similar to those

**Table 6.5.   Measured Values of Root Cell P.D. for
Atriplex hastata Plants Grown on the Solutions
Shown[a]**

Solution		Cell P.D. (mV)		Root
K^+	Na^+	Resting	+ mM KCN	$S_{K,Na}$
1	101	−137	−73	55
1	201	−138	−54	79
1	301	−140	−45	100
1	401	−132	−64	86
1	501	−124	−70	102

[a]From ref. 38.

Table 6.6. Ratios of Na^+/K^+ in Cytoplasm and Vacuoles of Roots of *Triglochin*, Together with Resting Potentials of Root Cells[a]

Solution (mM)		Cytoplasm		Vacuole	ψ_{vo}
K^+	Na^+	Na^+/K^+	$S_{K,Na}$	Na^+/K^+	(mV)
0.2	1	1.25	4.0	1.25	—
2	10	1.55	3.2	1.05	−65
2	100	3.45	14.5	1.65	−83
4	100	1.30	19	0.70	—
8	500	2.10	30	10	−74

[a]From ref. 34.

found for *Atriplex* (Table 6.5) and also imply a low value of P_{Na}/P_K, which is perhaps lower than that estimated by Jefferies (34) at about 1.

3.2. Uptake by Whole Plants

At more extreme salinities there are many published results showing increased ratios of Na^+/K^+ in the shoots of halophytes as external NaCl concentration was increased. Commonly, the ratio of Na^+/K^+ in the shoot is greater than the average ratio in the root (Figure 6.6) though, as has been seen (Table 6.4), the ratio of Na^+/K^+ in the cytoplasm is higher than this average value. Figure 6.7

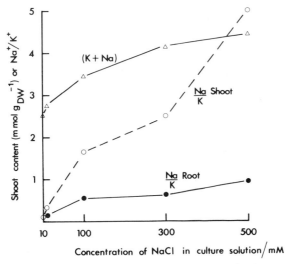

Figure 6.7. Concentration of $(K^+ + Na^+)$ in shoots and ratios of Na^+/K^+ in shoots and roots of *Aster tripolium* at 23 days growth on solutions of varied $(K^+ + Na^+)$. Data from ref. 36.

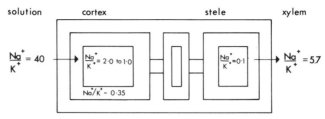

Figure 6.8. Distribution of K and Na in roots of *Atriplex spongiosa* grown in solution containing 400 m*M* NaCl. Based on data from refs. 33 and 37.

gives data for *Aster tripolium* (36) grown on solutions that contained constant Na^+/K^+ above 100 m*M* NaCl, showing the difference between root and shoot, but, nonetheless, a high selectivity for K^+ relative to Na^+. In these experiments Shennan measured growth and ionic content over a period of several weeks, allowing her to estimate the rate of transport of K^+ and Na^+ to the shoot (see Section 4).

Figure 6.8 summarizes the results of microprobe and growth studies. The symplast of halophyte roots and the steles (based on *Atriplex* and *Suaeda*) seem to have a moderately high selectivity for K^+ relative to Na^+ which is higher than found for the shoot. There is the implication that membrane permeability for Na^+ may be low, to account for the low ratios of P_{Na}/P_K. Measurements of shoot content, however, show that there is a relatively low selectivity for K^+ in transport from root to shoot, despite the high selectivity in the root. In general the major part of Na^+ in the plant (90%) can be in the shoot, and though the root may be only 15–25% of the plant it can contain more than 50% of the K^+.

The low selectivity for K^+ in transport to the shoot could involve flow "bypassing" the symplast, though this could lead to buildup of salt within the root (see Section 4). More probably, it is due to reabsorption of K^+ from the xylem and/or extrusion of Na^+ by xylem parenchyma cells from the xylem.

The low level of Na^+ in the cytoplasm, taken together with evident transport of Na^+ across the root implies that there must be effective means of excreting Na^+ from the xylem parenchyma to the vessels. Under these circumstances the low permeability of membranes to Na^+ would be an advantage at the vessel/parenchyma boundary by reducing the back-diffusion to the cells. At the entrance to the symplast, high Na^+ concentration in solution would tend to compensate for the low permeability coefficient, while the large negative potential would facilitate diffusive fluxes of cations *into* the cells, and restrict Cl^- influxes but facilitate outward fluxes.

Table 6.7 compares diffusive fluxes of K^+, Na^+, and Cl^- into and out of the cytoplasm of a "barley" cell and a "halophyte" cell with different membrane permeability to Na^+ but otherwise similar contents. The table shows the effect of a 60 mV change in cell P.D. on the fluxes; depolarization increased K^+ and Na^+ influx and reduced Cl^- efflux from the cell. This aspect of control of cell fluxes could be one reason why halophyte cells maintain large, metabolically

Table 6.7. Calculated Diffusive Fluxes into and out of Hypothetical "Cells" with Different P_{Na} using the Goldman Equation for Flux[a]

	"Barley" Cell			"Halophyte" Cell		
	K^+	Na^+	Cl^-	K^+	Na^+	Cl^-
P_j (x 10^{10} m s^{-1})	33	10	1.2	33	1.0	1.2
C_i	75	7	30	75	25	30
(a)						
$C_0 =$ (mM)	2.5	7.5	10	2.5	7.5	10
PD = −140 mV						
J_{in} (nmol m^{-2}s^{-1})	45	42	0.03	45	4.2	0.03
J_{out} (nmol m^{-2}s^{-1})	5.4	0.15	20	5.4	0.01	20
(b)						
PD = −80 mV						
J_{in}	27	25	0.18	27	2.5	0.18
J_{out}	34	1.0	11.8	34	0.35	11.8
(c)						
PD = −140 mV						
$C_0 =$ (mM)	5	100	100	5	100	100
J_{in}	92	556	0.26	92	56	0.26
J_{out}	5.4	0.15	20	5.4	0.01	20
(d)						
PD = −80 mV						
J_{in}	54	330	1.7	54	33	1.7
J_{out}	34	1.0	11.8	34	1.0	11.8

[a]The Goldman equation is

$$J = \frac{-P_j z_j FE/RT[C_0 - C_i \exp(z_j FE/RT)]}{1-\exp(z_j FE/RT)}$$

where 1 nmol m^{-2} s^{-1} = 0.36 μmol g$_{FW}^{-1}$ hr^{-1}.

dependent (negative) potentials at the cell membranes, since a combination of low P_{Na} and hyperpolarization offers control of Na$^+$ entry to the cell while efflux pumps can maintain lower Na$^+$ content in the cytoplasm operating to the solution and/or to the apoplast in the stele.

4. TRANSPORT IN THE APOPLAST

Estimates of the amounts of water flowing in the symplast compared with apoplast vary and range from locating the majority in the symplast to attributing most of the flow to the apoplast, with the proportion depending on the rate of water use by the plant. There is one consequence of water flow in the apoplast that could be most important for plants under saline conditions, namely, the buildup of NaCl in the cortical apoplast.

Flow of solution through the cell walls will carry with it salts from the external solution. Some of these salts may be accumulated by cells in the cortex, another part may be absorbed at the endodermis, but a gradient of concentration should develop in the cortex so that diffusion of salt out of the root balances the difference between input in water flow and absorption. Similar flow-induced profiles are produced in the soil around the root (6). The cortex of the root can be thought of as an internal, unstirred layer analogous to the unstirred layer at the surface of the root in the soil.

The differential equation relating flow, back-diffusion, and absorption along the apoplast pathway (assuming a linear rather than radial system) is

$$\frac{dC}{dt} = D\frac{d^2C}{dx^2} - \frac{v}{a}\frac{dC}{dx} - \frac{\alpha' l}{a}C$$

where C is concentration; D is the diffusion coefficient in the apoplast including the tortuosity of cell walls, but not the area of apoplast per root surface area (a); v is the water flux in the apoplast relative to root surface (m^3 per m^2 root surface); α' is the rate of absorption averaged over the surface area of cortical cells, and l is a term to relate this surface to the external surface area of the root. At steady state, when $dC/dt = 0$, this equation can be solved analytically to yield estimates of concentration in the apoplast at varied distances from the root surface. This has been done in Figure 6.9 for two external concentrations, 10 and 100 mM, and with varied diffusion coefficients. Both water flow rate and solute diffusion coefficient (D) affect the results; generally, a 10-fold increase in flow rate is equivalent to a 10-fold decrease in D. Values for D based on univalent cation tracer exchange are about 4×10^{-11} m^2 s^{-1} but there are various reasons why this is uncertain, and it could range from 10^{-10} to 10^{-11} m^2 s^{-1}.

Thicknesses of roots show no consistent relationship with halophyte or nonhalophyte life-style, although roots of *Atriplex* are notably narrow with only three cells along the radius of the cortex. The root seems to be on the edge of a physical dilemma concerning the relative importance of water flow in the apoplast or symplast, but an alternative solution to the problem would be for the epidermis or hypodermis to form a barrier to ion, but not water diffusion. The importance of the hypodermis has been considered in a number of publications (39, 40), but further investigation is warranted for plants in saline conditions. An alternative barrier would be the Donnan phase of the cell walls which would tend to exclude and so restrict net salt flow, but this in turn would set up electrical potentials across the root.

J. Passioura (personal communication) has pointed out that there is a simple solution to the differential equation for flow with back-diffusion when written in radial coordinates and with no absorption along the path. In this case the

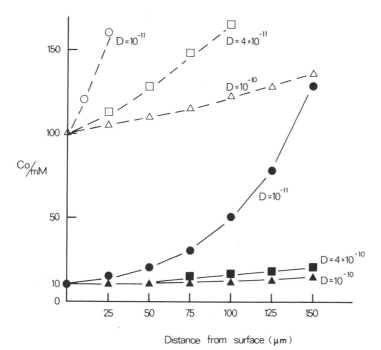

Figure 6.9. Calculated concentration of Na at increasing distances from the surface of the root into the cortex for a water flow of 10 mm^3 m^{-2} s^{-1} in 10 mM (closed symbols) or 100 mM (open symbols) NaCl. Diffusion coefficient in the apoplast varied from 10^{-11} to 10^{-10} m^2 s^{-1}.

concentration at the endodermis (radius x) is given by

$$C_x = C_0 \frac{b}{x}^{vb/Da}$$

where b is the outer radius of the root.

5. PLANT GROWTH AND RATE OF TRANSPORT FROM THE ROOT

The processes described above show where control of uptake can occur, due to selectivity in the symplast and release of solutes at the symplast–xylem interface. This section considers the extent of the "demand" from the shoot to the root for solutes, related to the need for solutes to fill developing cells or for osmotic adjustment. There are many examples in the literature that imply regulation of ionic concentrations in leaf tissues as part of the plant's adjustment to saline or water stress. The question, then, is whether regulation of concentration at the level of the leaf cell leads to regulation of solute output from root to shoot.

5.1. Relationships Between Concentration in the Leaf and Export from the Root

The rate of export from the root is J_R [also referred to as specific root transport (SRT) (36)], which is defined as

$$J_R = \frac{1}{W_R} \frac{dM_S}{dt} = \frac{1}{W_R} M_S \frac{d(\ln M_S)}{dt}$$

where M_S is the amount in the root. One approach would be to calculate functions fitting K_S (or $\ln M_S$) and W_R as a function of time and use them to determine J_R also as a function of time. However, the question of regulation of export involves plant growth and the *concentration* of solute in particular cells of the plant. The average "concentration" in the shoot can be defined as the amount of solute in the shoot divided by the shoot dry weight, or water content, that is, $C_S = M_S/W_S$. It then can be shown that

$$\frac{dC_S}{dt} + C_S R_S - J_R \frac{W_R}{W_S} = 0 \tag{1}$$

where R_S is the relative growth rate of the shoot and W_R is the root weight. It would be less ambiguous to use separate terms for content/dry weight and content/water content. The latter approximates best to an actual concentration and will be referred to as such, but "relative content" will be used for M_S (shoot dry weight).

This equation leads to two extreme situations, which are not uncommon. When the concentration in the shoot is constant, dC_S/dt is zero and

$$J_R = \frac{W_S}{W_R} C_S R_S \tag{2}$$

However, when growth is zero and solute is being used in osmotic adjustment,

$$J_R = \frac{W_S}{W_R} \frac{dC_S}{dt} \tag{3}$$

More commonly, the situation is intermediate with C_S showing some drift with time and R_S not being zero.

A consequence of constant relative content in the shoot, implied from Equation (2), is that the rate of transport from root to shoot is proportional to relative growth rate. Measurements of rates of transport of K^+ in barley seedlings grown on solutions of moderate (10 mM) K^+ concentration showed good proportionality to relative growth rate (30), and fluxes of K^+ into the root were also related to relative growth rate. This relationship between fluxes and growth has been established in more detail by Jeschke (41). Analysis of other data for barley (e.g., 10) shows that the average rate of transport from root to shoot is independent of external concentration to about 100 mM (K + Na)Cl as is also the average concentration of (K$^+$ + Na$^+$) in the shoot. However, these analyses with barley owe their apparent success to the small changes in fresh weight/dry weight ratio over the range of conditions studied. In many other plant species, both glycophyte and halophyte, changes in fresh weight/dry weight ratio are part of the plant's adjustment to water stress and then the simple relationship of Equation (1) needs modification.

If C_S in Equation (1) is expressed as the concentration relative to tissue water (mM), R then would be the relative increase in tissue water and W_S/W_R is (shoot water content/root fresh weight) or $(W_{S,F} - W_{S,D})/W_{R,F}$ where F and D refer to fresh and dry weights. If R_F and R_D are relative growth rates relative to fresh and dry weight, and W_F, and W_D refer to the shoot, then from Equation (1):

$$\frac{dC_S}{dt} + C_S \frac{R_F W_F - R_D W_D}{W_F - W_D} - J_{R,F} \frac{W_{R,F}}{W_F - W_D} = 0 \tag{4}$$

when C_S (mM) is regulated in the shoot so that $dC_S/dt = 0$, a more useful equation is

$$J_{R,F} = C_S \frac{R_F W_F - R_D W_D}{W_{R,F}} \tag{5}$$

Which would show how $J_{R,F}$ should change with changes in shoot relative water content, in order to maintain a particular concentration (C_S).

In examples where experimental interest is in C_S in a particular tissue, such as the growing region of the shoot (e.g., 26, 27) the overall transport from the root can be treated as the sum of a series of such terms as Equation (2) or Equation (5), and hence the input to a specific region can be calculated (see 26).

5.2. Interactions of Growth and Content

While the hypothetical "demand" of the shoot may set limits on rate of export from the root, inability of the root to meet this demand may lead to reduction in growth due to the relationship between turgor and stomatal conductance. Since turgor is equal to the difference between water potential of the leaf and vacuolar osmotic pressure, lower osmotic concentration will result in stomatal closure at higher water potentials (less negative). In practice, the balance between process controls in the leaf often leads to a compromise between osmotic concentration and size of leaf. The effect of atmospheric humidity on growth of *Atriplex halimus* at low salt concentrations (42) is an excellent example of this interaction.

Under saline conditions, the concentration of solutes in the shoot may be an indication of the plant's attempts to regulate its osmotic content. In certain

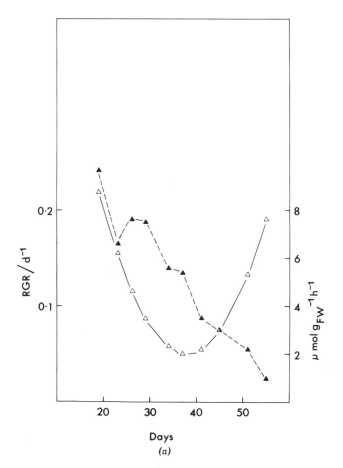

Figure 6.10. (*a*) Relative growth rate (day^{-1}) ($\triangle$) and rate of transport of K + Na from root to shoot (μmol g$_{FW}^{-1}$ root hr^{-1}) ($\blacktriangle$). (*b*) Concentration of K + Na in the shoot relative to tissue water (mM) ($\blacktriangle$) or to shoot dry weight (μmol g$_{DW}^{-1}$) ($\bigcirc$). Ratio of shoot fresh to dry weights ($\bullet$). Data from ref. 36.

species, the regulated osmotic component may be sugars (e.g., cereals), whereas in other species, particularly in halophytes, the concentrations may be ions (K$^+$, Na$^+$, Cl$^-$).

Shennan (36, 43) presented detailed analyses of the growth response and ionic content of *Aster tripolium* plants growing on varied levels of salinity (Fig. 6.7). From curves fitted to weight, ionic content, and water content, it was possible to estimate R and rates of solute transport as continuous functions, a technique that has much promise in studies of transport in whole plants (e.g., 44; see methods in 45).

The results showed characteristic responses of growth to salinity (42, 46) but there was no clear correlation between rate of transport and relative growth rate

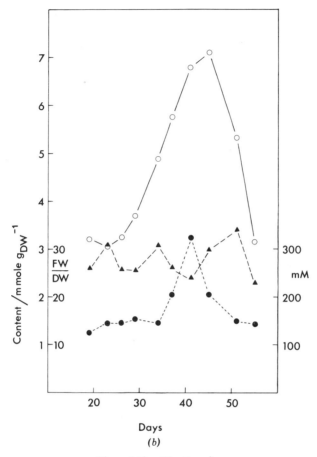

Figure 6.10. (*Continued*)

for all concentrations. Figure 6.10 shows more detailed data for *A. tripolium* plants growing on the 100 solution (containing 100 m*M* NaCl). Relative growth rate of the plants decreased from 19 days but increased again from about 40 days onward. However, ion transport from root to shoot (relative to root fresh weight) was unchanged, at least up to 29 days, and then fell gradually throughout the experiment. During the period of falling relative growth rate, the ratio of fresh to dry weight of the shoot increased and the concentration of $(K^+ + Na^+)$ relative to tissue water was essentially constant, though the relative content $(\mu mol/g_{DW})$ of the shoot increased considerably, falling again at the time that relative growth rate increased. It appears that the concentration in the leaf was being maintained at about 270 m*M* but that over the experimental period large changes in the ratio of fresh to dry weights resulted in changes in ionic content relative to dry weight. Although Equation (1) yields data for rate of transport relative to R and to C_S (dry weight basis), the more meaningful relationship

Table 6.8.   Effect of Transferring *Aster tripolium* from Low (CS10 Contains 10 m*M* NaCl) to High Salinity (CS100 Contains 100 m*M* NaCl) and the Reverse on Transport of Tracer K$^+$ and Na$^+$ from Root to Shoot[a]

	Rate of Transport	(pmol g_{FW}^{-1} hr^{-1})
Salinity Treatment	^{42}K	^{22}Na
CS10	1.5 ± 0.3	1.1 × 0.2
CS100	0.9 ± 0.2	7.3 × 1.6
CS10 → CS100	1.1 ± 0.3	7.3 ± 1.7
CS100 → CS10	0.9 ± 0.2	0.39 ± 0.1

[a]From ref. 36.

might be with C_S expressed on tissue water basis, since this seems to be the set point for control of leaf ionic concentrations [Equation (4)].

Other evidence for control of export from root to shoot was shown by transfer of *Aster* plants from high to lower salinity solutions which led to reduction in rate of export from root to shoot (Table 6.8).

5.3.   Interaction of Humidity and Salinity

Increase in ambient humidity to near saturation has two possible effects on the plant. One is to reduce water flow in the plant and the other is to reduce water stress, for example as in *A. halimus* (42). Plant species respond to varied humidity and salinity in different ways as shown by the following examples.

Figure 6.11 and Table 6.9 give results of a series of measurements using various plant species grown at two different humidities (30–35% and 95%) and on a range of salinities (50 m*M* and 100 m*M* NaCl for all except tomato) for up to 20 days. Measurements were based on two harvests about 10 days apart.

Table 6.9 shows the concentrations of (K$^+$ + Na$^+$) relative to tissue water at the second harvest. There are two responses that might be expected: one being an increase in concentration with external NaCl concentration and the other being lower concentrations at the lower humidity. The responses to external concentration and to humidity show particularly well for *Atriplex* and mung bean. The other species show some increase with external NaCl, but much less sensitivity to humidity.

Figure 6.11 summarizes the growth responses, water use, and rates of transport of (K$^+$ + Na$^+$) from the root to the shoot. Generally, the relative growth rates were reduced by increased external salinity and in the lower humidity, though there was less effect of salinity on *Atriplex* growth (dry weight). Water use (relative to root weight) was also reduced with increased salinity and as expected was lower in the high humidity.

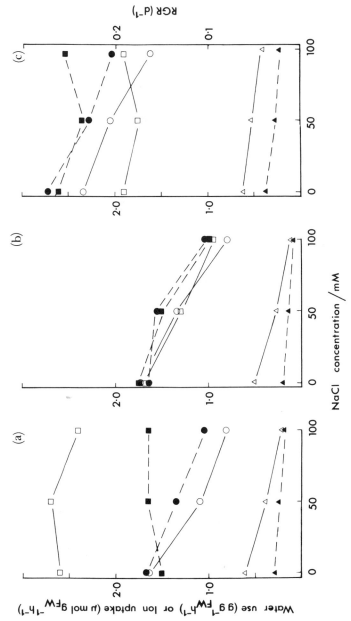

Figure 6-11 Measurements at higher humidity (closed symbols) and lower humidity (open symbols) of relative growth rates (day^{-1}; ●, ○) water use (g g$_{FW}$root^{-1}hr^{-1}; ▲, △) and (K + Na) transport from root to shoot (μmol g$_{FW}$root^{-1}hr^{-1}; ■, □) for (*a*) mung bean, (*b*) red kidney bean, (*c*) sunflower, (*d*) *Triticale*, (*e*) tomato, and (*f*) *Atriplex*. Data from ref. 51.

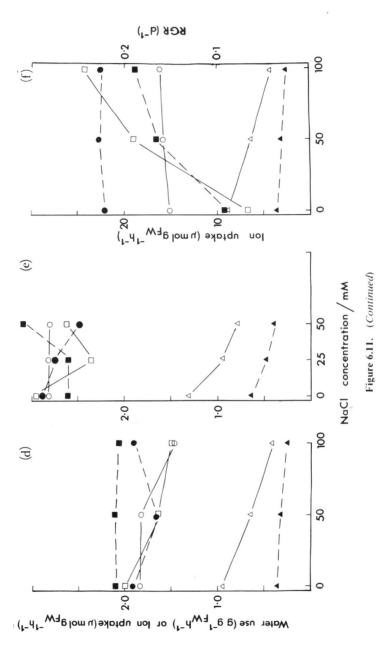

Figure 6.11. (*Continued*)

Table 6.9.   Concentrations of $(K^+ + Na^+)$ Relative to Tissue Water in the Shoot of Plants Grown on ½ Hoaglands Solution with Added NaCl and at Two Humidities

Plants	Control	Control + 25 mM NaCl	Control + 50 mM NaCl	Control + 100 mM NaCl
Mung bean				
Low humidity	195		300	385
High humidity	128		172	214
Red kidney bean				
Low humidity	127		139	165
High humidity	112		133	159
Sunflower				
Low humidity	150		166	227
High humidity	117		155	216
Tomato				
Low humidity	91	83	97	—
High humidity	60	64	83	—
Triticale				
Low humidity	232		257	313
High humidity	199		261	291
Atriplex				
Low humidity	300		367	495
High humidity	222		327	333

There was much more variability in the rates of transport from root to shoot. In mung bean and *Atriplex* plants the rate of transport was greater at the lower humidity, and was then associated with higher concentrations in the shoot (Table 6.9). In red kidney bean there was little effect of humidity (and hence water flow) on export from the root, but a reduction with increased salinity proportional to relative growth rate. For *Triticale*, sunflower, and tomato, the rate of transport was more closely related to relative growth rates, being higher at the high humidity than at the low humidity in certain cases.

Though not shown in Table 6.9 and Figure 6.11, *Atriplex* also showed changes in fresh to dry weight ratio during the experiment, as already described for *Aster*. For example, $(K^+ + Na^+)$ expressed on a dry weight basis was 30% higher at the second than at the first harvest on the 50 mM NaCl solution, but there was no significant difference in $(K^+ + Na^+)$ expressed relative to tissue water (mM).

These results illustrate that the response of transport to salinity can have a complex relationship to growth, to water flow, and to the nature of osmotic adjustment in the plant.

5.4. Adjustment to Salinity

It is well known that sudden salinization can lead to massive, nonselective uptake of Na^+ and Cl^- to the root and shoot, compared with gradual increases in salinity. For example, Storey and Wyn Jones (24) studied uptake of Na^+ and Cl^- to the barley variety California Mariout over a period of 96 hr after transfer to culture solution containing 250 mM NaCl and found there was rapid uptake of Na^+ and Cl^- to root and shoot and loss of K^+. After 48 hr the contents of the root and shoot were

Shoot: $K^+ = 1.78$; $Na^+ = 3.2$; $Cl^- = 3.25$ mmol/g_{DW}
Root: $K^+ = 0.2$; $Na^+ = 1.1$; $Cl^- = 0.55$ mmol/g_{DW}

Plants taken gradually to the same NaCl concentration and held at pseudo-steady state contained:

Shoot: $K^+ = 0.8$; $Na^+ = 1.8$; $Cl^- = 1.0$ mmol/g_{DW}
Root: $K^+ = 0.2$; $Na^+ = 2.6$; $Cl^- = 0.8$ mmol/g_{DW}

During the abrupt transfer to NaCl there was a large and nonselective transport of Na^+ and Cl^- to the shoot lasting nearly 48 hr, with less salt accumulation in the roots. The large nonselective uptake could be due to plasmolysis of root cells, destroying the barrier set up by the symplast, but it could also be due to a change in membrane permeability. Examples of changed membrane transport following transfer from low to higher concentration solutions have been found for *Lamprothamnion* (48), and could be due to changes in K^+ or Na^+ channels in the cell membranes as found for *Chara* (49) and *Hydrodictyon* (50).

5.5. Plant Death and Osmoregulation

Accumulation of solutes may lead to increased solute concentration in the leaves as part of osmotic adjustment, but this same accumulation may go beyond the

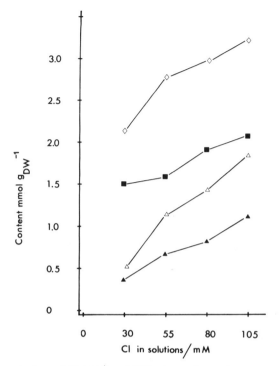

Figure 6.12.   Concentrations of $(K + Na^+)$ and Cl^- in shoots of mung beans grown on solutions containing varied NaCl or KCl, and Cl concentration as shown. NaCl plants: $K^+ + Na^+$ (■), Cl^- (▲). KCl plants: K^+ (◇), Cl^- (△). Data from ref. 51.

limits of regulation of cytoplasmic content with associated impairment of growth.

Figure 6.12 shows $(K^+ + Na^+)$ and Cl^- in shoots of mung beans grown on solutions containing either NaCl or KCl added to a basic nutrient solution (51). Plants grown on K^+ solutions without Na^+ took up more Cl^- from solution than the other series. Initially, this was accompanied by greater growth and there was little difference between controls and the plants growing on solutions with added KCl, but eventually growth was reduced as Cl^- accumulated in the leaves. There was good correlation between reduction in relative growth rate and either Cl^- or $(K^+ + Na^+)$ content of the shoot. There was no protection to the plant from effective feedback between the level in the shoot and the rate of export from the root in mung beans.

Previous examples have emphasized the need for regulation of export as a positive feedback so that the root can meet the demand of the shoot for $K^+ + Na^+$ for growth and osmotic adjustment. The example in this section shows the need for there to be effective negative feedback control reducing export if the plant is to avoid damage due to excess uptake. This aspect has been explored less than the positive feedback situation, but clearly warrants attention, particularly for plants growing in saline conditions.

6. CONCLUSIONS

Salinity appears to affect growth due to either toxic effects of Na^+ or Cl^- accumulation or the low osmotic potential of the soil or solution. The plant can avoid or minimize toxic effects by excluding salt from the plant, excreting it from glands, or translocating it to leaves that then drop from the plant, but in excluding salt, the plant may lose the opportunity of using NaCl as an osmotic solute in the leaves (1). Alternatively, plants may accumulate salts in the leaves, providing lower osmotic potentials but then may need to exclude salts from the cytoplasm to avoid ionic interactions with enzymatic reactions.

Studies of the kind discussed in this chapter show that export from the roots and the concentration of solutes in the leaves are related to growth and osmotic adjustment or regulation. An equally important approach, that cannot be included here, is the effect of endogenous salt content on rates of photosynthesis. Some progress has been made in determining whether stomatal conductance or photosynthetic efficiency limits growth by using gas exchange techniques. This approach can also give information about the efficiency of carboxylation by measuring the relationship between photosynthesis and internal CO_2 concentration, but here an open question is whether reduction in photosynthesis results from reduction in growth, due to the small sink for photosynthetic products or whether growth reduction is due to inhibitory effects of NaCl on photosynthesis directly. It is likely that this area of research will be particularly valuable in the next few years and have potential for better assessment of what is meant by salt tolerance in intact plants.

This chapter emphasizes the complex interaction in the plant between such processes as ion uptake to the root, transport to the shoot, osmotic solute regeneration in leaves, transpiration, phloem translocation, and growth. Truly, the whole is more than the sum of its parts.

ACKNOWLEDGMENTS

I am most grateful to C. Shennan, R. Storey, and M. Salim for use of their material and for valuable discussions.

REFERENCES

1. H. Greenway and R. Munns. Mechanisms of salt tolerance in non-halophytes. *Annu. Rev. Plant Physiol.* **31**, 149–190 (1980).
2. M. K. Smith and J. A. McComb. Effect of NaCl on the growth of whole plants and their corresponding callus cultures. *Aust. J. Plant Physiol.* **8**, 267–275 (1981).
3. D. J. F. Bowling. *Uptake of Ions by Plant Roots.* Chapman and Hall, London, 1976.

4. D. T. Clarkson. *Ion Transport and Cell Structure in Plants.* McGraw-Hill, London, 1974.

5. U. Lüttge and N. Higinbotham. *Transport in Plants.* Springer-Verlag, New York, 1979.

6. P. H. Nye and P. B. Tinker. *Solute Movement in the Soil-Root System,* Blackwell, Oxford, 1977.

7. D. T. Clarkson and J. B. Hanson. The mineral nutrition of higher plants. *Annu. Rev. Plant Physiol.* **31**, 239–298. (1980).

8. A. Läuchli. Translocation of Inorganic Solutes. *Annu. Rev. Plant Physiol.* **23**, 197–218 (1972).

9. U. Lüttge. Co-operation of organs in intact higher plants: A review. In U. Zimmerman and J. Dainty, eds., *Membrane Transport in Plants.* Springer-Verlag, Berlin, 1974, pp. 353–362.

10. M. G. Pitman. Ion uptake by plant roots. In U. Lüttge and M. G. Pitman, eds., *Transport in Plant II.* Springer-Verlag, Berlin, 1976, pp. 95–128.

11. M. G. Pitman. Ion transport into the xylem. *Annu. Rev. Plant Physiol.* **28**, 71–88 (1977).

12. M. G. Pitman. Transport across roots. *Q. Rev. Biophys.* **15**, 481–554 (1982).

13. D. J. F. Bowling. Release of ions to the xylem in roots. *Physiol. Plant.* **53**, 392–397 (1981).

14. M. G. Pitman. Sodium and potassium uptake by seedlings of *Hordeum vulgare. Aust. J. Biol. Sci.* **18**, 10–24 (1965).

15. M. G. Pitman and H. D. W. Saddler. Active sodium and potassium transport in cells of barley roots. *Proc. Natl. Acad. Sci. USA* **57**, 44–49 (1967).

16. M. G. Pitman, A. Lauchli, and R. Stelzer. Ion distribution in roots of barley seedlings measured by electron probe X-ray microanalysis. *Plant Physiol.* **60**, 673–679 (1981).

17. R. F. M. Van Steveninck, M. E. Van Steveninck, R. Stelzer, and A. Läuchli. Electron probe X-ray microanalysis of ion distribution in *Lupinus Luteus:* I. Seedlings exposed to salinity stress. In R. M. Spanswick, W. J. Lucas, and J. Dainty, eds., *Plant Membrane Transport: Current Conceptual Issues.* Elsevier/North-Holland, Amsterdam, 1980, pp. 489–492.

18. D. Kramer, W. P. Anderson, and J. Preston. Transfer cells in the root epidermis of *Atriplex hastata* L. as a response to salinity: A comparative cytological and X-ray microprobe investigation. *Aust. J. Plant Physiol.* **5**, 739–747 (1978).

19. W. D. Jeschke. Roots: Cation selectivity and compartmentation, involvement of protons and regulation. In R. M. Spanswick, W. J. Lucas, and J. Dainty, eds., *Plant Membrane Transport: Current Conceptual Issues.* Elsevier/North-Holland, Amsterdam, 1980, pp. 17–28.

20. M. G. Pitman, M. Schaefer, and R. A. Wildes. Relation between permeability to potassium and sodium ions and fusicoccin-stimulated hydrogen-ion efflux in barley roots. *Planta* **126**, 61–73 (1975).

21. M. G. Pitman. Uptake of potassium and sodium by seedlings of *Sinapis alba. Aust. J. Biol. Sci.* **19**, 257–269 (1966). F. W. Smith. The effect of sodium on potassium nutrition and ionic relations in rhodes grass. *Aust. J. Agric. Res.* **25**, 407–414 (1974).

22. M. G. Pitman. Transpiration and the selective uptake of potassium by barley seedlings. *Hordeum vulgare* cv. bolivia. *Aust. J. Biol. Sci.* **18**, 987–999 (1965).

23. H. Greenway. Plant response to saline substrates: VII. Growth and ion uptake throughout plant development in two varieties of *Hordeum vulgare. Aust. J. Biol. Sci.* **18**, 763–779 (1974).

24. R. Storey and R. G. Wyn Jones. Salt stress and comparative physiology in the Gramineae: I. Ion relations of two salt- and water-stressed barley cultivars, California mariout and arimar. *Aust. J. Plant Physiol.* **5**, 801–816 (1978).

25. R. G. Wyn Jones and R. Storey. Salt stress and comparative physiology in the Gramineae: IV. Comparison of salt stress in *Spartina X townsendii* and three barley cultivars. *Aust. J. Plant Physiol.* **5**, 839–850 (1978).

26. R. Delane, H. Greenway, R. Munns, and J. Gibbs. Ion concentration and carbohydrate status of the elongating leaf tissue of *Hordeum vulgare* growing at high external NaCl: I. Relationship between solute concentrations and growth. *J. Exp. Bot.* **33**, 557–573 (1982).

27. R. Munns, H. Greenway, R. Delane, and J. Gibbs. Ion concentration and carbohydrate status of the elongating leaf tissue of *Hordeum vulgare* growing at high external NaCl: II. Cause of the growth reduction. *J. Exp. Bot.* **33**, 574–583 (1982).

28. H. Greenway and D. A. Thomas. Plant response to saline substrates. *Aust. J. Biol. Sci.* **18**, 505–24 (1965). H. Greenway, A. Gunn, M. G. Pitman, and D. A. Thomas. Plant response to saline substrates: VI. Chloride, sodium and potassium uptake and distribution within the plant during ontogenesis of *Hordeum vulgare. Aust. J. Biol. Sci.* **18**, 525–540 (1965).

29. H. Lessani and H. Marschner. Relation between salt tolerance and long-distance transport of sodium and chloride in various crop species. *Aust. J. Plant Physiol.* **5**, 27–37 (1978).

30. M. G. Pitman. Uptake and transport of ions in barley seedlings: III. Correlation of potassium transport to the shoot with plant growth. *Aust. J. Biol. Sci.* **25**, 905–919 (1972).

31. B. Jacoby. Function of bean roots and stems in sodium retention. *Plant Physiol.* **39**, 445–449 (1964).

32. D. Kramer, A. Läuchli, A. R. Yeo, and J. Gullasch, Transfer cells in roots of *Phaseolus coccineus:* Ultrastructure and possible function in exclusion of sodium from the shoot. *Ann. Bot.* **41**, 1031–1040 (1977).

33. R. Storey, M. G. Pitman, R. Stelzer, and C. Carter. X-ray microanalysis of cells and cell compartments of *Atriplex spongiosa:* 1. Leaves. *J. Exp. Bot. 34:778–794* (1983).

34. R. L. Jefferies. The ionic relations of seedlings of the halophyte *Triglochin maritima* L. In W. P. Anderson, ed., *Ion Transport in Plants.* Academic Press, London, 1973, pp. 297–321.

35. D. M. R. Hervey, J. L. Hall, T. J. Flowers, and B. Kent. Quantitative ion localisation within Suaeda maritima leaf mesophyll cells. *Planta,* **151**, 555–560 (1981).

36. C. Shennan, Salt tolerance in *Aster tripolium* L, Ph.D. thesis. University of Cambridge, 1981.

37. R. Storey, M. G. Pitman, and R. Stelzer. X-ray microanalysis of cells and cell compartments of *Atriplex spongiosa:* 2. Roots. *J. Exp. Bot.* 34:1196–1206 (1983).

38. W. P. Anderson, D. A. Willcocks, and B. J. Wright. Electrophysiological measurements on the root of *Atriplex hastata. J. Exp. Bot.* **28**, 894–901 (1977).

39. D. T. Clarkson, A. W. Robards, J. Sanderson, and C. A. Peterson. Permeability studies on epidermal—Hypodermal sleeves isolated from roots of *Allium cepa* (onion). *Can. J. Bot.* **56**, 1526–1532 (1978).

40. C. A. Peterson, M. E. Emanuel, and G. B. Humphreys. Pathway of movement of apoplastic fluorescent dye tracers through the endodermis at the site of secondary root formation in corn (*Zea mays*) and broad bean *Vicia faba. Can. J. Bot.* **59**, 618–625 (1981).

41. W. D. Jeschke. Shoot-dependent regulation of sodium and potassium fluxes in roots of whole barley seedlings. *J. Exp. Bot.* **33**, 601–618 (1982).

42. J. Gale, R. Naaman, and A. Poljakoff-Mayber. Growth of *Atriplex halimus* L. in sodium chloride salinated culture solutions as affected by the relative humidity of the air. *Aust. J. Biol. Sci.* **23**, 947–952 (1970).

43. C. Shennan, R. Hunt, and E. A. C. Macrobbie. Salt tolerance in Aster tripolium. 1. The effect of salinity on growth. *J. Exp. Bot.* submitted (1983).

44. N. J. Halse, E. A. N. Greenwood, P. Lapins, and C. A. P. Boundy. An analysis of the effects of nitrogen deficiency on the growth and yield of a Western Australian wheat crop. *Aust. J. Agric. Res.* **20**, 987–998 (1969).

45. R. Hunt and G. C. Evans. Classical data on the growth of maize: curve fitting with statistical analysis. *New Phytol.* **86**, 155–180 (1980). R. Hunt. *Plant Growth Curves.* Arnold, London, 1982.

46. T. J. Flowers, P. F. Troke, and A. R. Yeo. The mechanism of salt tolerance in halophytes. *Annu. Rev. Plant Physiol.* **28**, 89–112 (1977).

47. M. Salim and M. G. Pitman. Interaction of water-flow, growth and humidity on salt uptake. Unpublished information (1983).

48. R. J. Reid, R. L. Jefferies, and M. G. Pitman. *Lamprothamnion*, a euryhaline charophyte: IV. Membrane potential, ionic fluxes and metabolic activity during turgor adjustment. *J. Exp. Bot.* submitted (1983).

49. D. W. Keifer and W. J. Lucas. Potassium channels in *Chara corallina* control and interaction with the electrogenic H. pump. *Plant Physiol.* **69**, 781–788 (1982).

50. G. P. Findlay. Electrogenic and diffusive components of the membrane of *Hydrodictyon africanum. J. Membr. Biol.* **68**, 179–189 (1982).

51. M. Salim and M. G. Pitman. Comparison of KCl and NaCl in salinity on growth of mung beans. *Aust. J. Plant Physiol.* in press (1983).

7

STRUCTURAL, BIOPHYSICAL, AND BIOCHEMICAL ASPECTS OF THE ROLE OF LEAVES IN PLANT ADAPTATION TO SALINITY AND WATER STRESS

Ulrich Lüttge and J. Andrew C. Smith

Institut für Botanik
Technische Hochschule Darmstadt
Darmstadt, Federal Republic of Germany

The stress effects of salinity are generally summarized as water stress, salt stress, and ion-imbalance stress. In the order cited we are dealing with increasingly specific problems of salinity. *Water stress* is a general condition, and we can expect that adaptations of plants to salinity may show features similar to those characteristic of adaptations to drought. *Salt stress* arises from excessive uptake of salt by the plants. This is a specific and unavoidable consequence of high ion concentrations in the medium, but at the same time it is part of the adaptive mechanism to the water stress associated with salinity. Here, salt is accumulated as a readily available osmoticum to maintain a downhill water-potential gradient ($\Delta \Psi$) between the medium and the various plant organs, thus allowing continued water uptake. *Ion imbalance* specifically results from disturbed ionic ratios in the cells after predominant accumulation of ions of the salt responsible for the salinity of the medium, NaCl being the most widely encountered example.

The integrated function of various organs in the plant as a whole (1; 2, section III; 3, chap. 13) allows adaptations at various levels of the plant. Thus, ion selectivity at the level of the roots (4) and specific reabsorption of ions from the xylem during transport to the shoot (5) can modify the degree of salt and ion-imbalance stress in the leaves. Resistance to water transport may be involved in water stress adaptation both at the level of the roots [increased root resistance in response to salinity (6)] and leaves [increased mesophyll and stomatal resistances in response to salinity (e.g., 7, 8)].

Leaves constitute the major sites of transpiration and photosynthesis in higher plants. In relation to salinity-induced water stress, one might thus expect the principal structural and metabolic modifications in leaves to be associated with a tendency to minimize transpiration rate and the occurrence of photosynthetic pathways with high water use efficiency (WUE). Two specific aspects are of importance here. First, rather than considering the absolute amounts of water lost in transpiration, we can more usefully discuss phenotypic variation in terms of the relationship between water loss and carbon gain under different environmental conditions; this is embodied in the optimization theory of Cowan and Farquhar (9, 10). Second, there is no implicit correlation between WUE and resistance to water stress: WUE is simply a measure of the efficiency with which *available* water is used during the production of dry matter, but says nothing about plant responses to *changes* in water availability (11).

What forms of structural and metabolic adaptation are found in leaves under conditions of salinity and water stress? Following Jennings (12), we will use the terms "xerophyte" and "halophyte" in this chapter in a broad sense to designate species that inhabit arid or saline environments, respectively. We note that aridity is not necessarily associated with salinity, but that in many deserts the salt factor may be of larger ecological significance than the water factor (13, 14). Thus, no attempt is made to distinguish between the multitude of putative "physiotypes" (cf. 13). We do, however, recognize the distinction made by van Eijk (15) in describing species that not only tolerate, but actually grow better, in saline environments: these are referred to as "halophilic".

One feature typical of halophytic species of Poaceae, Cyperaceae, and Junaceae is sclerophylly, which is characteristic of many xerophytes. Among dicotyledonous halophytes the dominant structural feature shared with many xerophytes is succulence. As regards leaf metabolism, many halophytes carry out either C_4 photosynthesis or crassulacean acid metabolism (CAM) (16–18) and as a consequence show high values of WUE. There is also a measure of interdependence between these two aspects of leaf adaptation. For example, some species of Aizoaceae shift from C_3 photosynthesis to CAM in response to salinity (19), and CAM is characteristically associated with a succulent leaf morphology (20). Leaf succulence is also highly developed in an number of C_4 halophytes, e.g., species of *Suaeda* and *Seidlitzia* (13). The nature of these relationships will be the main theme of this chapter.

The other principal structural adaptation found in the leaves of many halophytes is the presence of salt-eliminating glands and hairs. Their structure and function have been reviewed previously (21, 22) and will not be considered further in this chapter. We shall first discuss the biochemical and biophysical characteristics of succulent leaves, with emphasis on the properties of the cells and tissues. In the latter part we shall consider how these characteristics are related to plant performance in saline and arid environments.

1. THE FUNCTION OF SUCCULENCE

1.1. Definitions of Succulence

Succulence is essentially a morphological term referring to thick and fleshy plants organs. These have reduced surface to volume or surface to fresh weight ratios. One convenient measure of succulence is the ratio of water content to surface area for a tissue at equilibrium with a water-saturated atmosphere (23). Expressed in these ways, the distinction between "succulent" and "fleshy" is obviously a gradual one (24).

In anatomical terms succulent tissues are usually rather homogeneous and consist of large, isodiametrical (often spherical) cells. The cells have large central vacuoles and the cytoplasm makes up only a very small proportion of the total cell volume. This leads to low dry weight (DW) to fresh weight (FW) ratios in succulent tissues. For CAM, which is found only in succulent tissues whose cells contain chloroplasts, a definition of succulence has been proposed based on the ratio of chloroplasts (or chlorophyll) to vacuolar volume (25).

Although a morphological feature, succulence is clearly tied up with leaf function. A physiological definition of succulence is needed but hard to formulate. Functionally, succulence requires the storage of both water and osmotically active solutes like inorganic salts or organic metabolites, or both. A physiological definition may thus emerge from biophysical evaluations of water relations and membrane transport of solutes in succulent plants (26).

1.2. Xerophytes and Halophytes

Succulent organs of xerophytes possess internal or peripheral water-storage tissue (27), though this tissue is nonphotosynthetic. The high cuticular and stomatal resistances typical of xerophytes prevent rapid water loss from the organs to the atmosphere under conditions of drought. However, osmotically active solutes are required in addition for the establishment and maintenance of cell turgor pressure (P), which is given by the equation

$$\Psi = P - \pi \tag{1}$$

where π is the osmotic pressure of the cell sap and Ψ is the cell water potential.

In most higher plants the principal osmoticum is K^+ associated mainly with organic acid anions as counterions (28). In halophytes, organic acid content decreases in response to salinity (29, 30) and inorganic salts, mostly NaCl taken up from the saline medium, become the major osmotica (28). Although the role of organic acids in adaptation to salinity is restricted in this way, they still are at least partially important as osmotica, and organic acid metabolism remains essential (13). This applies especially to the Chenopodiaceae, where oxalate levels can actually increase in response to salinity (31).

In contrast to the xerophytes, the succulent tissues of halophytes are usually photosynthetic. This structural specialization in the leaves can be superimposed on any one of the three major metabolic pathways associated with photosynthetic carbon assimilation, viz., C_3- or C_4-photosynthesis or CAM (e.g., 13). One interesting morphological type is represented by the halophilic annual *Mesembryanthemum crystallinum*, which has succulent photosynthetic tissue but, in addition, large epidermal bladder cells (without chloroplasts) that form a peripheral water-storage tissue. These cells are covered by a cuticle that is almost impermeable to water and solutes (32). The bladders conceal the stomata between them, and this may also serve to increase transpirational resistance (33). When plants are grown under saline conditions the bladders become larger: they can reach a size of 2.7×10^{-9} m^3 and cover all aerial parts of the plant rather densely, making up as much as 25% of the shoot volume (34). In response to decreasing Ψ in the external medium brought about by salinity, the bladder-cell sap π increases markedly and the cell P (as measured directly with the Jülich pressure probe) decreases somewhat. At both of these salinities a downhill gradient for water transport is maintained between the medium and the bladders

Table 7.1.   Water-Relation Parameters of Epidermal Bladders of *Mesembryanthemum crystallinum* at Various Degrees of NaCl Salinity[a]

Medium	Water Potential (Ψ)	Turgor Pressure (P)	Pressure Osmotic (π)	Water Potential Difference Between Bladder and Medium ($\Delta \Psi$)
Soil culture without NaCl	-1.29	$+0.35 \pm 0.08$ (24)	$+1.64 \pm 0.40$ (7)	
After 5 weeks in hydroculture with 100 mol m^{-3} NaCl ($\Psi = -0.49$ MPa)	-1.98	$+0.22 \pm 0.09$ (5)	$+2.20 \pm 0.03$ (2)	-1.49
After 1–5 weeks in hydroculture with 400 mol m^{-3} NaCl ($\Psi = -1.85$ MPa)	-3.24	$+0.15 \pm 0.07$ (15)	$+3.39 \pm 0.68$ (5)	-1.39

[a] Values given in MPa with SD and number of replicates. After refs. 17 and 26.

(Table 7.1). Although NaCl is the major osmoticum in the *M. crystallinum* bladders, analyses of Na^+, Cl^-, and K^+ show that for reasons of charge balance other anions, presumably mainly oxalate, must be involved in addition (34).

Table 7.1 thus serves to emphasize an important point, which is that the development of tissue succulence (and the maintenance of positive cell *P*) requires the uptake and storage of both water *and* solutes.

1.3. CAM Plants

Malic Acid Storage

The succulence of CAM plants is usually considered to be a prerequisite for the accumulation of large amounts of organic acids. In CAM malic acid is synthesized during the night by CO_2 dark fixation via the cytosolic enzyme PEP carboxylase. The total tissue concentration can reach 220 mmol malic acid kg^{-1} FW in the CAM plant *Kalanchoë daigremontiana*. During the day malic acid is decarboxylated and the CO_2 refixed in the light via the Calvin cycle. Because PEP carboxylase is feedback inhibited by malate and in order to maintain the slightly alkaline cytoplasmic pH, the malic acid accumulated during the night must be sequestered in the vacuoles. Cytoplasmic malic acid concentrations must presumably be kept at <10 mM(35). Succulence and a high ratio of vacuolar volume in relation to total cell volume seem to be required to accommodate the large amounts of malic acid produced during the night. Indeed, in mature leaves of *K. daigremontiana* the large and uniform mesophyll cells (diameter $90.9 \pm 14.1 \mu m$, mean $\pm$ SD, $n = 266$ cells) have only 0.5–1.0% cytoplasm and 1–2% cell wall, the rest of the cell volume being occupied by the vacuole (26).

Considering these cell dimensions exclusively in relation to malic acid storage, is however, an oversimplification. Let us assume, for example, that cytoplasmic malic acid content is negligible and that all the malic acid is localized in the vacuole. Then at 220 mmol kg^{-1} FW and a cytoplasm to vacuole ratio of 1:99 ($\equiv$ 1% cytoplasm) the vacuolar malic acid concentration would remain close to this value (222 mol m^{-3}). But even at ratios of 1:9 ($\equiv$ 10% cytoplasm) or 1:4 ($\equiv$ 20% cytoplasm) vacuolar malic acid concentrations would not be excessively high (245 and 278 mol m^{-3}, respectively). Therefore, the capacity of the vacuoles for storage of malic acid is probably not limited by the malic acid concentration *per se*.

The driving force for malic acid accumulation in the vacuole of CAM cells has been suggested to be provided by a proton-pumping ATPase at the tonoplast (36, 37). Two protons are always accumulated together with one malate anion during the dark phase (35), and the protons must be actively transported into the vacuole (36). It is thus possible that the upper limit for malic acid accumulation is determined by the magnitude of the electrochemical proton difference ($\Delta\mu_{H^+}$) at the tonoplast against which the protons have to be pumped.

From measured values of vacuolar pH and the electrical potential difference across the tonoplast, together with assumed values of cytosolic pH, it has been calculated that the $\Delta\mu_{H^+}$ across the tonoplast in *Kalanchoë* species increases to about 26 kJ mol^{-1} at the end of the dark phase (37). Computed estimates of the free energy available from hydrolysis of ATP in the cytosol give a value about twice this (38), suggesting that the postulated proton pump at the tonoplast might be operating as a 2H$^+$-ATPase. The possibility that the proton pump could function as a 1H$^+$-ATPase under strong "kinetic" control seems to be excluded on metabolic grounds, as the ATP available from respiration and glycolysis during the night is not sufficient to support such a mechanism (37). Hence, bioenergetic considerations may set an upper limit to the concentration to which malic acid can be accumulated in the vacuole. This would be one factor favoring a low cytoplasm to vacuole ratio.

Of course, the buffering capacity of the vacuole will influence the magnitude of $\Delta\mu_{H^+}$ across the tonoplast for a given vacuolar concentration of malic acid. This varies in different CAM plants. For example, for a malic acid concentration of 150 mmol kg^{-1} FW the vacuolar pH in *K. daigremontiana* is 4.0, whereas in *K. tubiflora* it is 3.3 (37). The higher buffering capacity in *K. daigremontiana* seems to permit higher malic acid concentrations: *K. tubiflora* only accumulates up to 150 mmol kg^{-1} FW, as compared with the 220 mmol kg^{-1} FW attainable by *K. daigremontiana*.

Water Storage

A second aspect of the relationship between succulence and CAM is found in the particular water relations of CAM plants. Very few data are available for water-relation parameters of succulent organs at the tissue and cell levels. The Scholander pressure chamber technique was long considered to be inapplicable to fleshy tissues, and a wider use of the Jülich pressure probe in the investigation of ecophysiological problems is only just beginning. Some results obtained with the prototype of the pressure probe on the epidermal water-storage tissue of the annual halophyte and facultative CAM-plant *M. crystallinum* have been discussed above (Table 7.1). More recently, the pressure chamber has been used to determine the bulk water-relation parameters of CAM leaves at the tissue level (39) and the miniaturized pressure probe has provided data at the cellular level (26).

With CAM species it is possible to construct "pressure–volume curves" using the pressure chamber (40, 41), providing that a pressure seal is used to protect the relatively soft petioles during the course of the experiment. This is an extremely useful technique for obtaining information on water-relation parameters at the bulk-tissue level (42). Table 7.2 presents some data for three CAM species (all Crassulaceae). The ratio of the volume of osmotic (=symplastic) water at zero turgor to the volume at full turgor (V_p / V_o) shows that the cells retain a large proportion of the osmotic water at incipient plasmolysis. The osmotic water contributes close to 90% of the total fresh weight ($V_o/$FW) or

Table 7.2.   Relative Water Contents[a] of the Leaves of Three CAM Species[b] Compared with Average Values of Mesophytic C₃ Plants[c]

Plant	V_p/V_o	V_o/FW	$V_o/(FW-DW)$	n
CAM plants				
Kalanchoë	92.4 ± 1.6	88.6 ± 2.0	93.0 ± 2.2	5
daigremontiana				
Sedum praealtum	94.0 ± 1.6	91.1 ± 6.6	98.3 ± 7.2	5
Crassula obliqua	94.0 ± 2.2	88.8 ± 5.7	96.7 ± 6.4	6
C₃ plants	60 to 85	30 to 80	60 to 95	

[a] Values are mean $\% \pm$ SD. See text for explanation of symbols.
[b] From ref. 39 and unpublished.
[c] From ref. 40.

more than 95% of the total leaf water [$V_o/(FW-DW)$]. These values are higher than those generally observed in mesophytic C₃ plants (Table 7.2) and provide a quantitative index of the succulence of the CAM species.

The pressure probe, on the other hand, allows very precise measurements to be made of the water-relation characteristics of individual cells in a tissue (43). One parameter that can be determined using the pressure probe technique is the cell volumetric elastic modulus, ϵ. This relates a fractional change in cell volume (V) to the corresponding change in cell P, and is higher the less elastic the cells. For *K. daigremontiana* $\epsilon = 4.24 \pm 2.77$ MPa (mean $\pm$ SD, $n = 21$ cells) for an average cell P is 0.170 MPa (26). This value is similar to those obtained for cells of nonsucculent higher plants (43). However, in at least some plants there is a demonstrable relationship between cell volume and ϵ, the larger cells having higher ϵ values (43). If this relationship holds for interspecific comparisons, it may turn out (when more data are available) that *K. daigremontiana* cells have a relatively low ϵ considering their large volume (average 0.42×10^{-12} m³). This would presumably be a reflection of the thin cell walls characteristic of the CAM tissue.

The value of ϵ is also important in so far as it determines the water-storage capacity of the cells (C_c), which is given by the equation

$$C_c = \frac{V}{\epsilon + \pi^i} \tag{2}$$

where π^i is the osmotic pressure of the cell sap. For *K. daigremontiana* C_c is 9.1×10^{-20} m³ Pa⁻¹ cell⁻¹, and the capacity for the bulk tissue is 2.2×10^{-10} m³ Pa⁻¹ Kg⁻¹ FW (26). Although comparative data for C_c from other species are almost entirely lacking, we might expect the value for *K. daigremontiana* to be relatively high. This follows from the fact [see Equation (2)] that V is large and that ϵ is (probably) relatively small; π^i is also relatively small but is of less significance, because $\epsilon \gg \pi^i$. The value of C_c is likely to be of considerable importance for the water economy of CAM plants. For example, it can be shown for *K. daigremontiana* that the amount of water that could be stored or released for relatively small changes in cell P (say 0.1–0.2 MPa) is comparable to the amount lost during the entire dark period in transpiration (26).

Vacuolar Osmotica

The malic acid stored in CAM vacuoles in large amounts constitutes a major cellular osmoticum, and the large day–night changes in malic acid levels are associated with marked changes in leaf water relations. Measurements with *K. daigremontiana* using equilibration of leaf slices with mannitol solutions of varied water potentials or the pressure chamber to determine Ψ, and cryoscopy to measure π, have shown that Ψ decreases and π increases during the phase of nocturnal malic acid accumulation (35, 44). Cell P, calculated according to Equation (1), tends to remain low during the dark period, but it increases after a short lag phase at the beginning of the light period. This increase in cell P can be explained by the stomatal closure characteristic of CAM plants during malic acid remobilization in the light period. At the very beginning of the light phase there is a big reduction in the transpiration rate (and increase in leaf Ψ) but only a slight decrease in malic acid content (and cell sap π) (20). As a consequence cell P increases [see Equation (1)]. Thus, the malic acid plays an important role as a vacuolar osmoticum, providing the osmotic driving force for water uptake by the leaf tissue at the start of the light period. We are not assuming that the cell sap with these high solute concentrations behaves osmotically like an ideal solution. However, it can be shown that a large proportion of the day-night change of malic acid is osmotically effective (35).

Of course, cell turgor and the succulence of the leaf must be maintained throughout the day–night rhythm, even at times when the malic acid levels are low. Other organic acid anions are found in the vacuole (20), for example, citrate and isocitrate, as well as a component of the total malate that does not participate in the CAM rhythm (45). These anions are charge balanced by inorganic cations, of which Ca^{2+} and Mg^{2+} are by far the most important (13, 45, 46).

1.4.   Osmoregulation between Cytoplasm and Vacuole

Because the tonoplast is not mechanically supported by the cell wall, no significant difference in hydrostatic pressure can be maintained across this membrane. Hence, the $\Delta\Psi$ across the tonoplast must be equivalent to $\Delta\pi$, and at the steady state the osmotic pressure in the cytoplasm and vacuole must be identical. As a consequence, there must be osmoregulation between the cytoplasm and vacuole to match the accumulation of osmotically active solutes in the vacuole that are potentially harmful to cytoplasmic enzymes (e.g., NaCl in halophytes and malic acid in CAM plants). This has led to the concept that certain organic solutes synthetized by the cells and accumulated in the cytoplasm serve specifically to balance vacuolar π. Solutes such as proline, "betaines" (proline-, glycine-, and β-alanine-betaine), polyols (glycerol, sorbitol, mannitol), and sulfonium derivatives are known to have the potential to serve this function (47). The bulk tissue levels of such "compatible solutes" are not necessarily very high. Especially in succulents such as the halophyte *Suaeda maritima*, with low cytoplasm to vacuole ratios, low tissue levels in fact correspond to quite high cytoplasmic concentrations if the solute is restricted to the cytoplasm (48, 49).

For CAM plants it must be assumed that the levels of some cytoplasmic solutes oscillate in parallel with the malic acid levels in the vacuoles. However, attempts to demonstrate this have remained unsuccessful. It can be calculated that, owing to the very small relative volume of cytoplasm in CAM leaf cells, the absolute amounts involved in such changes would be below the resolution of standard analytical procedures for most of the substances that are possible candidates (J. Gorham, A.-M. Heun, R. G. Wyn Jones, and U. Lüttge, unpublished results).

1.5. Transport at the Tonoplast

Accumulation of solutes in the large vacuoles of succulents requires transport across the tonoplast. The only succulents for which transport processes at the tonoplast are currently being studied in detail are CAM plants (35, 50). From electrochemical and energetic considerations, it has been postulated that the driving force for malic-acid accumulation in CAM leaf cells is provided by a proton-pumping ATPase at the tonoplast (see Section 1.3) (35–38).

It can be argued that the transport activity across the tonoplast of CAM plants is probably not unique, but is rather just one example of very similar processes that are responsible for vacuolar solute accumulation in many plant types (35). Questions concerning the mechanism of solute transport across the tonoplast and the nature of energy coupling at this membrane, such as whether only secondary flux coupling to primary active H^+ transport establishing a $\Delta\mu_{H+}$ is involved or whether additional pumps are required [e.g., for NaCl accumulation in halophytes, (51)], must be tackled in experiments with isolated vacuoles and tonoplast vesicles. There is, in fact, already evidence for the existence of tonoplast ATPases in a number of species (52–55), and recent experiments suggest that they can translocate protons (56–59). In *Saccharum officinarum* a correlation between vacuolar sugar accumulation and $\Delta\mu_{H+}$ at the tonoplast has been demonstrated (56). Moreover in tonoplast vesicles from *Hevea brasiliensis* there is a relationship between H^+ pump activity, $\Delta\mu_{H+}$, and the rate of citrate uptake (58). Work is needed with vacuoles from a wider variety of plant types to determine whether generalizations can be made about the mechanism of vacuolar solute accumulation.

1.6.   Assessment of Similarities in the Characteristics of Succulent Tissues

The physiological evaluation of succulence outlined above draws heavily on investigations with CAM plants. This is largely due to the fact that few biophysical measurements have as yet been made on succulent halophytes. Nonetheless, we might expect there to be similarities between these succulent tissues, particularly in view of some of the ecological characteristics of CAM plants. For example, CAM can be regarded as a metabolic adaptation to restricted water availability not only under drought conditions but also under salinity stress. Halophilic members of the Mesembryanthemaceae are known to be CAM plants (17, 18, 60, 61), and salinity can induce a change from C_3 photosynthesis to CAM (19). Among the constitutive CAM plants there are species that inhabit saline environments, for example, *Cereus validus* (62) and other cactaceae, which grow on salt flats in South America. In these cacti the roots apparently die back in the dry season when salinity builds up in the enviornment, and neither salt nor water uptake occurs (E. Medina, personal communication). Roots redevelop in the rainy season. CAM and succulence allow these plants to survive just as in the CAM-idling *Opuntia* spp. in the Colorado desert (108).

In many CAM species there are actually two types of succulent tissue, namely, the green mesophyll cells with functional chloroplasts (chlorenchyma) and the colorless, chloroplast-free cells constituting the water parenchyma. This is very characteristic, for example, of the Cactaceae (24) and is found in many members of the Bromeliaceae (63), as well as in *Aloë* spp. [Liliaceae (64)]. Here, it is only the chlorophyll-containing cells that show the day–night changes in malic acid levels, and the water parenchyma seems just to function as a water reservoir (64). A very similar condition is found in the CAM form of *Mesembryanthemum crystallinum* (Section 1.2), in which the epidermal bladder cells do not participate in the malic acid rhythm shown by the mesophyll cells (65). These plants could be said to combine the features of xerophyte, CAM, and salinity-induced succulence.

Although the effects of salinity and water stress are species dependent, similarities undoubtedly exist in the characteristics of succulent tissues. This can be assessed on the basis of leaf morphology and the biophysics of water relations and solute accumulation.

Leaf Morphology

An effect observed in a wide range of halophytes, xerophytes, and CAM plants is the increase in succulence with leaf age (13, 51). In halophytes this is usually interpreted (as is the increase in succulence in response to an increase in salinity) as a means by which the absorbed salt is "diluted", thereby preventing intracellular NaCl concentrations becoming excessively high (13, 51, 66, 67). Examples that might be consistent with this view are provided by the mangrove

Laguncularia racemosa, in which leaf thickness increases greatly during ontogeny due to cell expansion but Cl content on a FW basis increases little (68), and by *Atriplex triangularis* (= *A. hastata*), where increases in the NaCl levels in the root medium bring about an increase in leaf thickness, this being associated with a decrease in cell number but a large increase in cell volume (23). Nonetheless, the widely held view that salinity inhibits cell division more than it does cell expansion may not be applicable to all species (12).

The effects of increased salinity stress on leaf morphology are diverse (69, 70). Besides the increased succulence, thickening of the cuticle and changes in number and dimensions of the xylem vessels observed in some species could also have important repercussions for plant water relations. Studies are also needed to determine the effects of salinity stress at the cellular level, similar to those conducted on the effects of water stress. Here at least the effects can be quite complex. For example, Berlin et al. (71) observed that water-stressed leaves of *Gossypium hirsutum* have somewhat larger mesophyll cells than the irrigated controls and considerably larger vacuoles. However, the stressed leaves are of much smaller area and volume than the controls, so the total volume of "vacuole" is actually similar in the two types of leaves. This kind of approach, in which differences at the cellular level are related to differences in growth rate and productivity, could be usefully extended to the halophytes.

It is noteworthy that a relationship exists under some circumstances between succulence and the occurrence of CAM. In *Kalanchoë blossfeldiana* cv. Tom Thumb, plants cultivated under long days show daytime CO_2 fixation whereas short-day plants perform CAM (20), and the latter are markedly more succulent (72). This response may be analogous to that in halophytes, because the vacuoles in CAM plants serve as the (transient) storage site for malic acid. The increased succulence of *Mesembryanthemum crystallinum* under salinity stress (see Section 1.2) does not allow a clear distinction between salinity effects and the induction of CAM because both occur simultaneously.

Water Relations

The importance of succulent tissues in water storage is most obviously seen in the differentation of water parenchyma. It must be anticipated that such tissue can be readily charged with water or drawn upon (27). This can be quantified as the half-time $(T_{1/2})$ for water exchange of the individual cells, where $T_{1/2}$ is determined by the water-storage capacity of the cell (Section 1.3) and also by the hydraulic conductivity (Lp) of the limiting membrane(s) (73). Values for $T_{1/2}$ and Lp can be obtained using the pressure probe (43) and the measurements made so far on the relevant species (Table 7.3) indicate that cells of such water-storage tissues can exchange water readily. The Lp values are among the highest yet obtained for cells of higher plants using the pressure probe (43). It is important to ask whether these values obtained with a selection of large cell types are a consequence of the particular ecological adaptations involved or are related more directly to the large succulent cell structure. A contribution to this

Table 7.3. **Water-Relation Characteristics of Large Cell Types, and for *C. annuum* Fruit Wall a Comparison of Two Cell Types of Very Different Sizes**[a]

Species and Cell Type	V (m³)	P (MPa)	ϵ (MPa)	$T_{1/2}$ (s)	Lp (pm s⁻¹ Pa⁻¹)	Reference
Mesembryanthemum crystallinum epidermal bladders	$(0.3 \text{ to } 2.7) \times 10^{-9}$	0.06 to 0.51	1.0 to 5.0	300 to 1980	0.08 to 0.28	32
Kalanchoë daigremontiana mesophyll cells	$(4.2 \pm 1.8) \times 10^{-13}$ ($n = 266$)	0.18 ± 0.06 ($n = 157$)	4.2 ± 2.8 ($n = 21$)	5.1 ± 2.1 ($n = 11$)	0.69 ± 0.46 ($n = 8$)	26
Suaeda maritima leaf epidermis cells	$(0.4 \text{ to } 21.2) \times 10^{-13}$	0.03 to 0.49	1.4 to 20.4	2.5 to 22.3	0.03 to 0.43	74, 75
Oxalis carnosa bladders of upper epidermis	$(0.8 \text{ to } 12) \times 10^{-12}$	0.07 to 0.29	0.2 to 1.7	22 to 213	0.38 ± 0.24	76
Oxalis carnosa bladders of lower epidermis	$(0.9 \text{ to } 18) \times 10^{-12}$	0.13 to 0.37	1.8 to 17	7 to 38	2.3 ± 1.6	76
Capsicum annuum giant subepidermal cells of inner pericarp surface	$(5.3 \pm 3.5) \times 10^{-10}$ ($n = 23$)	0.27 ± 0.06 ($n = 23$)	1.5 to 27	2.7 ± 2.2 ($n = 46$)	5.8 ± 3.7 ($n = 46$)	77
Capsicum annuum mesocarp parenchyma cells	$(3.6 \pm 2.7) \times 10^{-13}$ ($n = 6$) $(2 \text{ to } 12) \times 10^{-12}$	0.33 ± 0.07 ($n = 6$) 0.2 to 0.4	0.1 to 0.6 0.2 to 0.5	33 ± 10 ($n = 12$) 240	0.21 ± 0.07 ($n = 12$) ≃ 0.05	77 78

[a]Values are given ± SD.

136

question is made by a comparison between the mesophyll cells and the much larger subepidermal cells of and the inner surface of the fruit wall of *Capsicum annuum* included in Table 7.3. It is evident that small and large cells within the same tissue can differ significantly in their water-relation parameters. In the large *C. annuum* cells ϵ and *Lp* are similar to those of the other succulent cells described above. Work is needed to compare a variety of succulent CAM plants and halophytes.

Solute Accumulation

Considering the solutes that serve as the osmotically active agents in succulent tissues, we have salts of organic acid in xerophytes, mainly but not exclusively inorganic salts in halophytes (where organic acid anions are involved to differing extents), and free organic acids plus Ca^{2+}, Mg^{2+}, K^+, and Na^+ salts of organic acids in CAM plants. There is thus considerable flexibility in the kind of solutes used. Organic acid anions are always involved to some extent, their relative importance being modulated by the participation of inorganic ions. This is similar to other instances in which organic acid anions play a role as vacuolar osmotica, that is, in charge balancing and in the control of cytoplasmic pH as part of the "biochemical pH-stat" [e.g., in unequal cation and anion uptake (79), in extension growth (80), in stomatal guard cells (81), in the relation between nitrate and ammonium nutrition (82) and in special mechanisms of heavy-metal tolerance such as the sequestration of Zn in the vacuoles (13, pp. 491–495)]. The possibility that the different patterns of ion accumulation in succulent tissues might all be related to the activity of a proton pump at the tonoplast (Section 1.5) supports the notion that basic similarities exist between all of the cell types.

The flexibility found in the ions accumulated in succulent tissues can be interpreted as a reflection of the overriding importance of turgor maintenance (28). Under conditions of salinity or water stress plant survival is dependent on some degree of osmotic adjustment (83, 84). Again, this subject has been more extensively investigated in relation to the effects of drought (85–87), but there is at least some evidence that solute accumulation permits the maintenance of positive cell *P* in halophytes (7). This also underlines the point that in succulence the two aspects of salt dilution and solute accumulation should not be regarded separately [see Equation (1)].

2. PHOTOSYNTHETIC PATHWAYS AND LEAF ADAPTATION TO WATER AND SALINITY STRESS

Table 7.4 shows that the amount of water transpired for the production of new biomass is much smaller in C_4 and CAM plants than in C_3 plants. The transpiration ratio (TR), defined as the ratio of the mass of H_2O lost to the mass of CO_2 gained (90), is higher, or the WUE (= $1/TR$) lower, in the C_3 plants. A similar picture emerges when a more detailed comparison is made between closely related C_3 and C_4 species, as in the genus *Atriplex* (review ref. 84).

Table 7.4. Some Data Characterizing the Efficiency and Productivity of CAM, C₄, and C₃ Plants

Characteristic	CAM	C$_4$	C$_3$
Transpiration ratio (mass H_2O lost:mass CO_2 gained)[a]	25–145 (light period only: 220–880)	360–440	660–1400
Maximum rate of net photosynthesis $10^4 \times$ (g CO_2 m^{-2} leaf surface s^{-1})[a]	30–110 (highest reported values: 306–361)	1100–2200	410–1100
Maximum growth rate (g DW m^{-2}d^{-1})[a]	1.5–1.8	400–500	50–200
Theoretical energy requirement per CO_2 fixed[b]	5.5–6.5 ATP 2 [NADPH + H$^+$]	4–5 ATP 2 [NADPH + H$^+$]	3 ATP 2 [NADPH + H$^+$]
Measured quantum requirement per CO_2 fixed[c]	During phase III: 16 During phase IV: 41 Day + night: 46–68	17 (in the light)	17 (in the light)

[a] Recalculated for w/w after refs. 20 and 88, see also refs. 89 and 90.
[b] From refs. 37, 91, and 92.
[c] CAM phases III and IV (93); CAM day + night values are long-term growth requirements from refs. 94 and 95.

Comparatively low transpiration rates imply not only economic use of water but also a lower supply of salt to the leaves via the transpiration stream (13). Thus, a consideration of C_4 photosynthesis and CAM is interesting with respect to both the water and the salt stress brought about by salinity.

2.1. C_4 Plants

There are many C_4 plants among halophytes (17, Table 18.3). Winter (17) concludes that "most coastal halophytes are characterized by the C_3 pathway whereas many inland halophytes, particularly those of dry, saline habitats, are characterized by the C_4 pathway". Among these C_4 halophytes are sclerophyllous forms (e.g., species of Poaceae) but also succulents like *Salsola, Seidlitzia Suaeda, Portulaca,* and *Zygophyllum* (17). In the latter the typical "Kranz anatomy" of concentric layers of bundle sheath and mesophyll cells is modified.

2.2. CAM Plants

Water Economy and Ecological Flexibility of CAM plants

In the day–night rhythm of CAM four phases can be distinguished (92): I, nocturnal CO_2 fixation and malic acid accumulation, open stomata and hence relatively low leaf resistance; II, transition from dark to light period; III, remobilization and decarboxylation of malic acid, high CO_2 concentrations in the intercellular spaces of the leaf [typically 0.5–2.0% (96, 97)], light-dependent fixation of this endogenous CO_2, high stomatal resistance; and IV, in well-irrigated plants stomatal opening after completion of malic acid remobilization and performance of normal C_3 photosynthesis.

The low transpiration ratio of CAM plants results primarily from the CO_2 fixation occurring during phase I, when evaporative demand is much smaller than during the light period, for most of which the stomata are closed (phase III). However, if stomata open again and CO_2 is fixed by ribulosebisphosphate carboxylase in phase IV, the trasnspiration ratio is closer to that of C_3 plants. Therefore, two TR values are given for CAM plants in Table 7.4.

CAM can thus be regarded primarily as an adaptation to limited water availability. However, CAM plants show large ecological flexibility. Other environmental factors mingle, and the schematic performance of CAM as described above is highly modulated by H_2O, light, temperature, and salinity. Here, we will consider only the relationship between CAM and salinity, as the other subjects have been discussed in detail elsewhere (20, 92).

CAM and Salinity

As already mentioned (Section 1.6), the annual halophilic plant *M. crystallinum* can be induced by high NaCl concentrations (400–600 mol m^{-3}) in its root medium to change from C_3 photosynthesis to CAM (17, 65, 98). This is not a specific salt effect, however, because Na_2SO_4, KCl, and K_2SO_4 have the same effect as NaCl (99). Other means of imposing water stress on the plants such as withholding O_2 from the root medium or cooling the root medium to $+10°C$ also lead to CAM induction (100). Thus, it appears that water stress is the decisive factor in this process. A decrease in water potential and an increase in osmotic pressure due to increases in Na^+ and Cl^- levels occur rather rapidly when NaCl concentrations in the root medium are raised. These processes are completed within a few days and precede the diurnal malate oscillations, which become discernible after about 2 weeks (101). Slower metabolic changes must be involved in the induction of CAM, one of which is the formation of the CAM-type isozyme of PEP-carboxylase (102).

Winter (17) has followed the change of CO_2 and H_2O vapor exchange characteristics in a *M. crystallinum* plant for 41 days after the onset of salinity treatment. Salinity rather rapidly induces a midday depression of CO_2 and H_2O exchange, which becomes very substantial after the seventh day of the NaCl treatment. However, even on the forty-first day there is still net CO_2 assimilation during the day. Net CO_2 fixation in the dark period is discernible on the eighth day and increases greatly toward day 41.

Figure 7.1 gives the total amount of net CO_2 exchange, transpirational water loss, and transpiration ratios for the plant on each day. CO_2 fixation in the light period is stimulated during the first few days of the salt treatment. This agrees with the general observation that a limited salt load can stimulate the growth of

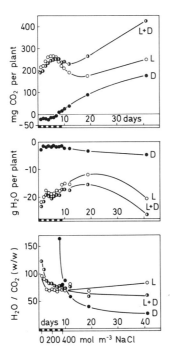

Figure 7.1. Net CO_2 and H_2O vapor (transpiration) exchange and transpiration ratio (H_2O/CO_2, w/w) for light (L) and dark (D) periods separately, and for the light + dark periods (L + D) of the shoot of one *Mesembryanthemum crystallinum* plant in hydroculture in response to increasing salinity (up to 400 mol m^{-3}) in the root medium. Positive and negative values correspond to a net gas uptake and release by the plant, respectively. Bars on the abscissa indicate increasing salinity of the root medium in daily steps of 50 mol m^{-3} up to 400 mol m^{-3} on day 9. Data are results of integration of the gas exchange curves of Figure 18.4 in ref. 17.

halophilic plants not only by gain of fresh weight due to salt accumulation and water uptake but also by carbon gain. CO_2 fixation in the light decreases as the salt load increases; the increase toward the end of the experiment can be explained by growth of the plant. The CAM-type dark fixation increases gradually from day 8 onward, and the total CO_2 fixation of the plant on days 19 and 41 reflects this large contribution from phase I of CAM. The respiratory CO_2 loss shows little response to the salt treatment at the beginning of the experiment. In contrast to net CO_2 uptake, transpirational water loss in the light initially decreases, and with the increasing CO_2 fixation this leads to a decrease in TR or an increase in WUE. The later increase in transpiration can again be explained by growth increments. Transpiration in the dark before addition of NaCl to the root medium is only 13% of transpirational H_2O loss in the light. It increases little during the experiment, at day 41 being only 1.7 times that observed at day 0 despite the considerable dark CO_2 fixation. As a consequence at day 41 the WUE of dark CO_2 fixation is three times that of daytime fixation; the WUE for the entire 24-hr period is twice as high when *M. crystallinum* performs CAM (day 41) as when it performs C_3 photosynthesis (day 0). Because CAM-performing *M. crystallinum* shows succulence related both to CAM and to salinity (see Sections 1.2 and 1.6), similar data would be desirable for non-CAM succulent halophytes to show the responses of WUE to salinity per se.

The initial responses of *M. crystallinum* to salinity that result in an increased WUE (Fig. 7.1) are indicative of a condition that may be widespread in plants, namely, that stomatal conductance is not the primary factor limiting CO_2 fixation (104). This follows from the fact that CO_2 fixation is actually increasing at the time that transpiration (and hence stomatal conductance) is decreasing (Fig. 7.1, days 1–3). One of the most detailed studies on the effects of salinity on these processes is that of Longstreth and Nobel (8, 105) on the C_3 plant *Gossypium hirsutum*. Here, it was shown that salinity caused (i) an increase in the ratio of mesophyll cell surface area to leaf surface area, A_{mes}/A (because of an increase in "succulence"), (ii) a decrease in the cell conductance to CO_2 and (iii) a decrease in water vapor conductance. The net result was an increase in WUE, and it is probable that similar effects occur in *M. crystallinum*.

The behavior of *M. crystallinum* in the field has been studied at a coastal site in Israel and compared with another species, *M. nodiflorum*, occurring both at the coast and in the Negev desert (106, 107). These annual species germinate after strong winter rains and grow lushly while performing C_3 photosynthesis. As the soil begins to dry out they shift to CAM. This enables them to survive longer than C_3 annuals and to bring a large number of seeds to maturity. *Mesembryanthemum nodiflorum* at the drier desert site changes earlier to CAM than it does at the les extreme coastal site (107).

Their high WUE notwithstanding, CAM plants do need the actual availability of water to survive either by seasonal or day–night patterns of water storage in their succulent photosynthetic and water-storing tissues. This is very evident in cacti of the Colorado desert. When no water is available from the

Table 7.5. Integration of the Net CO_2 and Net H_2O Vapor Exchange Curves of Gerwick and Williams (109) Showing CO_2 Uptake, H_2O Release, and Transpiration Ratios (H_2O/CO_2) for *Opuntia polyacantha* Plants Well Watered, Droughted, and Recovering from Drought at Two Different Temperature Regimes

Thermoperiod and Watering Regime	Dark Period $\sim$9 hr			Light Period $\sim$15 hr			Entire 24-hr Period		
	CO_2 (mg m^{-2})[1]	H_2O (mg m^{-2})[1]	H_2O/CO_2 (w/w)	CO_2 (mg m^{-2})[1]	H_2O (mg m^{-2})[1]	H_2O/CO_2 (w/w)	CO_2 (mg m^{-2})[1]	H_2O (mg m^{-2})[1]	H_2O/CO_2 (w/w)
D 15°C:L 35°C									
Well watered	0.20	−7.4	37	0.10	−28.0	280	0.30	−35.4	118
4 weeks droughted	0.02	−1.4	70	0.04	−12.9	322	0.06	−14.3	238
Recovery from drought: Day 1	0.03	−2.3	77	0.09	−33.7	374	0.12	−36.0	300
Day 2	0.12	−7.2	60	0.07	−34.6	494	0.19	−41.8	220
D 15°C:L 20°C									
Well watered	± 0	−1.4		0.15	−18.2	121	0.15	−19.6	130
4 weeks droughted	± 0	−1.3		0.05	−6.0	120	0.05	−7.3	146
Recovery from drought: Day 1	± 0	−1.1		0.14	−16.8	120	0.14	−17.9	128

[1] $10^{-4} \times$ (mg m^{-2}).

environment they can show "idling," where stomata remain closed and there is virtually no gas exchange day and night. A day–night malic acid rhythm is exhibited on account of refixation of respiratory CO_2. When water becomes available the plants return to normal CAM (20, 108).

In the shortgrass prairie cactus *Opuntia polyacantha* drought has been found to eliminate the nocturnal phase I CO_2 fixation but to inhibit only slightly phase IV CO_2 fixation in the late afternoon. The behavior of this cactus is highly modulated by temperature (Table 7.5). Well-watered plants at a thermoperiod of 15°C night/35°C day show both phase I (dark period) and phase IV (later light period) CO_2 fixation. At a thermoperiod of 15°C/20°C phase I CO_2 fixation disappears. Transpiration ratios of phase IV fixation are much higher than for phase I fixation and they do not appear to improve under conditions of drought (109). This response to water stress in *O. polyacantha* is unusual in that all other CAM plants studied so far show reduced phase IV CO_2 fixation as the first response to stress (92). It would be worth investigating other related CAM types to see how frequently this response if found.

CAM and Productivity

Table 7.4 implies that the productivity of CAM plants is very low. To some extent this is a result of the inherent energetic disadvantage of the metabolic pathway. Discussing malic acid storage, we have already mentioned (Section 1.3) that close to 100% of the glycolytic and respiratory energy in the dark is needed for malic acid transport into the vacuoles. This requires 1 ATP per malate, equivalent to 1 CO_2 stored. CAM plants largely gain their energy from photosynthetic electron flow in the light.

For a net gain of carbon the C_3 unit remaining after malate decarboxylation in the light must be reincorporated into glucans via gluconeogensis. This replenishes reserves for production of phosphoenolpyruvate (PEP) in the subsequent dark period.

If starch is synthezised and broken down by starch phosphorylase, an enzyme present in CAM plants (110), the energy requirement of the entire cycle is as follows. For a NADP-malic enzyme-type CAM plant, where malate decarboxylation results in CO_2 and pyruvate, which needs to be phosphorylated to PEP, an extra 2.5 ATP are required for the entire day-night cycle in addition to the 3 ATP + 2 [NADPH + H^+] necessary for the reduction of 1 CO_2 in the Calvin cycle. Extra reducing equivalents are not needed as there is internal stoichiometric balance between glycolysis, malate synthesis, and gluconeogenesis. In PEP—carboxykinase-type CAM plants the extra ATP requirement is only 1.5 ATP because the malate is reduced to oxaloacetate: this is decarboxylated with the consumption of only 1 ATP giving PEP directly, and the dikinase reaction is avoided (92).

However, the highly productive C_4 plants also have a higher ATP requirement than C_3 photosynthesis. These ATP and [NADPH + H^+] requirements are summarized in Table 7.4 together with the actual quantum yields of C_3, C_4, and

CAM photosynthesis. On the basis of ATP and $[NADPH + H^+]$ stoichiometries the quantum requirement of C_4 and CAM plants should not be excessively larger than that of C_3 plants. Thus, the observed high quantum requirement of CAM plants for growth must have other origins. Presumably, it is related to resistances involved in the particular CO_2 and H_2O exchange patterns of CAM plants (Section 1.3) (111).

With sufficient photosynthetically active radiation in semiarid environments CAM plants can, however, be quite successful at the community level (e.g., 103) and there are many successful CAM crops (20).

REFERENCES

1. M. G. Pitman. Whole plants. In D. A. Baker and J. L. Hall, eds., *Ion Transport in Plant Cells and Tissues.* North-Holland, Amsterdam, 1975, pp. 267–308.

2. U. Lüttge and M. G. Pitman, eds. Transport in Plants II, Part B, Tissues and Organs, *Encyclopedia of Plant Physiology, New Series*, Vol. 2. Springer-Verlag, Berlin, 1976, p. 456.

3. U. Lüttge and N. Higinbotham. *Transport in Plants*, Springer-Verlag, New York, 1979, 468 pp.

4. M. G. Pitman. Ion uptake by plant roots. In U. Lüttge and M. G. Pitman, eds., Transport in Plants II, Part B, Tissues and Organs, *Encyclopedia of Plant Physiology, New Series*, Vol. 2. Springer-Verlag, Berlin, 1976, pp. 95–128.

5. A. Läuchli. Regulation des Salztransportes und Salzausschließung in Glykophyten and Halophyten. *Ber. Deutsch. Bot. Ges.* **92**, 87–94 (1979).

6. J. W. O'Leary. Physiological basis of plant growth inhibition due to salinity. In W. G. McGinnies, B. J. Goldman, and P. Paylore, eds., *Food, Fiber and the Arid Lands.* University of Arizona Press, Tucson, 1971, pp. 331–336.

7. A. Kaplan and J. Gale. Effect of sodium chloride salinity on the water balance of *Atriplex halimus. Aust. J. Biol. Sci.* **25**, 895–903 (1972).

8. D. J. Longstreth and P. S. Nobel. Salinity effects on leaf anatomy. Consequences for photosynthesis. *Plant Physiol.* **63**, 700–703 (1979).

9. I. R. Cowan. Stomatal behaviour and environment. *Adv. Bot. Res.* **4**, 117–228 (1977).

10. I. R. Cowan and G. D. Farquhar. Stomatal function in relation to leaf metabolism and environment. *Symp. Soc. Exp. Biol.* **31**, 471–505 (1977).

11. A. D. Hanson and C. E. Nelsen. Water: Adaptation of crops to drought-prone environments. In P. S. Carlson, ed., *The Biology of Crop Productivity.* Academic Press, New York, 1980, pp. 77–152.

12. D. H. Jennings. The effects of sodium chloride on higher plants. *Biol. Rev.* **51**, 453–486 (1976).

13. H. Kinzel. *Pflanzenökologie und Mineralstoffwechsel.* Verlag Eugen Ulmer, Stuttgart, 1982, 534 pp.

14. O. Stocker. Das Halophytenproblem. *Erg. Biol.* **3**, 265–353 (1928).

15. M. van Eijk. Analyse der Wirkung des NaCl auf die Entwicklung der Sukkulenz

und Transpiration bei *Salicornia herbacea*; sowie Untersuchungen über den Einfluß der Salzaufnahme auf die Wurzelatmung bei *Aster tripolium*. *Rec. Trav. Bot. Neerl.* **36**, 559–657 (1939).

16. D. J. von Willert, E. Brinckmann, and E.-D. Schulze. Ecophysiological investigations of plants in the coastal desert of Southern Africa. Ion content and crassulacean acid metabolism. In R. L. Jefferies and A. J. Davy, eds., *Ecological Processes in Coastal Environments*, Blackwell, Oxford, 1979, pp. 321–331.

17. K. Winter. Photosynthetic and water relationships of higher plants in a saline environment. In R. L. Jefferies and A. J. Davy, eds., *Ecological Processes in Coastal Environments*, Blackwell, Oxford, 1979, pp. 297–320.

18. K. Winter and J. H. Troughton. Photosynthetic pathways in plants of coastal and inland habitats of Israel and the Sinai. *Flora* **167**, 1–34 (1978).

19. K. Winter and U. Lüttge. C_3-Photosynthese und Crassulaceen-Säurestoffwechsel bei *Mesembryanthemum crystallinum* L. *Ber. Deutsch. Bot. Ges.* **92**, 117–132 (1979).

20. M. Kluge and I. P. Ting. *Crassulacean Acid Metabolism, Analysis of an Ecological Adaptation*, Springer-Verlag, Berlin, 1978, p. 209.

21. A. E. Hiill and B. S. Hill. Elimination processes by glands. Mineral ions. In U. Lüttge and M. G. Pitman, eds., Transport in Plants II, Part B, Tissues and Organs, *Encyclopedia of Plant Physiology, New Series*, Vol. 2. Springer-Verlag, Berlin, 1976, pp. 225–243.

22. U. Lüttge. Salt glands. In D. A. Baker and J. L. Hall, eds., *Ion Transport in Plant Cells and Tissues*. North-Holland, Amsterdam, 1975, pp. 335–376.

23. R. F. Black. The effect of sodium chloride on leaf succulence and area of *Atriplex hastata* L. *Aust. J. Bot.* **6**, 306–321 (1958).

24. A. C. Gibson. The anatomy of succulence. In I. P. Ting and M. Gibbs, eds., *Crassulacean Acid Metabolism*. American Society Plant Physiologists, Rockville, Maryland, 1982, pp. 1–17.

25. M. Kluge, O. L. Lange, M.v. Eichenauer, and R. Schmid. Diurnaler Säurerhythmus bei *Tillandsia usneoides*. Untersuchungen über den Weg des Kohlenstoffs sowie die Abhängigkeit des CO_2-Gaswechsels von Lichtintensität, Temperatur und Wassergehalt der Pflanze. *Planta* **112**, 357–372 (1973).

26. E. Steudle, J. A. C. Smith, and U. Lüttge. Water-relation parameters of individual mesophyll cells of the crassulacean acid metabolism plant *Kalanchoë daigremontiana*. *Plant Physiol.* **66**, 1155–1163 (1980).

27. G. Haberlandt. *Physiologische Pflanzenanatomie*, 5. Auflage. W. Engelmann, Leipzig, 1981, p. 670.

28. W. J. Cram. Negative feedback regulation of transport in cells. The maintenance of turgor, volume, and nutrient supply. In U. Lüttge and M. G. Pitman, eds., Transport in Plants II, Part A, Cells, *Encyclopedia of Plant Physiology, New Series*, Vol. 2. Springer-Verlag, Berlin, 1976, pp. 284–316.

29. F.-A. Austenfeld. Untersuchungen zum Ionengehalt von *Salicornia europaea* L. unter besonderer Berücksichtigung des Oxalats in Abhängigkeit von der Substratsalinität. *Biochem. Physiol. Pflanzen* **165**, 303–316 (1974).

30. J. Rozema. An ecophysiological investigation into the salt tolerance of *Glaux maritima* L. *Acta Bot. Neerl.* **24**, 407–416 (1975).

31. W. Baumeister and G. Kloos. Über die Salzsekretion bei *Halimione portulacoides* (L.) Aellen. *Flora* **163**, 310–326 (1974).

32. E. Steudle, U. Lüttge, and U. Zimmermann. Water Relations of the epidermal bladder cells of the halophytic species *Mesembryanthemum crystallinum*: Direct measurements of hydrostatic pressure and hydraulic conductivity. *Planta* **126**, 229–246 (1975).

33. H.-D. Ihlenfeldt and H. E. K. Hartmann. Leaf surfaces in Mesembryanthemaceae. In D. F. Cutler, K. L. Alvin and C. E. Price, eds., *The Plant Cuticle*, Academic Press, London, 1982, pp. 397–424.

34. U. Lüttge, E. Fischer, and E. Steudle. Membrane potentials and salt distribution in epidermal bladders and photosynthetic tissue of *Mesembryanthemum crystallinum* L. *Plant Cell Environ.* **1**, 121–129 (1978).

35. U. Lüttge, J. A. C. Smith, and G. Marigo, Membrane transport, osmoregulation and the control of CAM. In I. P. Ting and M. Gibbs, eds., *Crassulacean Acid Metabolism*. American Society Plant Physiologists, Rockville, Maryland, 1982, pp. 69–91.

36. U. Lüttge and E. Ball. Electrochemical investigation of active malic acid transport at the tonoplast into the vacuoles of the CAM plant *Kalanchoë daigremontiana. J. Memb. Biol.* **47**, 401–422 (1979).

37. U. Lüttge, J. A. C. Smith, G. Marigo, and C. B. Osmond. Energetics of malate accumulation in the vacuoles of *Kalanchoë tubiflora* cells. *FEBS Lett.* **126**, 81–84 (1981).

38. J. A. C. Smith, G. Marigo, U. Lüttge, and E. Ball. Adenine-nucleotide levels during CAM and the energetics of malate accumulation in *Kalanchoë tubiflora. Plant Sci. Lett.* **26**, 13–21 (1982).

39. J. A. C. Smith and U. Lüttge. Tissue water-relation characteristics of CAM leaves studied using the pressure bomb. In *Abstracts of Lectures and Poster Demonstrations*, Federation of European Societies of Plant Physiology II. Congress, Santiago de Compostela, 1980, pp. 636–637.

40. Y. N. S. Cheung, M. T. Tyree, and J. Dainty, Water relations parameters on single leaves obtained in a pressure bomb and some ecological interpretations. *Can. J. Bot.* **53**, 1342–1346 (1975).

41. M. T. Tyree and H. T. Hammel. The measurement of the turgor pressure and the water relations of plants by the pressure-bomb technique. *J. Exp. Bot.* **23**, 267–282 (1972).

42. M. T. Tyree and P. Jarvis. Water in tissues and cells. In O. L. Lange, P. S. Nobel, C. B. Osmond, and H. Ziegler, eds., Physiological Plant Ecology II, Water Relations and Carbon Assimilation, *Encyclopedia of Plant Physiology, New Series.* Vol. 12B. Springer-Verlag, Berlin, 1982, pp. 35–77.

43. U. Zimmermann and E. Steudle. Fundamental water relations parameters. In R. M. Spanswick, W. J. Lucas, and J. Dainty, eds., *Plant Membrane Transport: Current Conceptual Issues.* Elsevier/North-Holland, Amsterdam, 1980, pp. 113–127.

44. U. Luttge and E. Ball. Water relation parameters of the CAM plant *Kalanchoë daigremontiana* in relation to diurnal malate oscillations. *Oecologia* **31**, 85–94 (1977).

45. R. D. Phillips. Deacidification in a plant with crassulacean acid metabolism associated with anion–cation balance. *Nature* **287**, 727–728.

46. R. D. Phillips and D. H. Jennings. Succulence, cations and organic acids in leaves of *Kalanchoë daigremontiana* grown in long and short days in soil and water culture. *New Phytol.* **77**, 599–611 (1976).

47. R. G. Wyn Jones, R. Storey, R. A. Leigh, N. Ahmad, and A. Pollard. A hypothesis on cytoplasmic osmoregulation. In E. Marrè and O. Ciferri, eds., *Regulation of Cell Membrane Activities in Plants*, North-Holland, Amsterdam, 1977, pp. 121–136.

48. J. L. Hall, D. M. R. Harvey, and T. J. Flowers. Evidence for the cytoplasmic localization of betaine in leaf cells of *Suaeda maritima. Planta* **140**, 59–62 (1978).

49. D. M. R. Harvey, J. L. Hall, T. J. Flowers, and B. Kent. Quantitative ion localization within *Suaeda maritima* leaf mesophyll cells. *Planta* **151**, 555–560 (1981).

50. C. Buser-Suter, A. Wiemken, and P. Matile. A malic acid permease in isolated vacuoles of a crassulacean acid metabolism (CAM) plant. *Plant Physiol.* **69**, 456–459 (1982).

51. D. H. Jennings. Halophytes, succulence and sodium in plants—A unified theory. *New Phytol.* **67**, 899–911 (1968).

52. J. d'Auzac. Caractérisation d'une ATPase membranaire en présence d'une phosphatase acide dans les lutoïdes du latex d'*Hevea brasiliensis. Phytochemistry* **14**, 671–675 (1975).

53. S. Doll, F. Rodier, and J. Willenbrink. Accumulation of sucrose in vacuoles isolated from red beet tissue. *Planta* **144**, 407–411 (1979).

54. R. A. Leigh and R. R. Walker. ATPase and acid phosphatase activities associated with vacuoles isolated from storage roots of red beet (*Beta vulgaris* L.). *Planta* **150**, 222–229 (1980).

55. W. Lin, G. J. Wagner, H. W. Siegelman, and G. Hind. Membrane-bound ATPase of intact vacuoles and tonoplasts isolated from mature plant tissue. *Biochim. Biophys. Acta* **465**, 110–117 (1977).

56. E. Komor, M. Thom, and A. Maretzki. Vacuoles from sugar cane suspension cultures. III. Protonmotive potential difference. *Plant Physiol.* **69**, 1326–1330 (1982).

57. B. Marin, M. Marin-Lanza, and E. Komor. The proton motive potential difference across the vacuo-lysosomal membrane of *Hevea brasiliensis* (rubber-tree) and its modification by a membrane-bound adenosine-triphosphatase. *Biochem. J.* **198**, 365–372 (1981).

58. B. Marin, J. A. C. Smith, and U. Lüttge. The electrochemical proton gradient and its influence on citrate uptake in tonoplast vesicles of *Hevea brasiliensis. Planta* **153**, 486–493 (1981).

59. G. J. Wagner and W. Lin. An active proton pump of intact vacuoles isolated from *Tulipa* petals. *Biochim. Biophys. Acta* **689**, 261–266 (1982).

60. D. J. von Willert, D. A. Thomas, W. Lobin, and E. Curdts. Ecophysiologic investigations in the family of the Mesembryanthemaceae. *Oecologia* **29**, 67–76 (1977).

61. K. Winter, J. H. Troughton, M. Evenari, A. Läuchli, and U. Lüttge. Mineral ion composition and occurrence of CAM-like diurnal malate fluctuations in plants of coastal and desert habitats of Israel and the Sinai. *Oecologia* **25**, 125–143 (1976).

62. N. P. Yensen, M. R. Fontes, E. P. Glenn, and R. S. Felger. New salt tolerant crops for the Sonoran desert. *Desert Plants* **3**, 111–118 (1981).

63. E. Medina. Dark CO_2 fixation habitat preference and evolution within the Bromeliaceae. *Evolution* **28**, 677–686 (1974).

64. M. Kluge, I. Knapp, D. Kramer, I. Schwerdtner, and H. Ritter. Crassulacean acid metabolism (CAM) in leaves of *Aloë arborescens* Mill. Comparative studies of the carbon metabolism of chlorenchym and central hydrenchym. *Planta* **145**, 357–363 (1979).

65. K. Winter and U. Lüttge. Balance between C_3 and CAM pathway of photosynthesis. In O. L. Lange, L. Kappen, and E.-D, Schulze, eds., *Water and Plant Life. Problems and Modern Approaches.* Ecological Studies, Vol. 19. Springer-Verlag, Berlin, 1976, pp. 323–334.

66. V. J. Chapman. The new perspective in the halophytes. *Q. Rev. Biol.* **17**, 291–311 (1942).

67. M. Steiner. Zur Ökologie der Salzmarschen der nordöstlichen Vereinigten Staaten von Nordamerika *Jahrb. Wiss. Bot.* **81**, 94–202 (1935).

68. R. Biebl and H. Kinzel. Blattbau und Salzhaushalt von *Laguncularia racemosa* (L.) Gaertn. f. und anderer Mangrovenbäume auf Puerto Rico. *Österr. Bot. Z.* **112**, 56–93 (1965).

69. A. Poljakoff-Mayber. Morphological and anatomical changes in plants as a response to salinity stress. In A. Poljakoff-Mayber and J. Gale, eds., *Plants in Saline Environments.* Springer-Verlag, Berlin, 1975, pp. 97–117.

70. Y. Waisel. *Biology of Halophytes.* Academic Press, New York, 1972.

71. J. Berlin, J. E. Quisenberry, F. Bailey, M. Woodworth, and B. L. McMichael. Effect of water stress on cotton leaves. I. An electron microscopic stereological study of the palisade cells. *Plant Physiol.* **70**, 238–243 (1982).

72. R. Harder and H. von Witsch. Über den Einfluß der Tageslänge auf den Habitus, besonders die Blattsukkulenz, und den Wasserhaushalt von *Kalanchoe Bloßfeldiana. Jahrb. wiss. Bot.* **89**, 354–411 (1940).

73. J. Dainty. Water relations of plant cells. In U. Lüttge and M. G. Pitman, eds., Transport in Plants II, Part A, Cells, *Encyclopedia of Plant Physiology, New Series,* Vol. 2. Springer-Verlag, Berlin, 1976, pp. 12–35.

74. A. D. Tomos and R. G. Wyn Jones. Water relations in the epidermal cells of the halophyte *Suaeda maritima.* In F. Franks, ed., *Biophysics of Water.* Cambridge University Press, Cambridge, 1982.

75. A. D. Tomos and U. Zimmermann. Determination of water relation parameters of individual higher plant cells. In F. Franks, ed., *Biophysics of Water.* Cambridge University Press, Cambridge, 1982.

76. E. Steudle, H. Ziegler, and U. Zimmerman. Water relations of the epidermal bladder cells of *Oxalis carnosa* Molina. *Planta* in press(1983).

77. J. Rygol and U. Lüttge. Water relation parameters of giant and normal cells of *Capsicum annuum* pericarp. *Plant Cell Environ.* **6**, in press (1983).

78. D. Hüsken, E. Steudle, and U. Zimmermann. Pressure probe technique for

measuring water relations of cells in higher plants. *Plant Physiol.* **61,** 158–163 (1978).

79. C. B. Osmond. Ion absorption and carbon metabolism in cells of higher plants. In U. Lüttge and M. G. Pitman, eds., Transport in Plants II, Part A, Cells, *Encyclopedia of Plant Physiology, New Series*, Vol. 2. Springer-Verlag, Berlin, 1976, pp. 347–372.

80. H.-P. Haschke and U. Lüttge. Auxin action on K^+ -H^+-exchange and growth, $^{14}CO_2$-fixation and malate accumulation in *Avena* coleoptile segments. In E. Marrè and O. Ciferri, eds., *Regulation of Cell Membrane Activities in Plants*. North Holland, Amsterdam, 1977, pp. 243–248.

81. H. Schnabl. Der Anionenmetabolismus in starkehltigen und starkefreien Schließ-zellprotoplasten. *Ber. Deutsch. Bot. Ges.* **93,** 595–605 (1980).

82. J. A. Raven and F. A. Smith. Nitrogen assimilation and transport in vascular land plants in relation to intracellular pH regulation. *New Phytol.* **76,** 415–431 (1976).

83. T. C. Hsiao. Plant responses to water stress. *Annu. Rev. Plant Physiol.* **24,** 519–570 (1973).

84. C. B. Osmond, O. Björkman, and D. J. Anderson. *Physiologica Processes in Plant Ecology. Towards a Synthesis with Atriplex*, Ecological Studies, Vol. 36. Springer-Verlag, Berlin, 1980, 468 pp.

85. A. D. Hanson and W. D. Hitz. Metabolic responses of mesophytes to plant water deficits. *Annu. Rev. Plant Physiol.* **33,** 163–203 (1982).

86. T. C. Hsiao, E. Acevedo, E. Fereres, and D. W. Henderson. Water stress, growth and osmotic adjustment. *Phil. Trans. R. Soc. London, Ser. B* **273,** 479–500 (1976).

87. N. C. Turner and M. M. Jones. Turgor maintenance by osmotic adjustment: A review and evaluation. In N. C. Turner and P. J. Kramer, eds., *Adaptation of Plants to Water and High Temperature Stress*, Wiley-Interscience, New York, 1980, pp. 87–103.

88. C. C. Black. Photosynthetic carbon fixation in relation to net CO_2 uptake. *Annu. Rev. Plant Physiol.* **24,** 253–286 (1973).

89. T. F. Neales. The effect of night temperature on CO_2 assimilation, transpiration, and water use efficiency in *Agave americana* L. *Aust. J. Biol. Sci.* **26,** 705–714 (1973).

90. P. S. Nobel. Water relations and photosynthesis of a desert CAM plant, *Agave deserti. Plant Physiol.* **58,** 576–582 (1976).

91. M. D. Hatch and C. B. Osmond. Compartmentation and transport in C_4 photosynthesis. In C. R. Stocking and U. Heber, eds., Transport in Plants III, Intracellular Interactions and Transport Processes, *Encyclopedia of Plant Physiology, New Series*, Vol. 3. Springer-Verlag, Berlin, 1976, pp. 144–184.

92. C. B. Osmond. Crassulacean acid metabolism: A curiosity in context. *Annu. Rev. Plant Physiol.* **29,** 379–414 (1978).

93. M. H. Spalding, G. E. Edwards, and M. S. B. Ku. Quantum requirement for photosynthesis in *Sedum praealtum* during two phases of crassulacean acid metabolism. *Plant Physiol.* **66,** 463–465 (1980).

94. P. S. Nobel. Water relations and photosynthesis of a barrel cactus, *Ferocactus acanthodes*, in the Colorado desert. *Oecologia* **27,** 117–133 (1977).

95. P. S. Nobel and T. L. Hartsock. Resistance analysis of nocturnal carbon dioxide

uptake by a crassulacean acid metabolism succulent, *Agave deserti*. *Plant Physiol.* **61**, 510–514 (1978).

96. W. Cockburn, I. P. Ting, and L. O. Sternberg. Relationship between stomatal behavior and internal carbon dioxide concentration in crassulacean acid metabolism plants. *Plant Physiol.* **63**, 1029–1032 (1979).

97. M. Kluge, C. Böhlke, and O. Queiroz. Crassulacean acid metabolism (CAM) in *Kalanchoë*: Changes in intercellular CO_2 concentration during a normal CAM cycle and during cycles in continuous light or darkness. *Planta* **152**, 87–92 (1981).

98. K. Winter and D. J. von Willert. NaCl-induzierter Crassulaceensäurestoffwechsel bei *Mesembryanthemum crystallinum*. *Z. Pflanzenphysiol.* **67**, 166–170 (1972).

99. K. Winter. Zum Problem der Ausbildung des Crassulaceensäurestoffwechsels bei *Mesembryanthemum crystallinum* unter NaCl-Einfluß. *Planta* **109**, 135–145 (1973).

100. K. Winter. Enfluß von Wasserstreß auf die Aktivität der Phosphoenolpyruvat-Carboxylase bei *Mesembryanthemum crystallinum* L. *Planta* **121**, 147–153 (1974).

101. A.-M. Heun, J. Gorham, U. Lüttge, and R. G. WynJones. Changes of water-relation characteristics and levels of organic cytoplasmic solutes during salinity induced transition of *Mesembryanthemum crystallinum* from C_3-photosynthesis to crassulacean acid metabolism. *Oecologia* **50**, 66–72 (1981).

102. H. Greenway, K. Winter, and U. Lüttge. Phosphoenolpyruvate carboxylase during development of crassulacean acid metabolism and during a diurnal cycle in *Mesembryanthemum crystallinum*. *J. Exp. Bot.* **29**, 547–559 (1978).

103. K. Winter. $\delta^{13}C$ values of some succulent plants from Madagascar. *Oecologia* **40**, 103–112 (1979).

104. G. D. Farquhar and T. D. Sharkey. Stomatal conductance and photosynthesis *Annu. Rev. Plant Physiol.* **33**, 317–345 (1982).

105. P. S. Nobel. Leaf anatomy and water use efficiency. In N. C. Turner and P. J. Kramer, eds., *Adaptation of Plants to Water and High Temperature Stress*. Wiley-Interscience, New York, 1980, pp. 43–55.

106. K. Winter, U. Lüttge, E. Winter, and J. H. Troughton. Seasonal shift from C_3 photosynthesis to crassulacean acid metabolism in *Mesembryanthemum crystallinum* growing in its natural environment. *Oecologia* **34**, 225–237 (1978).

107. K. Winter and J. H. Troughton. Carbon assimilation pathways in *Mesembryanthemum nodiflorum* L. under natural conditions. *Z. Pflanzenphysiol.* **88**, 153–162 (1978).

108. I. P. Ting and L. Rayder. Regulation of C_3 to CAM shifts. In I. P. Ting and M. Gibbs, eds., *Crassulacean Acid Metabolism*. American Society Plant Physiologists, Rockville, Maryland, 1982, pp. 193–207.

109. B. C. Gerwick and G. J. Williams. Temperature and water regulation of gas exchange of *Opuntia polyacantha*. *Oecologia* **35**, 149–159 (1978).

110. N. Schilling and P. Dittrich. Interaction of hydrolytic and phosphorolytic enzymes of starch metabolism in *Kalanchoë diagremontiana*. *Planta* **147**, 210–215 (1979).

111. P. Nobel. Interaction between morphology, PAR interception, and nocturnal acid accumulation in cacti. In I. P. Ting and M. Gibbs, eds., *Crassulacean Acid Metabolism*, American Society Plant Physiologists, Rockville, Maryland, 1982, pp. 260–277.

8

MECHANISMS OF SALINITY RESISTANCE IN RICE AND THEIR ROLE AS PHYSIOLOGICAL CRITERIA IN PLANT BREEDING

A. R. Yeo and T. J. Flowers

Biology Building
The University of Sussex
Brighton, United Kingdom

The selection of plant material that is potentially suitable for improving the species response to adverse conditions is generally mediated by some type of mass screening of the phenotype as a whole. Most commonly, the condition to which resistance is sought is imposed (in the manner of a severe selection pressure) and those entries showing the lowest mortality, or least obvious damage, are selected. This approach has led to the majority of agricultural and horticultural developments that man has produced, though by its nature it is very consuming of time and resources where large numbers of accessions and hybrids must be screened for prolonged periods. There is an additional problem with physiological disorders such as salt toxicity in that the measured criterion (usually visual damage assessment) is a remote consequence of the actual lesions and is subject to considerable bias in its expression due to uncontrolled environmental conditions (1). The initial aim of the research program from which much of the presented data stems was simply to seek a more rapid and objective physiological parameter to assist mass screening programs.

The high residual variability in correlations involving even the most physiologically plausible single parameter examined leads to the general note of this chapter. The background is that whatever single parameter or criterion is chosen (be it survival, visible damage, or the most sophisticated physiological measurement) one assumption persists: the behavior of the single parameter can only fully describe the potential of the genotype if a single factor mediates the response of the plant to the applied conditions.

The data presented here and the other literature described point to this assumption being untrue. The response of plants to conditions like soil salinity emerge as complex whole plant phenomena not necessarily due to, or dominated by, any single factor. As a consequence the best response lies with the optimization of several, probably independent, physiological characteristics.

This does not prejudice the role of mass screening in producing varietal improvement; historical experience supports this. However, physiological work suggests a parallel "building block" approach to the incorporation of various recognized traits that singly contribute to the whole of salinity resistance.

The overview chapters in this volume, and indeed the existence of the conference itself, obviate the need to catalogue the many reviews and articles (e.g., 2–5) emphasizing the significance of soil salinity in restricting and limiting the production and productivity of many agricultural crops. In introduction it needs to be emphasized that *Oryza sativa* (rice) is widely documented as one of the most important agricultural crops most sensitive to salinity (6–15, summary in 14).

1. BASIS OF SALINITY DAMAGE

The literature frequently contrasts the central dilemma of ion toxicity versus water deficit in saline conditions (16, 17), and an important initial question is whether salinity damage in rice is attributable to excessive ion entry or to water deficit imposed by the coligative properties of dissolved salts in the environment. In this context, water deficit imposed internally due to high inorganic salt concentrations in the extracellular spaces of the leaves (18) is classed as a consequence of excessive ion entry.

Figure 8.1 compares the dry matter production of two rice varieties[*] with a salinity resistant member of the Graminae (*Festuca rubra*) showing the relative sensitivity of rice and that varietal differences tend to be manifest only at rather moderate (50 mol m^{-3} NaCl) salt concentrations. Almost all varieties are appreciably damaged by 50 mol m^{-3} NaCl imposed at the seedling stage (14 days); the time for 50% of individuals to die ranged with variety from 9 to 60 days (15). A concentration of 50 mol m^{-3} NaCl is thereby established as a useful working level, eliciting a wide range of varietal response.

Table 8.1 shows the effect on survival of adding PEG (polyethyleneglycol, a polymer of ethane 1:2 diol) to the 50 mol m^{-3} NaCl medium [throughout, all culture solutions contain a full nutrient solution with K at 1.25 mol m^{-3}; basically that of Yoshida et al. (19) though often with slight modifications to

---

[*]The word variety cannot always be equated with pure homozygosity and is used loosely for textual simplicity. The figure and table legends specify designated cultivars, land races, and breeding lines.

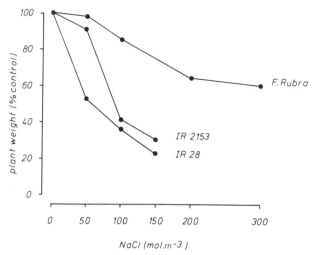

Figure 8.1.   Relative dry matter production of cv. IR28, breeding line IR2153-26-3-5-2 and *Festuca rubra* after 10 days in various NaCl additions imposed at 14 days.

buffer pH drifts]. Survival was greatly increased by adding PEG (ψ_w −220 kPa) to the 50 mol m^{-3} NaCl (ψ_w −220 kPa) although this clearly lowers the water potential of the medium (to approx. −440 kPa), double that of NaCl alone.

This benefit is attributed to the reduction of transpiration rate (volume flow J_v) and consequently the flux (J_s) of Na through the root. The shoot Na concentration was significantly reduced by adding PEG to the medium (20) as it is by other methods that reduce J_v (increasing ψ_w of the atmosphere or inducing stomatal closure by aerosol abscisic acid) (Table 8.2).

The interpretation is that salinity damage at this external concentration is due predominantly to excessive ion entry. Poorer performance on NaCl than PEG was considered persuasive evidence of this in a recent review (17), and simple arithmetic using data which follow later show that sodium accumulation in the leaves is 5 to 10 times that needed for osmotic adjustment to the changed solute potential of the external medium.

Table 8.1.   Survival of Seedlings of Breeding Line IR 2153-26-3-5-2 in 50 mol m^{-3} NaCl With and Without PEG 1540 (220 kPa) Imposed at 14 Days

Time from Salinization (days)	Plants Dead (%)	
	NaCl	NaCl + PEG
10	0	0
15	82	17
20	85	25

Table 8.2.   Effects of Conditions That Reduce the Transpirational Volume Flow on Average Shoot Sodium Content of Plants Growing in 50 mol m^{-3} NaCl[a]

Treatment		Leaf Sodium (mmol g^{-1} dry wt)	
a	b	a	b
60% RH	90% RH	2.28	0.82
Spray H$_2$O	Spray ABA[b]	2.25	1.42
-PEG	220 kPa PEG[c]	1.42	0.72

[a]Illustrative data from three unrelated experiments of differing durations.
[b]10 mmol m^{-3}.
[c]Mean MW = 1540.

2.   THE RELATIONSHIP BETWEEN SODIUM UPTAKE AND SURVIVAL

2.1.   Individuals

Most rice varieties of tropical origin are not homozygous lines (21). This means that although considerable uniformity exists for characters of long-term agronomic importance (growth rate and nutrient uptake) a uniform response to nonselected characters is not necessary. Table 8.3 illustrates this by the coefficients of variation for a number of characteristics among individuals of a breeding line, and it is particularly notable that variability in sodium uptake is seven times that for K uptake. This is understandable in that there has been strong selection for the K uptake, a nutrient response, and there are strong biochemical reasons why plants should demonstrate uniformity of internal K from disparate environments (22). As discussed by Flowers and Yeo (15) there

Table 8.3.   Variability in Different Parameters for Seedlings of Breeding Line IR2153-26-3-5-2 Grown for 7 Days after Imposition of NaCl (50 mol m^{-3} at 13 Days)

Parameter	Units	$\bar{x}$	SD	V^a(%)
Shoot dry wt	mg	67.5	11.0	16.3
Transpiration per plant	g. day^{-1}	4.06	0.98	24.1
Na	mmol g^{-1} dry wt	7.44	0.497	66.8
Cl	mmol g^{-1} dry wt	0.787	0.331	42.1
K	mmol g^{-1} dry wt	0.654	0.051	7.7

[a]Coefficient of variation (100 $\times$ SD/$\bar{X}$).

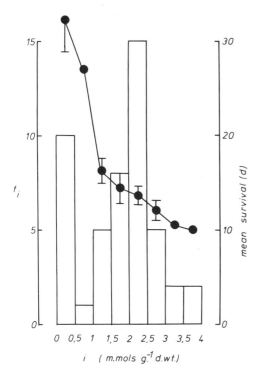

Figure 8.2. Frequency distribution of leaf 3 sodium concentration in class intervals of 0.5 mmol g^{-1} dry wt and mean survival period of each class. Plants of cv. Hwaizawi (Iraq) were salinized (50 mol m^{-3}) at 14 days, a small leaf sample taken after 6 days, and survival followed.

are few restrictions on sodium uptake except the rather wide limits of micronutrient necessity (23) and toxicity; there is insufficient selection history for a sodium response to impose any uniformity.

It is highly significant as a generality, and potentially important, that individual variability in sodium uptake and in survival are negatively correlated (15) (Fig. 8.2). In the figure initial sodium uptake (measured by a small leaf sample after 6 days) has been segregated into classes and the mean survival period for individuals falling into each class interval computed. There is a clear inverse relationship that is consistent with salt toxicity being due to excessive sodium uptake to the leaves.

2.2. Varieties

The experimental procedures are the same, but in deference to the degree of individual variability observed (15), not less than 50 internal replicates are used to assess a mean varietal Na uptake. There is again an overall negative relation between Na uptake and varietal survival in saline conditions, confirming the predictive value of initial Na uptake measurements, which is illustrated in Figure 8.3.

On both varietal bases, and on individual variation within varieties, sodium

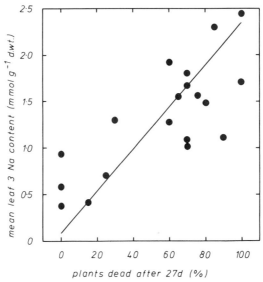

Figure 8.3.   Scatter diagrams of the mean sodium content of a leaf 3 sample ($N = 50$) and resistance (based on survival) for 19 cultivars and breeding lines. Salinity (50 mol m^{-3} NaCl) imposed at 14 days, leaf samples after further 6 days. A linear regression ($y = 8.85x - 0.62$, $r^2 = 63\%$) is illustrated.

uptake is negatively related to survival. This may be interpreted as fitting a general pattern of exclusion described for glycophytes though it must not be applied so simplistically as to infer strategy (24). The location of this relative "exclusion" in the continuum needs noting: even the most resistant varieties are being damaged by 50 mol m^{-3} NaCl and even they are showing Na net transport rates and internal salt concentrations rivalling those of a halophyte! It is arguable that the halophyte growing at several hundred mol m^{-3} NaCl is in fact the better excluder although it accumulates NaCl for strategic reasons.

The regressions of individual and varietal survival on sodium uptake show a major point: they account at best for some 60% of the total sums of squares and frequently less. The predictive value of Na uptake on survival (resistance) has a large uncertainty factor. There is an overall relationship with a sound physiological basis of causality, but it is a loose one. The most obvious explanation is that the line of causality is not due to a single factor dominating the relationship between sodium uptake and leaf death.

There are numerous possibilities for mitigating the consequences of excessive initial salt entry, which are summarized in the scheme of Figure 8.4. Ions may be extruded from the roots or retained in the cortical vacuoles in preference to long distance transport. If fluxed to the xylem they may be reabsorbed anywhere along the water pathway (mature roots, stems, petioles in dicots, leaf veins) and possibly retranslocated back to the medium. Saline ions entering the shoot may be involved in complex interactions with growth rate (diluted by high growth rates) or they may be compartmentalised on tissue (leaf to leaf) intercellular and

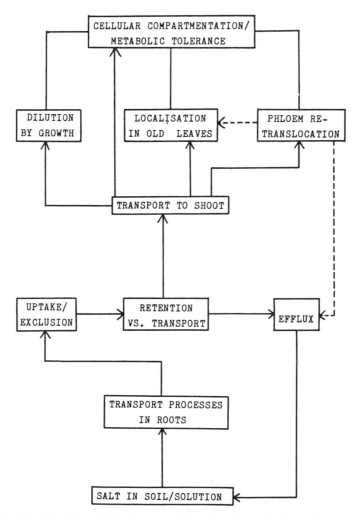

Figure 8.4. Physiological variables that may affect salt resistance in rice and schematic representation of their possible interaction.

intracellular bases. Evidence for constitutive changes in metabolic ion activation requirement of the type seen in holobacteria (25) is unknown in higher plants.

The existence and implications of many of these possibilities are explored in the ensuing sections.

3. SODIUM-POTASSIUM SELECTIVITY

NaCl salinity is used as an experimental system because it is the simplest starting point, without the more complex nutrient deficiency problems introduced, for example, by alkaline salinity. Nonetheless it comprises the questions of Na and

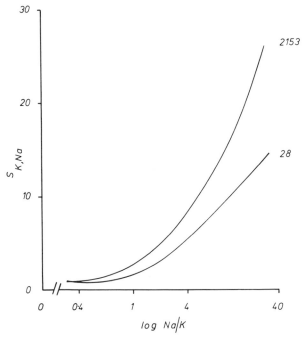

Figure 8.5.   Sodium–potassium selectivity in cv. IR28 and breeding line IR2153-26-3-5-2 at differing external Na:K ratios.

Cl toxicity per se, water deficit, and competitive interaction of the saline ions to induce deficiencies.

Emphasis is placed on sodium uptake and toxicity because chloride is regarded as essentially a neutral anion physiologically which is tolerated over a wide range of concentrations (26). There is more biochemical and chemical evidence that the disruptive effect of sodium on conformation of macromolecular structure and interference with cytoplasmic K roles will preempt Cl toxicity. There is, anyway, a very clear quantitative relationship between Na and Cl uptake by rice (27). $S_{K,Na}$ or the selectivity of the root for K versus Na is defined as K plant/Na plant: K medium/Na medium. Figure 8.5 illustrates the behavior of $S_{K,Na}$ at different external ratios (up to Na:K = 40, i.e., 50 mol m^{-3} NaCl) for IR2153 (a fairly resistant breeding line) and IR28 (a sensitive variety). There are differences in that the sensitive variety showed the poorer selectivity, but the most notable point is the very low values. They mean Na:K ratios of 2 or more and specific Na concentrations of 2 m mol g^{-1} dry wt. (almost 500 mol m^{-3} in rice) at 50 mol m^{-3} NaCl. By contrast, these values of selectivity (20–30) are much lower than seen for the resistant *F. rubra* (about 1000 in the same conditions) (Waite, Yeo, and Flowers, unpublished data).

This poor initial selectivity sums up most of the problems rice faces in a saline environment. It is possibly due in part to very poor carrier specificity but is more likely due to either passive membrane leakage or to a large membrane bypass flow [J_{vb} in the terminology of Pitman (28)]. Our data do not permit a conclusive statement at the present time. Selectivity may be aided by extrusion [shown for barley roots by Nassery and Baker (29)], but it is implausible that this would be an energetically viable option in rice.

4. MITIGATION OF EXCESSIVE SALT ENTRY

4.1. Reabsorption from the Xylem

Reabsorption has not been specifically investigated in rice roots, although it is hardly plausible that it could make a significant impact upon the massive sodium influx at 50 mol m^{-3} concentrations. Reabsorption is inferred for roots of maize based on measurements of root exudate concentration at different excision levels, and ^{22}Na autoradiography (30). Yeo et al. (31) used X-ray microanalysis to estimate ion ratios in the mature (nonabsorbing) root region of *Zea mays* and interpreted these results as showing a reabsorption of sodium from the xylem mediated by the xylem parenchyma cells. This was found to be consistent with ultrastructural observations (31). Xylem parenchyma has been observed to change into transfer cells in *Glycine max* varieties (32), and the xylem parenchyma was also implicated in *Phaseolus coccineus* roots (33). Observations had been made earlier for Na retention by bean stems when perfused with solution by Jacoby (34). Na retention in the leaf petiole of *Lycopersicon* was reported by Besford (35); in fact the whole xylem transport pathway may be bounded by backup reabsorption.

4.2. Retranslocation

Lessani and Marschner (36) considered that retranslocation was a salt resistance strategy based on a correlation between the appearance of foliarly applied ^{22}Na in the root medium and the resistance of the species determined by relative growth. To make an appreciable contribution in rice, it would require more sodium in the phloem than is usually considered likely, and Yeo and Flowers could find at most 1% of sodium retranslocated per day (27) when the sodium was applied in a pulse loading sequence from the medium to ensure so far as practical that the sodium really was in the leaf cells to start with. Of interest, retranslocation of Na was not observed in the salt tolerant halophyte, *Suaeda maritima* (37).

4.3. The Interaction with Growth Rate

The relationship between quantity of salt leaving the root and the resulting concentration in the shoot is determined by the growth rate of the shoot in relation to the net transport of ions out of the root. This value [J_s, cf Pitman (28)]

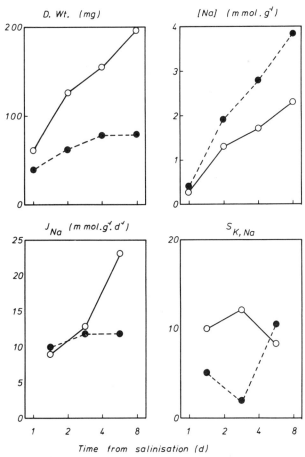

Figure 8.6. Time course of shoot dry weight, shoot average sodium content, net sodium transport, and Na/K selectivity for cv. IR22 (●) and land race Pokkali (○) following salinization (50 mol m⁻³ NaCl) at 14 days).

is net quantity of ions transported to the shoot per unit weight of roots in unit time.

Figure 8.6 shows the time course of shoot weight, shoot average sodium content, net transport of Na, and selectivity for two varieties: the nondwarfed Land Race 'Pokkali' and the dwarfed variety IR22. The dwarfing is reflected in the shoot weight graph where 'Pokkali' is clearly the larger plant. 'Pokkali' has a substantially lower average shoot Na content, and in the light of foregoing arguments, this will contribute substantially to its salinity resistance. However, J_{Na}, the net sodium transport by both varieties is initially similar and in fact 'Pokkali' later has a higher J_{Na}. This clearly indicates that the lower shoot Na content in 'Pokkali' is not due to any better control of Na transport by its roots but is directly attributable to the diluting effect of its rapid vegetative growth. This might be a significant point in planning varietal suitability for growth in saline conditions, a universal feature of glycophyte response to which is a reduced growth.

4.4.   Compartmentation within the Plant Body (Leaf to Leaf)

The pattern of leaf death in rice exposed to saline conditions is initially suggestive of a nonuniformity of sodium concentration in the different leaves at any one time. That is, the older leaves die while younger ones remain green and growing. This is a useful feature of the Graminae, and in fact the plant has been treated as a population of leaves where these are the units of modular construction and so in fact the real individuals (38). Yoshida (39) points out that the leaf/node/nodel roots assemblage is the unit of construction in rice, and tillers are sufficiently independent to be separated. This growth form makes parts of the monocot plant more expendable for strategic purposes than the dicot which must conserve its primary axis. Shedding of leaves is a common strategy in monocot halophytes (40), and gradients in salt content from leaf to leaf are often reported for nonhalophytes exposed to salinity (16).

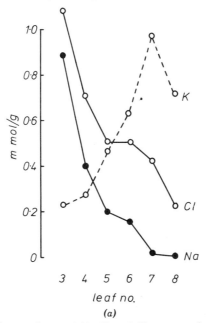

(a)

Figure 8.7.   Leaf to leaf ion gradients. (a) Na, K, and Cl concentrations in different leaves (8, youngest; 3, oldest surviving) in plants of breeding line IR2153-26-3-5-2 showing most resistance to salinity (survivors after 35 days, following 50 mol m^{-3} NaCl at age 14 days). (b) Profiles of leaf sodium concentration at five times following salinization (50 mol m^{-3} NaCl) at 14 days for breeding line IR4630-22-3-10-2. Solid line, 0 days; long dashes, 3 days; short dashes, 7 days; dots and dashes, 10 days; dots, 14 days.

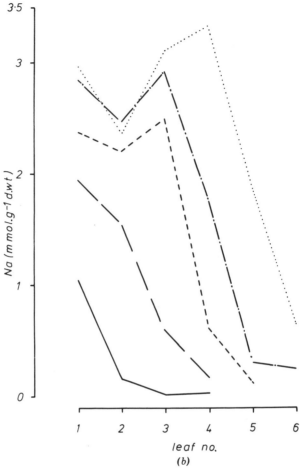

Figure 8.7. (*Continued*)

Leaf to leaf gradients of sodium content are very pronounced in rice (Fig. 8.7). The data show a dramatic range of sodium content from young to old leaves in individuals which have survived a long period of salinity. This gradient develops rapidly on exposure to salinity (27). The figure shows a profile of leaf Na contents from 1 (oldest) to the youngest at various periods after salinisation at 14 days. There is a progressive 'filling up' of leaves, but the younger leaves are protected. This permits some leaves to remain at sublethal salt concentrations.

Where such observations have been made in the literature they are normally explained as a mixture of (i) dilution by growth in the younger leaves and (ii) integrating time of exposure and xylem input leading to more salt in older leaves (17). This is not a sufficient explanation in rice (27) since a gradient develops rapidly upon a pulse exposure and it can be shown by tracer uptake that the net input of Na to the younger leaf is absolutely low (Table 8.4). Hence leaf to leaf distribution of sodium is a physiological mechanism of survival significance

Table 8.4.   The Timecourse of Radionuclide Content in Individual Leaves following Addition of ^{22}Na-Labeled NaCl (50 mol m^{-3} Chemical Concentration) to the Growing Medium (Breeding Line IR2153-26-3-5-2 Salinized at 14 Days)[a]

Time from Salinization (days)	Leaf ^{22}Na Content (Bq)		
	Leaf 3	Leaf 4	Leaf 5
2	255	160	25
4	350	151	30
6	440	205	60
8	510	340	61
10	570	650	110

[a]Calculated from ref. 27.

which differs from rice variety to variety (27). A result of this is that the average salt content of the rice shoot is an inadequate indicator of resistance potential.

4.5.   Intercellular and Intracellular Compartmentation

Suggestions that some level of compartmentation within the leaf is possible, if not probable, came from two converging lines of evidence. Sodium toxicity *in vivo* is difficult to estimate and the method adopted is a paraphrase of the LD$_{50}$ of toxicity using chlorophyll concentration as an indicator of viability (due to its ease of measurement). Figure 8.8 shows a scatter diagram of the Na and chlorophyll contents of a population of 50 individual leaves of a single variety sampled at a period after salinisation (50 mol m^{-3} at 14 days) when a range of damage was visibly evident. As the sodium content of the leaf increases, the chlorophyll content declines and this is used to estimate the concentration of sodium causing 50% loss of chlorophyll. Despite the residual variability, varietal differences are large enough to be significant. The estimated sodium concentration (on a tissue water basis), which brings about a 50% loss of chlorophyll, ranges from about 135 to 500 mol m^{-3} (Table 8.5) (41). The overall phenotypic resistance of the varieties is also shown (based on survival), and there is no correlation between leaf resistance and phenotype resistance. Presumably, differences in leaf tolerance are masked by performance in respect to other criteria.

A general consensus in the literature seems to be that the enzymes and metabolic systems of rice are very unlikely to function at sodium concentrations of over 300 mol m^{-3}; indeed, the optimal monovalent cation activation of NADH nitrate reductase and malate dehydrogenase was only 60–80 mol m^{-3} (27).

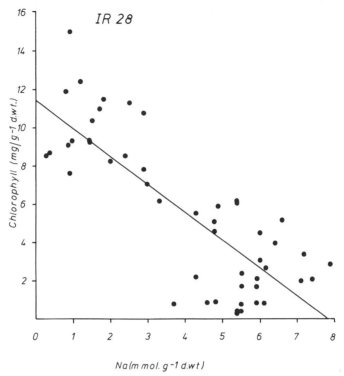

Figure 8.8. Scatter diagram of sodium and chlorophyll contents (ethanol-insoluble dry weight basis) for a population of leaves of cv. IR28 sampled 7 days after salinization (50 mol m^{-3} NaCl at 14 days). A linear regression is illustrated.

This is circumstantial evidence that there is some discontinuity of Na distribution within the leaf, the effectiveness of which may differ with variety. I am not aware of any data on the intracellular distribution of ions in rice, although this is the subject of investigation.

The role of intercellular compartmentation has been approached by X-ray microanalysis of leaves frozen in melting N$_2$ ($\sim$70° K) essentially as described for roots by Yeo et al. (31). The problems of standardization still limit the quantification, and the data (Table 8.6, Fig. 8.9) show only the Na:K ratio and an overall concentration estimate based on spectral intensity (net peak count rate ratio to background electron deceleration count rate). Even at a preliminary stage the data shows pronounced differences in the ion contents of different leaf cells. The highest Na:K ratios are seen in the bundle sheath cells. This fits the generalized pattern of Na reabsorption from the vascular system and with the observation of Stelzer (42) for the salinity tolerant grass *Puccinellia peisonis*.

The cryomicroanalysis technique also has uses in examining root barriers to Na movement. A profile across the root cortex of ion distribution is shown in Figure 8.10 and demonstrates a consistent trend across the cortical layers.

Table 8.5.   The Leaf Sodium
Concentrations (mol m^{-3} Tissue Water)
Causing 50% Loss of Chlorophyll (LC$_{50}$)
for Eight Varieties and One
Breeding Line.[a]

Variety/Line	LD$_{50}$	Resistance
SR26	133	R
Hwaizawi	166	S
Amber	230	M
Bhura Rata	367	R
MI48	410	M
Pokkali	434	R
IR28	434	S
IR2153.26.3.5.2	461	M
IR20	491	S

[a]The parameter is estimated from Na/Chlorophyll scatter diagrams. The approximate salinity resistance is included based on survival (R = resistant, M = moderate, S = sensitive; relative scale).

5. THE RELEVANCE OF PHYSIOLOGICAL CRITERIA TO PLANT SELECTION

It is difficult to advocate one approach to a problem without seeming critical of other approaches to the same problem. It is, however, the purpose of the concluding section of this chapter to outline the role of plant physiology in plant breeding, not to assert their antipathy.

The use of physiological criteria may simplify mass screening by virtue of

Table 8.6.   X-Ray Analysis of Cell Vacuoles in the Third Leaf of cv. Cigaron (France) after 24-hr Exposure to 25 mol m^{-3} NaCl Followed by 48-hr at 50 mol m^{-3} NaCl[a]

Cell Type	Na:K	Cl:(Na+K+Mg)	I[b]
Parenchyma	0.492 ± 0.054	0.743 ± 0.093	0.81 ± 0.10
Bulliform	0.243 ± 0.021	0.786 ± 0.035	0.83 ± 0.14
Mesophyll[c]	0.260	0.584	0.35
Xylem vessel	0.356 ± 0.114	0.575 ± 0.209	0.23 ± 0.11
Bundle sheath	1.22 ± 0.48	0.825 ± 0.136	0.58 ± 0.12

[a]Relative positions are shown in Figure 8.9.
[b]I is an estimate of spectral intensity as a guide to overall concentration. It is (all counts ascribed to elements)/(all background counts) in the 0–5.15 keV range.
[c]Too few fractures to permit standard errors.

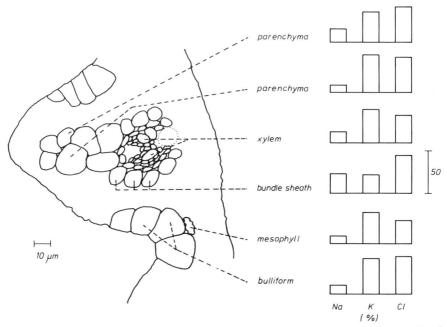

Figure 8.9. Schematic representation (traced from secondary electron image photograph) of the leaf fracture-surface showing the relative positions of the cells considered in Table 8.6. The bar diagrams show the relative composition of the vacuoles of the different cell types.

being more rapid and objective. For instance, the transport of sodium to the shoot has a substantial theoretical and experimental background for its predictive value for salinity damage to the shoot in glycophytes both generally (17) and for rice specifically (15). The fit of the Na uptake/survival regression is far from perfect, but the scatter appears to be of a similar order to that for visual damage on final yield. And it is very rapid. However, as discussed earlier, the use of a single parameter is subject to the same basic assumptions whatever that criterion is.

A physiological parameter provides more information about the system. As an example, a Na uptake measurement enables a mechanistic distinction to be made between ion toxicity damage (where uptake greatly exceeds osmotic needs) and water deficit (where uptake is inadequate). A knowledge of how a variety was resistant would be valuable in predicting its likely response to similar problems such as alkalinity. In the field of other toxicity problems, such as Fe, a knowledge of Fe transport would enable the basis of resistance to be attributed to improved exclusion or improved internal chelation of Fe which is important since cotolerance of heavy metals is widely reported in the ecological literature.

The third and major role of physiology comes in what can be termed the "building block" approach. This is the need to combine several characteristics,

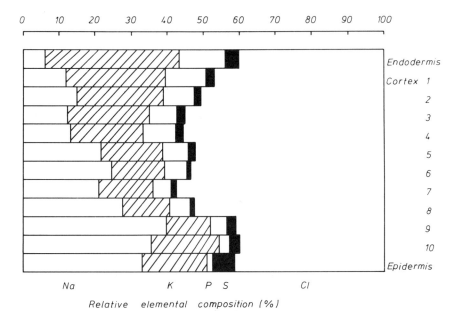

Figure 8.10.   Relative elemental composition of cell vacuoles in a transect across the root of a salt-grown plant (breeding line IR2153-26-3-5-2 after 3 days at 50 mol m^{-3} NaCl). The net counts acquired for each element in the atomic number range Na to Ca were proportioned; elements contributing 1% or less are excluded.

which do not on their own lead to improved survival in saline conditions, to achieve a resistant variety.

The data presented deal mainly with the response of rice to NaCl salinity in solution culture. They show differences in Na/K selectivity, interaction of net transport and growth rate, and sodium compartmentation from leaf to leaf and cell to cell. Other evidence is that extrusion and reabsorption would also be important in other species such as *Zea mays* though they would be masked by the sheer quantity of sodium transport in rice.

NaCl salinity is a simple experimental starting point (in comparison with soil salinity with the contribution of different salt types and accessory problems), but even this reveals the physiological complexity of something like higher plant salinity resistance.

The factors that influence the toxicity of sodium, both its initial uptake and any mitigating effects, are likely to be independent, incidental variation in a salt-sensitive glycophyte. There is no reason to suppose their simultaneous optimization in a genotype other than by chance (Table 8.5). It is fairly evident that an advantageous variation in one aspect (such as leaf to leaf distribution) may be confounded or completely masked by a poor performance in another (initial sodium uptake). Poor control of sodium uptake can be lethal to the phenotype even if it is better than many others at mitigating excessive uptake;

salt entry will simply swamp this capacity and the phenotype will be rejected both on grounds of survival and its overall sodium content.

A combination of all physiological traits assisting overall salinity resistance is logically a desirable long-term aim. The data reported, and the high residual variability in Na/survival correlations, show that salinity resistance is a multivariable whole plant phenomenon, each aspect of which needs to be separately optimized and combined, that is, a building block approach to a resistant variety.

It is not suggested that mass screening can, or should, be applied to all these variables. Physiological methods can identify potential donors for various traits which can be combined by conventional plant breeding. This could be a valuable parallel to mass screening for these difficult multivariate problems like salinity.

ACKNOWLEDGMENTS

Research described on rice is supported by the United Kingdom Overseas Development Administration. X-ray microanalysis measurements on rice were performed at the Tierarzliche Hochschule, Hannover, and we are deeply indebted to Dr. R. Stelzer for his help.

REFERENCES

1. J. Gale, R. Neuman, and A. Poljakoff-Mayber. Growth of *Atriplex halimus L.* in NaCl salinated culture solutions as affected by the relative humidity of the air. *Aust. J. Biol. Sci.* **23**, 947 (1970).

2. L. Bernstein. *Arid Zone Res.* (UNESCO) **18**, 139 (1962).

3. H. Boyko. *Saline Irrigation for Agriculture and Forestry.* Junk, The Hague, 1968.

4. E. Epstein. *Mineral Nutrition of Plants, Principles and Perspectives.* Wiley, New York, 1972.

5. IRRI *Annual Report for 1973.* International Rice Research Institute, Manila, 1974, p. 100.

6. L. C. Kapp. The effect of common salt on rice production. *Ark. Agric. Exp. St. Bull.* **465**, 3 (1947).

7. G. A. Pearson. Factors influencing salinity of submerged soils and growth of Caloro rice. *Soil Sci.* **87**, 198 (1959).

8. M. T. Kaddah and S. I. Fakhry. Tolerance of Egyptian rice to salt. *Soil Sci.* **91**, 113 (1961).

9. M. Akbar, T. Yabuno, and S. Nakao. Breeding for saline-resistant varieties of rice. I. Variability for salt tolerance among some rice varieties. *Jpn. J. Breeding* **22**, 277 (1972).

10. S. K. Datta. A study of salt tolerance of twelve varieties of rice. *Curr. Sci.* **41**, 456 (1972).

11. J. Venkateswarlu, M. Ramesam, and G. U. N. N. Rao. Salt tolerance in rice varieties. *Indian Soc. Soil. Sci. J.* **20,** 169 (1972).

12. S. A. Korkor and R. M. Abdel-Aal. Effect of total soil salinity and type of salts on rice crop. *Agric. Res. Rev. (Cairo)* **5,** 273 (1974).

13. R. K. Bhattacharyya. New salt tolerant rice varieties for coastal saline soils of Sunderban (West Bengal). *Sci. Culture* **42,** 122 (1976).

14. E. V. Maas and G. J. Hoffman. Crop salt tolerance: Evaluation of existing data. In H. E. Dregre, ed., *Managing Saline Water for Irrigation.* Proc. Int. Conf. pp. 187–198, Texas Technical Univ., 1976.

15. T. J. Flowers and A. R. Yeo. Variability in the resistance of sodium chloride salinity within rice (*Oryza sativa* L.) cultivars. *New Phytol.* **88,** 363 (1981).

16. T. J. Flowers, P. F. Troke, and A. R. Yeo. The mechanism of salt tolerance in halophytes. *Annu. Rev. Plant Physiol.* **28,** 89 (1977).

17. H. Greenway and R. Munns. Mechanisms of salt tolerance in non-halopytes. *Annu. Rev. Plant Physiol.* **31,** 149 (1980).

18. J. J. Oertli. Extracellular salt accumulation, a possible mechanism of salt injury in plants. *Agrochemica* **12,** 461 (1968).

19. S. Yoshida, D. A. Forna, J. H. Cock, and K. A. Gomez. *Laboratory Manual for Physiological Studies of Rice.* International Rice Research Institute, Manila, 1972.

20. A. R. Yeo and T. J. Flowers. Non-osmotic effects of polyethylene glycol on sodium transport and sodium/potassium selectivity. Plant Physiol. (Submitted).

21. T. T. Chang. Varietal variability and its relation to seed purity and maintenance of seed stocks. In *Third Workshop on Field Experiments,* International Rice Research Institute, Manila, 1974.

22. R. G. Wyn Jones, C. J. Brady, and J. Speirs. Ionic and osmotic relations of plant cells. In D. L. Laidman and R. G. Wyn Jones, eds., *Recent Advances in the Biochemistry of Cereals.* Academic Press, London, 1979, pp. 63–103.

23. P. F. Brownell. Sodium as an essential micronutrient element for a higher plant *(Atriplex vesicaria). Plant Physiol.* **40,** 460 (1965).

24. A. R. Yeo. Salinity resistance; physiologies and prices. *Physiol. Plant.* **58,** 214 (1983).

25. J. K. Lanyi. Salt dependent properties of proteins from extremely halophilic bacteria. *Bacteriol Rev.* **38,** 272 (1974).

26. D. T Clarkson and J. B. Hanson. The mineral nutrition of higher plants. *Annu. Rev. Plant Physiol.* **31,** 239 (1980).

27. A. R. Yeo and T. J. Flowers. Accumulation and localisation of sodium ions within the shoots of rice *(Oryza sativa)* varieties differing in salinity resistance. *Physiol. Plant.* **56,** 343 (1982).

28. M. G. Pitman and W. J. Cram. Regulation of ion content in whole plants. In D. H. Jennings, ed. *Integration of Activity in the Higher Plant.* Cambridge Univ. Press, Cambridge, 1977, pp. 391–424.

29. H. Nassery and D. A. Baker. Extrusion of sodium ions by barley roots III. The effect of high salinity on long-distance sodium ion transport. *Ann. Bot.* **38,** 141 (1974).

30. M. G. T. Shone, D. T. Clarkson and J. Sanderson. The absorption and translocation of sodium by maize seedlings. *Planta* **86,** 301 (1969).

31. A. R. Yeo, D. Kramer, A. Lauchli, and J. Gullasch. Ion distribution in salt-stressed mature *Zea mays* roots in relation to ultrastructure and retention of sodium. *J. Exp. Bot.* **28,** 17 (1977).

32. A. Lauchli, D. Kramer, and R. Stelzer. Ultrastructure and ion localisation in xylem parenchyma cells of roots. In U. Zimmerman and J. Dainty, eds., *Membrane Transport in Plants,* Springer-Verlag, Berlin, 1974, pp. 363–371.

33. D. Kramer, A. Lauchli, A. R. Yeo, and J. Gullasch. Transfer cells in roots of *Phaseolus coccineus:* Ultrastructure and possible function in exclusion of sodium from the shoot. *Ann. Bot.* **41,** 1031 (1977)

34. B. Jacoby. Sodium retention in excised bean stems. *Physiol. Plant.* **18,** 730 (1965).

35. R. T. Besford. Effect of replacing nutrient potassium by sodium on uptake and distribution of sodium in tomato plants. *Plant Soil* **58,** 399 (1978).

36. H. Lessani and H. Marschner. Relationship between salt tolerance and long distance transport of sodium and chloride in various crop species. *Aust. J. Plant Physiol.* **5,** 27 (1978).

37. A. R. Yeo. Salt tolerance in the halophyte *Suaeda maritima* L. Dum.: Intracellular compartmentation of ions. *J. Exp. Bot.* **32,** 487 (1981).

38. J. L. Harper and A. D. Bell. The population dynamics of growth form in organisms with modular construction. In R. M. Anderson, B. D. Turner, and L. R. Taylor, eds., *Symp. Br. Ecol. Soc. 20, Population Dynamics,* Blackwell, Oxford, 1979, pp. 29–52.

39. S. Yoshida. *Fundamentals of Rice Crop Science.* International Rice Research Institute, Manila, 1981, 269 pp.

40. R. Albert. Salt regulation in halophytes. *Oecologia* **21,** 57 (1975).

41. A. R. Yeo and T. J. Flowers. Varietal differences in the toxicity of sodium ions in rice leaves. *Physiol. Plant.* 59: 189–195 (1983).

42. R. Stelzer. Ion localisation in the leaves of *Puccinellia peisonis. Z. Pflanzenphysiol.* **103,** 27 (1981).

9

SALT EXCLUSION:
AN ADAPTATION OF LEGUMES
FOR CROPS AND PASTURES
UNDER SALINE CONDITIONS

André Läuchli

Department of Land, Air and Water Resources
University of California
Davis, California

Among the leguminous plants there are many important agricultural species. They are used as grain legumes, forage crops, or components of pastures, and extend from cool-temperate regions to the tropics. Soybean (*Glycine max*) is the most important species among the grain legumes; other significant grain legumes are pea (*Pisum sativum*), beans (*Phaseolus* spp.), broadbean (*Vicia faba*), peanut (*Arachis hypogaea*), chickpea (*Cicer arietinum*), and cowpea (*Vigna unguiculata*). Alfalfa (*Medicago sativa*) is the most outstanding forage legume, and clovers (*Trifolium* spp.) are the predominant legume components of pastures.

Legumes have long been recognized to be either sensitive or only moderately resistant to salinity (1). Salinity responses of crops in general are extremely complex, and an array of mechanisms appear to be involved in salt tolerance of plants (reviews: 2–5). Considering legumes, two features are relevant: there is variability in salt resistance among legumes, and most of them respond to saline conditions by *salt exclusion*, that is, exclusion of sodium and/or chloride from the leaves. Table 9.1 shows a compilation of comparative salt resistance of legumes. Except for *Lupinus luteus*, a legume of limited agricultural significance in Australia, all listed species responded to 50 mM NaCl salinity in solution culture by growth inhibition. Although fresh and dry weights of the foliage of *Lupinus luteus* were increased by this salinity level, 100 mM NaCl proved toxic to this species (6). Overall, this legume may be rated salt resistant, in contrast to

Table 9.1.   Comparative Salt Resistance of Legumes[a]

	% of Control (approximate values)	References
Resistant		
Lupinus luteus	150	6
Moderately resistant		
Lupinus angustifolius	85	6
Trifolium alexandrinum	85	7
Intermediate		
Glycine max, cv. Lee	75	8
Medicago sativa	70	11[b]
Phaseolus coccineus	75	9
Trifolium pratense	70	7
Vigna aureus (*V. radiata*)	70	10
Moderately sensitive		
Trifolium repens	60	12
Vigna sinensis (*V. unguiculata*)	60	10
Sensitive		
Arachis hypogaea	20–40	13[b,c]
Cicer arietinum	40	14
Glycine max, cv. Jackson	65[d]	8
Phaseolus vulgaris	55	15
Pisum sativum	53	15

[a]Estimated from leaf or shoot growth of plants grown in solution culture containing 50 mM NaCl for about 3 to 7 weeks.
[b]Grown in soil at comparable salinity.
[c]Harvested at maturity.
[d]At 10 mM NaCl.

the others that are considerably more salt sensitive. Peanut, chickpea, soybean cv. 'Jackson,' *Phaseolus vulgaris*, and pea are most sensitive to salinity. More salt resistant are particularly *Lupinus angustifolius*, the clovers, soybean cv. 'Lee,' alfalfa, and *Phaseolus coccineus*. Noteworthy are the species or cultivar differences in *Lupinus* (6), *Trifolium* (7, 12), soybean (8), and *Phaseolus* (9, 15).

Sodium is excluded from the leaves of many plant species (16, and earlier references quoted therein), including legumes (Table 9.2). Jacoby (16) demonstrated that bean plants exclude Na by retention in the basal parts of the plants but readily translocate Cl to the tops. Retention of Na in the stems became gradually saturated with increasing salt concentrations in the medium. Furthermore, Jacoby (17) concluded that Na retention in bean stems is energy

Table 9.2.   Salt Exclusion from Leaves of Legumes

	References
Na⁺ exclusion	
Cicer arietinum	
(depending on N supply)	14
Glycine max cv. Lee	8
Glycine wightii	20
Lupinus angustifolius	6
Medicago sativa	11
Phaseolus coccineus	9
Phaseolus vulgaris	16 (and others)
Trifolium alexandrinum	7
Cl⁻ exclusion	
Cicer arietinum cv. L550	Lauter and Munns, unpublished
Glycine max cv. Lee	18
Glycine wightii	20
Lupinus angustifolius	6
Trifolium alexandrinum	7

dependent and due to accumulation by cells in the vascular tissue. Chloride is also known to be excluded in some legumes (Table 9.2) and particularly in many fruit trees (see refs. in 18). Abel and MacKenzie (18) found that varietal differences in salt resistance of soybean were correlated with exclusion of Cl, the resistant varieties controlling Cl content in stems and leaves to a low level. Furthermore, Abel (19) showed that Cl transport to the tops of soybean varieties is under genetic control. In particular, Cl exclusion in soybean cv. 'Lee' was proposed to be controlled by a dominant, single gene pair, whereas Cl accumulation in the leaves of soybean cv. 'Jackson' appeared regulated by a recessive, single gene pair (19). In these studies, however, no attempt was made to unravel the mechanism of Cl exclusion in soybean.

On a comparative basis (Table 9.2), it is obvious that some legumes exclude both Na and Cl (e.g., soybean cv. 'Lee,' *Lupinus angustifolius, Trifolium alexandrinum*). Beans and alfalfa, however, are only Na excluders. More to the point is the question whether there is a widespread, positive correlation in legumes between Na or Cl exclusion and relative salt resistance. Such a positive correlation indeed exists in soybean varieties when Cl exclusion is considered (8, 18). Also, *Trifolium alexandrinum*, a major forage crop in the Middle East, India, and Pakistan was found by Winter and Läuchli (7) to be more salt resistant and a more efficient Na and Cl excluder than the salt-sensitive species *Trifolium pratense*. On the other hand, the two lupin species studied by Van Steveninck et al. (6) exhibited the opposite response, that is, *Lupinus luteus* appeared more salt resistant and less effective in excluding Na and Cl than *Lupinus angustifolius*. Thus, some legumes can adapt to moderate salinity levels by exclusion of Na and/ or Cl from the leaves, whereas the salt resistant *Lupinus luteus* may be termed a salt accumulator and thus features a halophytic response.

1. MECHANISMS OF SALT EXCLUSION IN LEGUMES

1.1. Sodium Exclusion

Since the groundbreaking work of Jacoby (16, 17), using *Phaseolus vulgaris*, the mechanisms of Na exclusion have been investigated extensively. The key control points in regulating Na transport in the whole plant appear to be a series of membrane transport processes in root, stem, and leaves (Fig. 9.1). In barley roots, the cortex cells are able to sequester Na predominantly in the vacuoles while maintaining the K/ Na ratio in the cytoplasm at a reasonably high level (21, 22). This important pattern of ion compartmentation appears to be brought about by selective K influx and Na efflux at the plasmalemma and by Na/ K exchange at the tonoplast (21). What is not well understood, however, is the capacity of Na sequestration in root cell vacuoles under saline conditions. In addition, a similar pattern of ion compartmentation in legume roots has not yet been elucidated.

In legumes, Na-specific events in the vascular tissues of the proximal region of the root and the base of the stem are of primary significance. Following up on Jacoby's work, Kramer et al. (9) revealed the existence of transfer cell-like xylem parenchyma cells in the proximal region of the root of *Phaseolus coccineus*. In the same region of the root, Na accumulated to levels greatly exceeding those in the apical root and the leaves (Table 9.3). Using X-ray microanalysis, these authors further demonstrated very high Na/ K ratios in the xylem parenchyma cells, in contrast to the xylem vessels where the ratio was approximately at unity (Table 9.4). Chloride did not accumulate in the xylem parenchyma cells (Table 9.4). Läuchli (23) concluyded that Na accumulation in the xylem parenchyma cells is due to Na reabsorption from the xylem sap in exchange for K, possibly by a Na/ K exchange process operating at the plasmalemma of these transfer cells. In view of evidence by Jeschke (21), on Na/ K exchange at the tonoplast of root cells, it is feasible that the tonoplast of the xylem parenchyma cells in bean roots is also important in Na reabsorption from the xylem vessels and controls Na accumulation in the vacuole of the xylem parenchyma cells. Whatever the mechanism of Na reabsorption, this process contributes effectively to Na exclusion from the leaves (Fig. 9.1). Its practical importance for conferring salt tolerance, however, is limited to low degrees or short duration of salinity stress, because the capacity of the xylem parenchyma cells to accumulate Na must become rapidly exhausted, as was discussed by Lüttge (24).

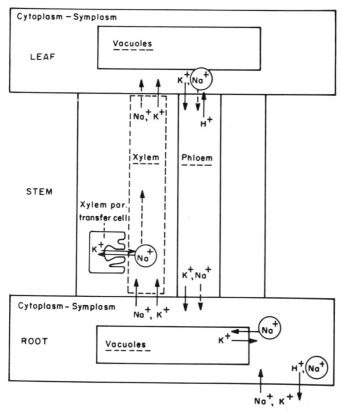

Figure 9.1. Model of the regulation of Na⁺ transport in a Na⁺-excluding glycophyte (relatively salt sensitive).

Table 9.3 Distribution of Na⁺ and Cl⁻ in Three Regions of the Root and in the Leaves of *Phaseolus coccineus*[a]

	Root						Leaves	
0–10 cm		10–30 cm		>30 cm			Leaves	
Na⁺	Cl⁻	Na⁺	Cl⁻	Na⁺	Cl⁻		Na⁺	Cl⁻
31	33	37	35	95	95		11	162

[a]Data from ref. 9. Nutrient solution (containing $2 \, mM \, K^+$) + $50 \, mM$ NaCl, 4 weeks. Total plant fresh weight inhibited by $50 \, mM$ NaCl to 73% of the control. Ion contents in μmol g^{-1} fresh weight.

Table 9.4.   Electron Probe X-Ray Microanalysis of the Proximal Region of a *Phaseolus coccineus* Root[a]

	Na/K	Na/Cl
Medium	25	1.0
Cortex	1.4	1.2
Xylem parenchyma	32	32
Xylem vessel	0.8	0.8

[a]Data from ref. 9. See Table 9.3 for conditions.

Sodium retranslocation from the leaves also contributes to maintaining Na at low levels in the leaves (Fig. 9.1). Phloem loading of Na may not be very significant, however, if the maintenance of low Na concentrations in the root symplasm (22) also applies to sieve tubes, which are considered a symplasmic pathway. By applying ^{22}Na to the tip of a primary bean leaf, Marschner and Ossenberg-Neuhaus (25) demonstrated significant Na retranslocation through the phloem to the root followed by Na efflux from the proximal root region to the medium, but the Na amounts applied and transported were low. More recently, Jacoby (26) applied Na to the upper roots of bean and found that Na moved upward in the stem, then downward to the main roots (probably through the phloem), from where most of it was lost to the medium. As in the previous study (25), the conditions in Jacoby's experiments were nonsaline, and comparable experiments using salt-stressed plants are lacking. In addition, Na lost to the medium may be reabsorbed by other regions of the root and, thus, net loss of Na to the root environment may be small.

1.2.   Chloride Exclusion

Following the study by Abel and MacKenzie (18) on Cl exclusion in soybean, Läuchli and Wieneke (8) studied ion distribution in the soybean varieties 'Lee' and 'Jackson' and suggested that Cl exclusion from the leaves of 'Lee' is regulated by the root. In subsequent papers the mechanism of Cl exclusion in 'Lee' was studied in more detail. It was shown (27) that Cl influx in roots of 'Lee' over a wide concentration range was much lower than in the salt sensitive, nonexcluding variety 'Jackson.' Furthermore, 'Lee' only transported considerable amounts of Cl to the shoot immediately upon onset of the salinity treatment and then effectively controlled Cl transport to the shoot, contrary to the behavior of 'Jackson.' X-ray microanalysis revealed (28) Cl accumulation in the cortex of the apical root of 'Lee,' suggesting that Cl accumulation in this root is mediated by sequestration of Cl in the vacuoles of the cortex. A model of the possible regulation of Cl transport in the roots of soybean varieties 'Lee' and

'Jackson' is presented in Figure 9.2. It indicates that the low Cl flux to the xylem (Φ_{cx}) and the shoot of the Cl excluding 'Lee' is mainly the consequence of low influx Φ_{oc} and high vacuolar transport Φ_{cv}. In contrast, 'Jackson' has an inefficient vacuolar transport Φ_{cv} and, thus, the high Φ_{cx} mirrors the high Φ_{oc} leading to uncontrolled Cl transport to the shoot. Chloride efflux (Φ_{co}) does not appear to differ much between 'Lee' and 'Jackson' (27 and unpublished results by H. Eggers and A. Läuchli).

Additional data shown in Figures 9.3 and 9.4 are in support of this model. The time course of Cl transport to the xylem of excised soybean roots indicates that in 'Lee' Cl transport rates are only significant during the first 4 hr of the experiment and then drop to very low levels. In 'Jackson,' however, Cl transport rates rise continuously after 7 hr (Fig. 9.3). The experiment exhibited in Figure 9.4 demonstrates the effects of low temperature. Chloride transport to the xylem of 'Jackson' is very temperature sensitive and almost completely blocked at 3°C. Surprisingly, the temperature response in 'Lee' is more complex: 12°C is inhibitory, but Cl transport to the xylem is stimulated at 3°C. A tentative explanation is that Φ_{cv} in 'Lee' (see Fig. 9.2) is a metabolically driven transport process that requires physiological temperature; at 3°C this condition is not met and there is rapid transport to the xylem (Φ_{cx}). Additional experiments using isolated protoplasts and vacuoles from the cortex of soybean roots 'Lee' and 'Jackson' are needed to lend definitive support to the proposed model in Figure 9.2. Furthermore, some of the data pertaining to the model were obtained under low-salt conditions; they need to be supplemented by similar studies using saline conditions.

The only Cl excluding legume system studied in detail concerns the soybean variety 'Lee,' in comparison with the Cl including 'Jackson.' There are now X-

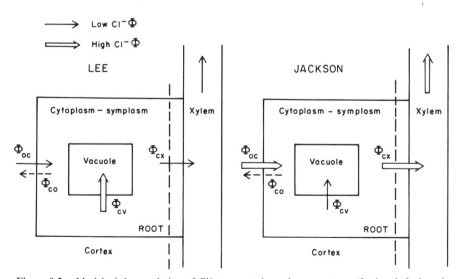

Figure 9.2.   Model of the regulation of Cl⁻ transport in soybean roots, cv. 'Lee' and 'Jackson.'

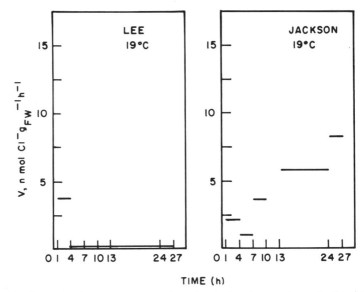

Figure 9.3. Rates of Cl⁻ transport to the xylem of excised soybean roots, cv. 'Lee' and 'Jackson': time course. Uptake medium: 1 mM KCl + 0.5 mM CaSO₄. Original data by H. Eggers and A. Läuchli.

ray microanalysis data from *Lupinus luteus* (6) which indicate that the apparent lack of Cl exclusion in this species is due to low rates of Cl accumulation in the vacuoles of the cortex cells, similar to the situation in soybean cv. 'Jackson.'

2.   REGULATION OF SALT CONCENTRATION IN LEAVES OF SALT EXCLUDERS

The concentrations of Na and Cl in leaves of salt excluding plants are regulated by a number of membrane transport processes, as discussed above. In addition, they also depend on the growth rates of the plant. For salt excluders without salt glands, such as many legumes, the concentration of Na and Cl in a leaf is given by the following equations (modified after H. Greenway, personal communication):

$$Na_{leaf} = \frac{net\ import}{RGR} = \frac{\Phi_{cx,\ root} - (\Phi_{xt,\ root,\ stem} + \Phi_{lp})}{RGR} \tag{1}$$

$$Cl_{leaf} = \frac{net\ import}{RGR} = \frac{\Phi_{cx,\ root} - \Phi_{lp}}{RGR} \tag{2}$$

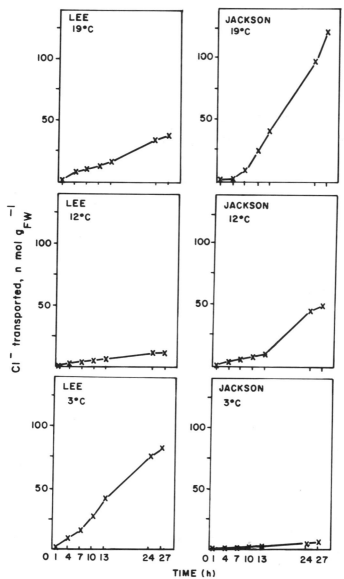

Figure 9.4. Cl⁻ transport to the xylem of excised soybean roots, cv. 'Lee' and 'Jackson': effect of temperature. Conditions as in Figure 9.3. Original data by H. Eggers and A. Läuchli.

179

where RGR = relative growth rate, $\Phi_{cx, root}$ = ion flux from root symplasm to xylem, $\Phi_{xt, root, stem}$ = ion flux from xylem vessels to xylem parenchyma-transfer cells in root and stem, and Φ_{lp} = ion flux from leaf mesophyll to phloem (retranslocation). Chloride reabsorption from the xylem vessels in root and stem is negligible and therefore not included in Equation (2).

Lessani and Marschner (29) examined the significance of retranslocation for regulation of salt concentration in the leaves and its relation to salt resistance in several species. They found no correlation between the extent of Cl retranslocation and growth depression caused by salinity. Wieneke and Läuchli (27) arrived at a similar conclusion for the soybean varieties 'Lee' and 'Jackson.' With regard to Na, however, there were significant correlations between decrease in dry matter production by 100 mM NaCl in the medium and Na retranslocation from leaves (Fig. 9.5a) and, in particular, efflux of Na from the roots (Fig. 9.5b). Efflux from the roots increased with decreasing salt resistance indicating that Φ_{lp} in Equation (1) does not contribute effectively to salt resistance of the plant. Rather, the regulation of net salt import to the leaves primarily controls salt exclusion and salt resistance. Indirect support of this conclusion comes from recent work by Winter (30, 31). This author demonstrated that in leaves of *Trifolium alexandrinum* grown at 50 mM NaCl, there was gradual destruction of the phloem transfer cells prior to the development of leaf burn due to salt toxicity (30). The destruction of phloem transfer cells coincided with very high Na/K ratios in these cells (31). Thus, the buildup of Na in the leaves probably caused the breakdown of Na export through the phloem.

3.   RECENT PROGRESS IN PERFORMANCE OF LEGUMES UNDER SALINE CONDITIONS

The exploitation of genetic variability in cultivated species offers the possibility of developing salt-tolerant crops [for a recent review see Epstein et al. (32)]. Such genetic variability exists among legumes. Croughan et al. (33) selected a salt-tolerant line of alfalfa using cell culture techniques (see also Chapter 18). Growth of the alfalfa line W75RS was not affected by a salt level of 62.5 mM NaCl in nutrient solution, no matter whether callus cultures or whole plants were examined (12). Cerdá et al. (34) found the pea cultivar Durana to be moderately salt resistant, in contrast to most pea varieties, but salt resistance in this cultivar was not correlated with salt exclusion. Among 150 cultivars of chickpea, only L550 survived at 50 mM NaCl and was found to be a Cl excluder, and *Phaseolus lunatus* (lima bean) produced pods at 200 mM NaCl in solution culture (35, 36, Lauter and Munns, personal communication). Thus, there is considerable salt resistance in some legume genotypes but no consistent correlation emerges between salt resistance and exclusion of salt from the leaves.

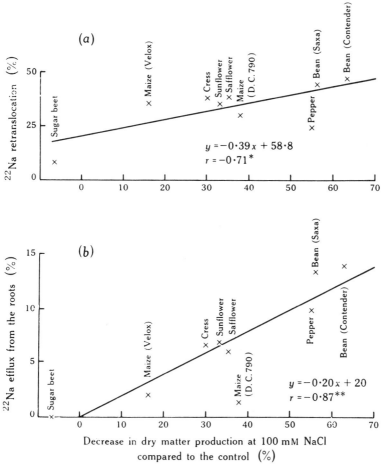

Figure 9.5. Correlation between decrease in dry matter production of species differing in salt resistance and (a) retranslocation and (b) efflux of ^{22}Na from the roots. Plants grown at 100 mM NaCl. From ref. 29. Reproduced with permission from H. Marschner and *Australian Journal of Plant Physiology.*

4. PRIORITIES FOR FUTURE RESEARCH

4.1. Basic Research

Serious gaps exist in our knowledge of the mechanisms of salt exclusion and its significance in adaptation of legumes to saline conditions. First and foremost, salt exclusion can only function if the leaves have developed some means of *osmotic adjustment*. In Na exclusion, for example, selective K transport to the leaves may be a means for achieving K accumulation, charge balance, and osmotic adjustment. Evidence from *Phaseolus coccineus* (9) indicates that the sum of K + Na does not balance Cl in leaves, in contrast to the halophilic grass *Puccinellia peisonis* (37). Hence, organic solutes may have to be synthesized and accumulated in salinized legumes. Glycinebetaine accumulates in many salt-tolerant and salt-resistant plants under saline conditions, but not in green tissues of legumes (38, 39). Proline-betaine, however, is present in alfalfa and has recently been demonstrated to accumulate under saline conditions much more in a salt-resistant cultivar than in a salt-sensitive one (40). The disadvantage of osmotic adjustment by means of organic compounds is its high energy cost which may have growth reduction as an inevitable consequence. A more efficient and energetically cheaper way is *subcellular compartmentation* of inorganic ions coupled with salt sequestration in the leaf cell vacuoles (41). Compartmentation of Na in leaves may not be developed efficiently in legumes; in line with this suggestion, Jacoby (42) found that metabolic Na absorption in bean leaves is restricted to certain tissues associated with the veins. It would be very revealing to investigate the tonoplast properties in leaf cells of legumes, because this membrane must be efficient in maintaining a very asymmetric ionic distribution between cytoplasm and vacuole of halophyte leaves (43).

Salinity may have adverse effects on *K and Ca nutrition* (44). It will be important to study the interactions between nutrition and salinity in salt excluding legumes. Increasing external Ca concentrations were shown to inhibit Na absorption in beans (45) and soybeans (46) and thus may be an important factor in controlling salinity response of legumes. Calcium in the medium also appears significant in relation to salinity-induced loss of K from roots (47–49).

There are also plant developmental problems that need to be addressed. On a cellular level, the challenge lies in whether and by what mechanism salinity induces the *development of xylem parenchyma transfer cells* in root and stem of salt excluding legumes. The significance of salt exclusion in contributing to salt resistance must also be evaluated in relation to *plant ontogeny*, as salt resistance can change dramatically with ontogenetic development (e.g., 13, 50–52).

4.2.   Applied Research

Probably the most significant question is that of the capacity of salt exclusion under saline conditions. A quantitative assessment of *salt exclusion capacity* is required and may be particularly relevant using genotypes of salt excluding legumes that differ in salt resistance. Only when we better understand the capacity of salt exclusion as a means for coping with a saline environment, can we make recommendations as to whether screening, selection, and breeding for salt exclusion or for accumulation of salt in the leaves ought to be the method of choice for developing more salt-resistant legumes.

At the Boden Symposium on "Salinity and Plant Productivity" held in

Australia in February 1982, the recommendation was made that salinity tolerance of horticultural crops may be improved by *grafting* salt-excluding rootstocks with shoots resistant to moderate water stress. This recommendation appears applicable to legumes. Although little evidence is available on improved salt resistance by grafting legumes, unpublished data by J. Wieneke (Table 9.5) from a reciprocal grafting experiment with the soybean cultivars 'Lee' and 'Jackson' convincingly demonstrates that Cl exclusion in the cultivar 'Lee' is controlled by the root.

The agricultural importance of legumes is particularly related to their ability of *fixing* N_2 in their root nodules. Any assesment of the feasibility of growing legumes under saline conditions needs to take into consideration the relative salt resistance of legumes fixing N_2 in comparison with fixed N supply. There is mounting evidence that salt resistance of legumes may vary with the mode of N acquisition. Growth of soybean (53), *Glycine wightii* (54), and chickpea (14) was more reduced by salinity when grown symbiotically than under N fertilization. In alfalfa, however, relative growth inhibition by salinity was similar for N-fertilized and N_2-fixing plants (53). Furthermore, it appears that N_2 fixation is impaired by salinity in mung beans, but not in cowpea (10). The effects of salinity on symbiotic N_2 fixation need to be studied further, both as far as the mechanism of N_2 fixation is concerned and relative to agronomic implications. Recent advancements in genetic engineering of *Rhizobium* spp. with the aim of developing salt tolerant strains suitable to be paired with relatively salt-resistant legumes (55) may be a component of this line of research, although rhizobia appear to be more salt resistant than the host plants. Ultimately, the various avenues for genetic improvement of crop plants such as selection and breeding, cell culture techniques, and genetic engineering will all contribute to the

**Table 9.5. Reciprocal Grafting in Soybean Cultivars 'Lee' and 'Jackson':
Distribution of Na^+ and Cl^- in the Plant**[a]

	Na^+	Cl^-	Na^+	Cl^-
	root 'Lee' $\times$ shoot 'Jackson'		root 'Jackson' $\times$ shoot 'Lee'	
Root	0.8 ± 0.2	12.5 ± 0.6	0.7 ± 0.2	8.5 ± 1.0
Stem	0.6 ± 0.1	1.3 ± 0.5	0.6 ± 0.3	4.9 ± 0.7
Leaves	0.1 ± 0.06	0.8 ± 0.3	0.1 ± 0.05	3.1 ± 0.5
	'Lee' (not grafted)		'Jackson' (not grafted)	
Root	0.7 ± 0.06	10.1 ± 0.7	0.7 ± 0.05	7.0 ± 0.7
Stem	0.6 ± 0.2	1.1 ± 0.6	0.9 ± 0.2	4.3 ± 0.6
Leaves	0.1 ± 0.03	0.6 ± 0.3	0.1 ± 0.03	3.4 ± 0.4

[a]Plants grown at 25 mM NaCl for 1 week. Ion contents in mmol g^{-1} dry weight. From J. Wieneke, unpublished data. Reproduced with permission from the author.

development of legumes better adapted to saline environments. As Boyer (56) stated, better adaptation of plants to adverse environments could result in significant gains in productivity because there is a large but at present unrealized genetic potential for yield.

ACKNOWLEDGMENT

I thank D. N. Munns for reviewing the manuscript.

REFERENCES

1.   E. V. Maas and G. J. Hoffman. Crop salt tolerance—Current assessment. *J. Irrig. Drainage Div. ASCE* **103**, 115–134 (1977).

2.   E. V. Maas and R. H. Nieman. Physiology of plant tolerance to salinity, In G. A. Jung, ed., *Crop Tolerance to Suboptimal Land Conditions.* American Society of Agronomy, Madison, Wisconsin, 1978, pp. 277–299.

3.   A. Läuchli. Regulation des Salztransportes und Salzausschliessung in Glykophyten und Halophyten. *Ber. Deutsch. Bot. Ges.* **92**, 87–94 (1979).

4.   H. Greenway and R. Munns. Mechanisms of salt tolerance in nonhalophytes. *Annu. Rev. Plant Physiol.* **31**, 149–190 (1980).

5.   R. G. Wyn Jones. Salt tolerance. In C. B. Johnson, ed., *Physiological Processes Limiting Plant Productivity.* Butterworths, London, 1981, pp. 271–292.

6.   R. F. M. Van Steveninck, M. E. Van Steveninck, R. Stelzer, and A. Läuchli. Studies on the distribution of Na and Cl in two species of lupin (*Lupinus luteus* and *Lupinus angustifolius*) differing in salt tolerance. *Physiol. Plant.* **56**, 465–473.

7.   E. Winter and A. Läuchli. Salt tolerance of *Trifolium alexandrinum* L. I. Comparison of the salt response of *T. alexandrinum* and *T. pratense. Aust. J. Plant Physiol.* **9**, 221–226 (1982).

8.   A. Läuchli and J. Wieneke. Studies on growth and distribution of Na^+, K^+ and Cl^- in soybean varieties differing in salt tolerance. *Z. Pflanzenernaehr. Bodenkd.* **142**, 3–13 (1979).

9.   D. Kramer, A. Läuchli, A. R. Yeo, and J. Gullasch. Transfer cells in roots of *Phaseolus coccineus:* Ultrastructure and possible function in exclusion of sodium from the shoot. *Ann. Bot.* **41**, 1031–1040 (1977).

10. V. Balasubramanian and S. K. Sinha. Effects of salt stress on growth, nodulation and nitrogen fixation in cowpea and mung beans, *Physiol. Plant.* **36**, 197–200 (1976).

11. J. W. Brown and H. E. Hayward. Salt tolerance of alfalfa varieties, *Agron. J.* **48**, 18–20 (1956).

12. M. K. Smith and J. A. McComb. Use of callus cultures to detect NaCl tolerance in cultivars of three species of pasture legumes. *Aust. J. Plant Physiol.* **8**, 437–442 (1981).

13. J. Shalhevet, P. Reiniger, and D. Shimshi. Peanut response to uniform and non-uniform soil salinity. *Agron. J.* **61**, 384–387 (1969).

14. D. J. Lauter, D. N. Munns, and K. L. Clarkin. Salt response of chickpea as influenced by N supply. *Agron. J.* **73**, 961–966 (1981).

15. R. H. Nieman. Some effects of sodium chloride on growth, photosynthesis, and respiration of twelve crop plants. *Bot. Gaz.* **123**, 279–285 (1962).

16. B. Jacoby. Function of bean roots and stems in sodium retention. *Plant Physiol.* **39**, 445–449 (1964).

17. B. Jacoby. Sodium retention in excised bean stems. *Physiol. Plant.* **18**, 730–739 (1965).

18. G. H. Abel and A. J. MacKenzie. Salt tolerance of soybean varieties (*Glycine max* L. Merrill) during germination and later growth. *Crop Sci.* **4**, 157–161 (1964).

19. G. H. Abel. Inheritance of the capacity for chloride inclusion and chloride exclusion by soybeans. *Crop. Sci.* **9**, 697–698 (1969).

20. C. T. Gates, K. P. Haydock, and M. F. Robins. Salt concentration and the content of phosphorus, potassium, sodium and chloride in cultivars of *G. wightii* (*G. javanica*). *Aust. J. Exp. Agric. Anim. Husb.* **10**, 98–110 (1970).

21. W. D. Jeschke. Roots: Cation selectivity and compartmentation, involvement of protons and regulation. In R. M. Spanswick, W. J. Lucas, and J. Dainty, eds., *Plant Membrane Transport: Current Conceptual Issues.* Elsevier/North-Holland, Amsterdam, 1980, pp. 17–28

22. M. G. Pitman, A. Läuchli, and R. Stelzer. Ion Distribution in roots of barley seedlings measured by electron probe X-ray microanalysis. *Plant Physiol.* **68**, 673–679 (1981).

23. A. Läuchli. Symplasmic transport and ion release to the xylem. In I. F. Wardlaw and J. B. Passioura, eds., *Transport and Transfer Processes in Plants.* Academic Press, New York, 1976, pp. 101–112.

24. U. Lüttge. Import and export of mineral nutrients in plant roots. In A. Läuchli and R. L. Bieleski, eds., *Encyclopedia of Plant Physiology,* New Series, Vol. 15. Springer-Verlag, Berlin, 1983, pp. 181–211.

25. H. Marschner and O. Ossenberg-Neuhaus. Langstreckenstransport von Natrium in Bohnenpflanzen. *Z. Pflanzenernaehr. Bodenk.* **139**, 129–142 (1976).

26. B. Jacoby. Sodium recirculation and loss from *Phaseolus vulgaris* L. *Ann. Bot.* **43**, 741–744 (1979).

27. J. Wieneke and A. Läuchli. Short-term studies on the uptake and transport of Cl⁻ by soybean cultivars differing in salt tolerance. *Z. Pflanzenernaehr. Bodenk.* **142**, 799–814 (1979).

28. A. Läuchli and J. Wieneke. Salt relations of soybean mutants differing in salt tolerance: Distribution of ions and localization by X-ray microanalysis. In *Plant Nutrition 1978*, Proc. 8th Int. Colloq. Plant Analysis and Fertilizer Problems, Auckland, New Zealand. Wellington, Government Printer, 1978, pp. 275–282.

29. H. Lessani and H. Marschner. Relation between salt tolerance and long-distance transport of sodium and chloride in various crop species. *Aust. J. Plant Physiol.* **5**, 27–37 (1978).

30. E. Winter. Salt tolerance of *Trifolium alexandrinum* L. III. Effects of salt on ultrastructure of phloem and xylem transfer cells in petioles and leaves. *Aust. J. Plant Phyisol.* **9**, 239–250 (1980).

31. E. Winter and J. Preston. Salt tolerance of *Trifolium alexandrinum* L. IV. Ion

measurements by X-ray microanalysis in unfixed, frozen hydrated leaf cells at various stages of salt treatment. *Aust. J. Plant Physiol.* **9**, 251–259 (1982).

32. E. Epstein, J. D. Norlyn, D. W. Rush, R. W. Kingsbury, D. B. Kelley, G. A. Cunningham, and A. F. Wrona. Saline culture of crops: A genetic approach. *Science* **210**, 399–404 (1980).

33. T. P. Croughan, S. J. Stavarek, and D. W. Rains. Selection of a NaCl tolerant line of cultured alfalfa cells. *Crop Sci.* **18**, 959–963 (1978).

34. A. Cerda, M. Caro, and F. G. Fernandez. Salt tolerance of two pea cultivars. *Agron. J.* **74**, 796–798 (1982).

35. D. N. Munns. Effect of Salinity on Legumes. In D. W. Rains, ed., *A Conference on Biosalinity: The Problem of Salinity in Agriculture.* University of California, Davis, 1981, pp. 79–80.

36. D. J. Lauter and D. N. Munns. Salt response of chickpea and cowpea genotypes to different ionic compositions. *Agron. Abstr.,* **213** (1982).

37. R. Stelzer and A. Läuchli. Salt- and flooding tolerance of *Puccinellia peisonis* III. Distribution and localization of ions in the plant. *Z. Pflanzenphysiol.* **88**, 437–448 (1978).

38. R. Storey and R. G. Wyn Jones. Quaternary ammonium compounds in plants in relation to salt resistance. *Phytochemistry* **16**, 447–453 (1977).

39. R. G. Wyn Jones. An assessment of quaternary ammonium and related compounds as osmotic effectors in crop plants. In D. W. Rains, R. C. Valentine, and A. Hollaender, eds., *Genetic Engineering of Osmoregulation.* Plenum, New York, 1980, pp. 155–170.

40. R. G. Wyn Jones and R. Storey. Betaines. In L. G. Paleg and D. Aspinall, eds., *The Physiology and Biochemistry of Drought Resistance in Plants.* Academic Press, New York, 1981, pp. 171–204.

41. T. J. Flowers and A. Läuchli. Sodium versus potassium: Substitution and compartmentation. In A. Läuchli and R. L. Bieleski, eds., *Encyclopedia of Plant Physiology*, New Series, Vol. 15. Springer-Verlag, Berlin, 1983, pp. 651–681.

42. B. Jacoby. Light sensitivity of ^{22}Na, ^{86}Rb, and ^{42}K absorption by different tissues of bean leaves. *Plant Physiol.* **55**, 978–981 (1975).

43. J. Dainty. The ionic and water relations of plants which adjust to a fluctuating saline environment. In R. L. Jefferies and A. J. Davy, eds., *Ecological Processes in Coastal Environments*, British Ecology Society Symposium No. 19, Blackwell, Oxford, 1979, pp. 201–209.

44. A. Läuchli. Mineral nutrition of plants exposed to salinity stress. In D. W. Rains, eds., *A Conference on Biosalinity: The Problem of Salinity in Agriculture.* University of California, Davis, 1981, pp. 63–69.

45. P. A. LaHaye and E. Epstein. Calcium and salt toleration by bean plants. *Physiol. Plant.* **25**, 213–218 (1971).

46. J. Wieneke and A. Läuchli. Effects of salt stress on distribution of Na^+ and some other cations in two soybean varieties differing in salt tolerance. *Z. Pflanzenernaehr. Bodenk.* **143**, 55–67 (1980).

47. L. C. Campbell and M. G. Pitman. Salinity and Plant Cells. In T. Talsma and J. R. Philip, eds., *Salinity and Water Use.* Macmillan, London, 1971, pp. 207–224.

48. H. Nassery. The effect of salt and osmotic stress on the retention of potassium by excised barley and bean roots. *New Phytol.* **75,** 63–67 (1975).

49. H. Nassery. Salt-induced loss of potassium from plant roots. *New Phytol.* **83,** 23–27 (1979).

50. D. Pasternak, M. Twersky, and Y. De Malach. Salt resistance in agricultural crops. In H. Mussell and R. C. Staples, eds., *Stress Physiology in Crop Plants.* Wiley, New York, 1979, pp. 127–142.

51. D. W. West and J. A. Taylor. Germination and growth of cultivars of *Trifolium subterraneum* L. in the presence of sodium chloride salinity. *Plant Soil* **62,** 221–230 (1981).

52. J. Lynch, E. Epstein, and A. Läuchli. Na^+ - K^+ Relationships in Salt-Stressed Barley. In A. Scaife, ed., *Plant Nutrition 1982*, Proc. 9th Int. Plant Nutr. Colloq., Vol. 1. Commonwealth Agricultural Bureaux, Slough, United Kingdom, 1982, pp. 347–352.

53. L. Bernstein and G. Ogata. Effects of salinity on nodulation, nitrogen fixation, and growth of soybeans and alfalfa. *Agron. J.* **58,** 201–203 (1966).

54. J. R. Wilson. Response to salinity in *Glycine*. VI. Some effects of a range of short-term salt stresses on the growth, nodulation, and nitrogen fixation of *Glycine Wightii* (formerly *Javanica*). *Aust. J. Agric. Res.* **21,** 571–582 (1970).

55. J. Mielenz, K. Andersen, R. Tait, and R. C. Valentine. Potential for Genetic Engineering for Salt Tolerance. In A. Hollaender, J. C. Aller, E. Epstein, A. San Pietro, and O. R. Zaborsky, eds. *The Biosaline Concept: An Approach to the Utilization of Underexploited Resources.* Plenum, New York, 1979, pp. 361–370.

56. J. S. Boyer. Plant productivity and environment. *Science* **218,** 443–448 (1982).

10

ORGANIC AND INORGANIC SOLUTE CONTENTS AS SELECTION CRITERIA FOR SALT TOLERANCE IN THE TRITICEAE

R. G. Wyn Jones, J. Gorham, and E. McDonnell

Department of Biochemistry and Soil Science
University College of North Wales
Bangor, Wales, United Kingdom

The conference at which this chapter was presented was a further indication of the growing international concern at the progressive salinization and alkalinization of many irrigated lands. Although good agronomic practices and water management are essential to combat the problem, the urgent need to compensate, at least in part, for the degeneration in soil conditions and the paucity of good quality water by breeding crop plants of greater tolerance is also being appreciated (1). Closely allied to this aspect of irrigation agriculture are the twin problems of drought and salt tolerance in arid zones where a combination of water and salt stress generally limit productivity.

In no group of plants is the need to understand and manipulate the characters conferring tolerance more apparent than in the graminaceous tribe, the Triticeae. Within this tribe are found several of the world's most important crop plants: various forms of wheat, barley, rye, and the increasingly important hybrid, *Triticale*. Wheat alone accounts for approximately a quarter of the world's food production, and barley, rye, and *Triticale* together account for an additional 15–18%. Barley and rye are commonly regarded as robust crop species resistant to a number of environmental stresses including salt. Bread wheats are, in contrast, generally thought to be less resistant to salt stress (2). Nevertheless, as wheats are the more important sources of human nutrition, the study and ultimately the improvement of salt tolerance in this crop deserves the highest priority.

There are cogent reasons to expect significant progress in the breeding of wheat for enhanced tolerance without recourse to the unproved, though exciting, techniques of genetic engineering. Within the existing 20,000 or so wheat cultivars there is a range of tolerance (3–5) and, by selection and breeding from within this gene pool, it is possible that some improvement could be achieved. However, it might be more advantageous to cast the net wider as there are members of the Triticeae with relatively high salt and drought tolerances. Cytogeneticists are now able to hybridize many of these "alien" relatives with *T. aestivum* or *T. turgidum* (see Fig. 10.1) and bring about the transfer of chromosome fragments (6). Thus, there is an expectation that this genetic variability can in time be exploited to produce highly tolerant wheats. The problems of achieving such transfers should not be underestimated and it will undoubtedly take time to produce agronomically valuable cultivars in which characters from an alien grass are expressed in a genetic background giving other desirable agronomic traits. Unfortunately, we know remarkably little about the nature, inheritance, segregation, and so on, of the character (almost certainly several different characters) that confer tolerance.

An essential first step in any attempt to exploit the potential variability in alien members of the Triticeae is a survey to establish the degree and mechanisms of salt tolerance in these species. Factors such as chromosome number must also be recorded as they will influence the ease with which subsequent cytogenetic manipulations can be carried out. Such data should then allow the selection of aliens for a wide crossing program with wheat. It would also be advantageous to devise simple selection criteria in order to follow the fate of the salt tolerance characters through the extensive backcrossing and commercial breeding programs that will be required. A purely pragmatic approach could and perhaps ought to be followed in which the tolerance of progeny from a wide crossing program are simply selected in field trials without any biochemical or physiological input. Such empirical methods have served breeders well over many decades (7). Nevertheless, as we hope will become apparent in the course of this chapter, there are good reasons to expect biochemical and physiological data to be of value in tackling this problem and that collaboration between biochemists and the cytogeneticists and breeders would be highly desirable.

In this chapter we describe some experiments which contribute to laying the groundwork for a long-term investigation of this problem. Two principal approaches have been used. First, since the ancestry of the modern hexaploid bread wheats is now established with some confidence (8), we have compared the salt tolerance and patterns of ion accumulation found in these ancestors and in a selection of hexaploid cultivars. Any attempt at screening the formidable numbers of wheat cutivars now available would be quite beyond our resources. Second, we have followed the earlier studies of Dewey (9) and Shannon (10) and

Subtribe	Genus	Number of Species	Ploidy Level ($X = 7$)	Growth Habit: Perennial	Growth Habit: Perennial Annual	Growth Habit: Annual	Mode of Pollination: Cross	Mode of Pollination: Cross, Self	Mode of Pollination: Self	Successful Hybrids with Triticum
Hordeinae	Horedeum	25	$2X-6X$		■			■		■
	Elymus	60	$2X-12X$	■				■		■
	Asperella	7		■			■			
	Sitanion	1	$4X$	■			■			
	Psathyrostachys	6		■			■			
	Crithopsis	1	$2X$			■			■	
	Taeniatherum	2	$2X$			■			■	■
Triticinae	Agropyrum	100	$2X-10X$	■			■			■
	Haynaldia	2	$2X, 4X$		■		■			■
	Secale	6	$2X$		■		■			■
	Heteranthelium	1	$2X$			■			■	
	Henrardia	2	$2X$			■			■	
	Eremopyrum	5	$2X, 4X$			■			■	■
	Triticum	27	$2X-6X$			■			■	■

Figure 10.1. Details of growth habit and genetics for the different genera within the Triticeae (redrawn from Feldman and Sears, (8)).

examined accessions of both *Agropyrum elongatum*, which was shown by these authors to be a potential source of salt tolerance, and of other, previously unexamined, species of *Agropyrum*. Since *Elymus* (*Leymus*) species can now be hybridized with wheat (11) and are often found in saline and arid environments, they have also been investigated.

There has been considerable interest in the possible role of proline and glycinebetaine in both drought and salt tolerance and in the possibilty that the ability to accumulate these solutes might confer tolerance (cf. 12, 13). It was, therefore, of particular interest to examine the levels of these solutes in a variety of wheats and alien species which showed differences in tolerance and ion accumulation patterns.

1. ION ACCUMULATION PATTERNS AND SALT TOLERANCE IN HEXAPLOID WHEATS AND THEIR ANCESTORS

The cytogenetic studies of Sears, Feldman, and their colleagues, now supplemented by more biochemical data on DNA sequences, have established a probable evolutionary pathway for the hexaploid bread wheats (6, 8; but see 14 for reservations) (see Fig. 10.2). Paleobotanical studies have also partially illuminated the occurrence and distribution of the various primitive wheats (15). By historic times the hexaploid wheats had become dominant although the tetraploid durum wheats derived from *T. turgidum* are still grown extensively in drier regions. Some primitive emmer and einkorn wheats may still be found in remote areas.

A number of trials have been carried out to compare the salt tolerance and ion accumulation patterns of the wheat ancestors (Tables 10.1 and 10.2). The

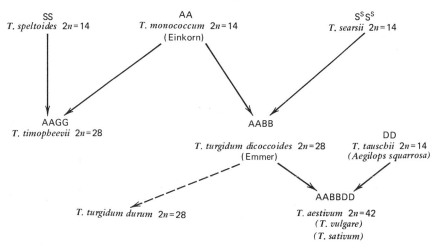

Figure 10.2. Probable evolutionary pathways of hexaploid wheat (from Feldman and Sears, (8); Sears, (6)).

Table 10.1.   Influence of Salt Stress on the Growth (Fresh Weight) of a Representative Bread Wheat and Some Ancestors of Modern Bread Wheats

Species[a]	Shoot Fresh Weight Yield as % of Low Salt Control. Plants Grown with Additions of NaCl (mol m^{-3})[b]		
	0	75	150
T. monococcum	100	22	12
A. searsii	100	40	30
T. dicoccoides	100	52	33
A. squarrosa	100	52	42
T. aestivum cv. Ciano 79	100	62	40

[a]Ancestral species kindly supplied by Dr. C. N. Law, P. B. I., Cambridge, and Ciano 79 by C.I.M.M.Y.T.
[b]Plants grown in hydroponic culture in $^{1}/_{2}$ strength Hoagland's medium in a glasshouse with supplementary light. Salt was added in 25 mol m increments per day until the final concentration in the medium had been reached. The plants were harvested after 2 weeks at the final salinity.

diploid *T. monococcum* was found to be very salt sensitive whereas *Aegilops searsii*, the putative source of the B genome, and the tetraploid *T. dicoccoides* were somewhat more tolerant. *Aegilops squarrosa* exhibited marginally greater tolerance as did the hexaploid bread wheat used in this experiment, *T. aestivum* cv. Ciano 67. (Variations within the hexaploid wheats will be alluded to later.) The ion accumulation patterns suggested interesting differences between the various species (Table 10.2; Fig. 10.3). Both *A. searsii* and *T. dicoccoides* had

Table 10.2.   Major Ion Contents of Wheat Ancestors Subjected to Salt (NaCl) Stress[a]

Species	Shoot Ion Contents (mmol kg^{-1} plant water)				
	Low Salt Controls		Salt-Treated Plants (100 mol m^{-3})		
	K$^+$	Na$^+$	K$^+$	Na$^+$	K/Na
T. monococcum	183	6	160	301	0.54
A. searsii	198	4	92	377	0.24
T. dicoccoides	179	4	84	332	0.25
A. squarrosa	166	4	158	154	1.03
T. aestivum cv. Flanders	199	3	169	114	1.48

[a]Plants grown as outlined in Table 10.1.

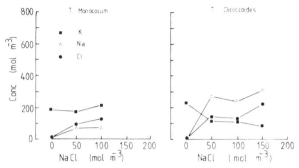

Figure 10.3. Inorganic ion contents of *T. monococcum* and *T. dicoccoides* grown at different NaCl levels (for growth conditions see Table 10.1).

very low K^+/Na^+ ratios even at modest salinities and appear to tolerate lower leaf K^+ levels than *T. monococcum*. In a direct comparison of *T. monococcum* and *T. dicoccoides* carried out under conditions of low transpirational demand (Fig. 10.3), the differences between these species were illustrated dramatically. It should be noted that growth of *T. monococcum* was severely inhibited by 100 mol m^{-3} NaCl in this trial despite the low leaf levels of Na^+ and Cl^-. However, in experiments carried out in the summer, presumably with higher transpiration rates, high Na^+ and Cl^- levels were found even in *T. monococcum*. These data indicate tentatively that the B genome from *A. searsii* or a close relative contributed a modestly enhanced salt tolerance in comparison with *T. monococcum* and the ability to withstand a higher leaf salt load. In contrast, *A. squarrosa* is a more efficient salt excluder maintaining a higher K^+/Na^+ ratio. This same character is found in many of the hexaploid wheats (see also Fig. 10.4) and may be associated with the D genome from *A. squarrosa*. Since genotypic plasticity within these grasses is very great any generalizations are dangerous and any correlation of ion transport characters with evolution must be regarded as extremely tentative.

Further data on four bread wheats, one *T. turgidum* cultivar and one *Triticale*, are presented in Table 10.3 and Figure 10.4. Both with regard to the salt tolerance and ion accumulation patterns the cultivar Cheyenne is typical of a number of the hexaploid wheats. Data for Kogo II was presented since this was one of the most salt-sensitive cultivars tested and appeared to be a less efficient excluder at salinities greater than 75 mol m^{-3} NaCl. By way of contrast Atou was the most tolerant cultivar tested. Few tetraploid wheats have yet been screened but the data from Cocorit 71 suggest that it has some characteristics similar to *T. dicoccoides*. The *Triticale* cultivar, on the other hand, is an efficient salt excluder. Jeschke and collaborators (16, 17) have shown that there is an efficient K^+/Na^+ exchange system at the plasmalemma of cortical cells in *T. aestivum* cv. Carstacht, and in rye. It would be of great interest if differences in the efficiency of such exchange systems were to occur in the wheat ancestors and could be related to the ion accumulation patterns presented here. Certainly, if

Table 10.3.   Effect of Salt (NaCl) Stress on the Fresh Weight Yield of Shoots of Four Cultivars of *T. aestivum*, One Tetraploid Wheat, and One Triticale Cultivar[a]

| | | Shoot Fresh Weight as % of Low Salt Controls at Different NaCl Treatments (mol m^{-3}) | | | |
		75	150	200	250
T. aestivum cv.					
Chinese spring	Used in cytogenetic research	49	41	25	15
Capelle-Desprez	French winter wheat	45	40	38	15
Kogo II	European spring wheat	72	45	17	[b]
Cheyenne	North American frost-resistant winter wheat	50	43	35	15
Atou	High-vigor winter wheat	77	54	50	29
T. turgidum cv. Corcorit 71		70	27	27	38
Triticale cv. GLT 176		(80)[c]	71	64	N.D.[d]

[a]Growth conditions as outlined in Table 10.1.
[b]No survivors.
[c]Extrapolated from another experiment.
[d]Not determined.

particular ion transport characters can be associated with specific genomes, and ultimately with single chromosomes, it could open up a productive line of physiological and biochemical studies as well as being of potential practical importance.

In these experiments it might be objected that the Na^+/Ca^{2+} ratio is not kept constant (cf. 18). However, neither in *T. aestivum* cv. Flanders nor in *T. dicoccoides* (Table 10.4) were significant differences detected in the salt tolerance between plants subjected to NaCl or to NaCl with $CaCl_2$ added so as to maintain a constant Na:Ca ratio of 10:1. More detailed experimentation on the effects of different salt mixtures at different relative humidities, and so on, on the wheat ancestors is now in progress.

2.   SALT TOLERANCE IN *AGROPYRUM* AND *ELYMUS* (*LEYMUS*) SPECIES

With the possible exception of *Triticale* GLT 176 the degree of salt tolerance exhibited by the wheat cultivars and ancestors was modest. As already mentioned *Agropyrum* and *Elymus* species are possible sources of far greater

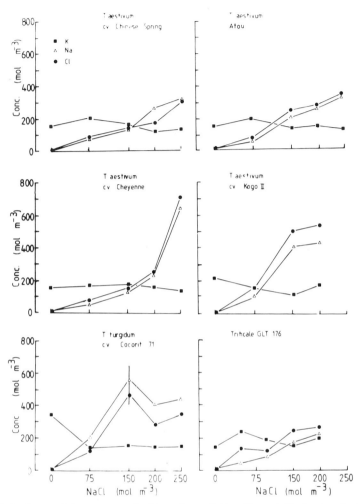

Figure 10.4. Inorganic ion contents of cultivated tetraploid and hexaploid wheats and a Triticale grown at different levels of NaCl (for growth conditions see Table 10.1).

tolerance. However, the earlier work also indicated great variation even within a single species. This was confirmed by a study of five different accessions of *Agropyrum elongatum* (Table 10.5). Great diversity was found even within this species and no simple correlation was observed with any of the ion uptake characteristics measured. The accession from China (20014) was considerably more tolerant than any of the hexaploid wheats tested so far and was apparently able to withstand a substantial leaf salt load. This accession could be regarded as a potential candidate for a crossing programme but unfortunately it has a high level of ploidy (C. Law, personal communication).

With the generous cooperation of Dr. Dewey we have been able to test a variety of other *Agropyrum* species, which again showed a wide range of

Table 10.4. Comparison of the Effects of Constant and Variable Na$^+$:Ca^{2+} Ratios on the Salt Tolerance of *T. aestivum* cv. Flanders and *T. dicoccoides*

	Shoot Yield (g fresh weight) of Plants Grown at Different NaCl Levels (mol m^{-3})^a				
	0	25	75	150	225
T. aestivum cv. Flanders					
Changing ratio	62 ± 7	—	48 ± 7	22 ± 2	14 ± 1
Constant Na:Ca^b	—	64 ± 9	56 ± 7	20 ± 4	15 ± 3
T. dicoccoides					
Changing ratio	31 ± 2	—	9 ± 3	8 (2 reps)	^{c}
Constant Na:Ca^b	—	30 ± 4	14 ± 2	6 (1 rep)	^{c}

^{a}Values are means of five replicates except where stated.
^{b}A constant ratio of Na$^+$ to Ca^{2+} (10:1) was maintained by adding CaCl$_2$.
^{c}No survivors.

tolerance from the relatively sensitive *A. intermedium* (cf. 19) to the highly tolerant *A. junceum* (Table 10.6). The latter also appears to exclude Na and Cl from its leaves very efficiently, thus exhibiting a character frequently associated with tolerance in glycophytic crop species (for references: see 18, 20). This species is also of particular interest since it has recently been reported to be very drought tolerant (21). Furthermore, its low chromosome number ($2n = 14$) makes it a suitable subject for cytogenetic work and Mujeeb-Kazi and Rodriquez (11) have reported intergeneric hybrids between a hexaploid *A. junceum* ($2n = 6x = 42$) and *T. aestivum* and *T. turgidum*.

Table 10.5. Influence of Salinity on the Growth (Fresh Weight) and Ion Content of Various Accessions of *Agropyrum elongatum*^{a} and *T. aestivum* cv. Flanders (accession no. 10030)

Accession Number	Origin	Fresh Weight at 200 mol m^{-3} as % of Controls	Ion Contents (mmol kg^{-1} plant water) of Plants Grown in 200 mol m^{-3} NaCl			
			K$^+$	Na$^+$	Cl$^-$	K/Na
20003	U.S.S.R.	41	177 (182)^b	161	184	1.1
20004	Turkey	16	152 (157)	140	214	1.1
20014	China	80	222 (159)	213	230	1.0
20007	Iran	44	—	—	—	—
20010	Israel	34	186 (278)	466	494	0.4
10030	U.K.	23	133 (193)	164	219	018

^{a}Kindly supplied by the Germplasm Resources Laboratory.
^{b}Figures in parentheses are low-salt control potassium contents.

Table 10.6.   Influence of Salinity on the Growth (Fresh Weight) and Ion Contents of Various Agropyrum Species

Species, Origin, and Chromosome Number	Fresh Weight at 200 mol m^{-3} NaCl as % of Controls	Ion Contents (mmol kg^{-1} plant water) of Plants Grown at 200 mol m^{-3} NaCl			
		K$^+$	Na$^+$	Cl$^-$	K/Na
A. intermedium, Latvia, 2n = 42	26	186 (180)[a]	215	377	0.9
A. scirpeum, Aegean, 2n = 48	60	156 (212)	209	235	0.7
A. junceum, U.S.S.R., 2n = 14	106	173 (185)	102	184	1.7
A. curvifolium, Spain, 2n = 28	64	262 (194)	128	252	2.1
A. dasystachyum, Canada, 2n = 28	31	159 (173)	156	225	1.0

[a]Figures in parentheses are low-salt control potassium contents.

A similar trial to that on the *Agropyrum* species has been conducted on a number of *Elymus* species (Table 10.7). A considerable degree of salt tolerance was observed in all the species tested except for *E. triticoides*. However, very different ion accumulation patterns were observed, as may be seen from a comparison of *E. dahuricus* and *E. sabulosus*. The former grew quite well with a low leaf K$^+$/Na$^+$ ratio and high leaf Na and Cl load, whereas the latter maintained a high K$^+$/Na$^+$ ratio and efficiently excluded Na and Cl. This salt exclusion trait was even more pronounced in *E. sabulosus* than in *A. junceum*. It should also be noted that the two *Elymus* species differed radically in their morphology. *E. sabulosus* has tough, gray-green, waxy leaves whereas *E. dahuricus* has green, apparently less robust leaves. On the basis of these crude data it is not apparent how the former species is able to adjust osmotically to 200 mol m^{-3} NaCl and it must be assumed that some other (organic?) solutes are accumulated in quantity.

3.   PROLINE AND GLYCINEBETAINE ACCUMULATION

One aim of these experiments was to consider whether the cytosolutes, proline and glycinebetaine, might be correlated with tolerance in the Triticeae. Some typical data for glycinebetaine and proline for a range of species, cultivars, and accessions are shown in Table 10.8. An inspection of this table and the salt tolerance data in Tables 10.1, 10.3, 10.5, 10.6, and 10.7 immediately shows that no correlation was found. Similarly in the *Elymus* species neither the constitu-

Table 10.7. Influence of Salinity on the Growth (Fresh Weight) and Ion Contents of Various Elymus Species

Species, Origin, and Chromosome Number	Fresh Weight at 200 mol m⁻³ NaCl as % of Controls	Ion Contents (mmol kg⁻¹ plant water) of Plants Grown at 200 mol m⁻³ NaCl			
		K^+	Na^+	Cl^-	K/Na
E. giganteus, U.S.S.R., $2n = 63$	63	192 (290)[a]	151	216	1.3
E. dahuricus, U.S.S.R., $2n = 67$	67	90 (176)	217	245	0.4
E. angustus, U.S.S.R., $2n = 78$	78	254 (301)	106	170	2.4
E. sabulosus, U.S.S.R., $2n = 70$	70	232 (275)	60	125	3.8
E. triticoides, unknown, $2n = 28$	46	220 (199)	72	157	3.1

[a]Figures in parentheses are low-salt control potassium contents.

Table 10.8. Glycinebetaine Contents of Some Wheat Cultivars and Agropyrum Species

Species	Glycinebetaine Content[a] (mmol kg⁻¹ fresh weight)	
	Low Salt Controls	Salt Treated (200 mol m⁻³ NaCl)
T. aestivum cv.		
Chinese spring	0.6	10.0
Flanders	1.7	2.8
Triticale	3.8	17.0
A. elongatum		
20003	6.2	8.8
20014	2.9	9.8
A. intermedium	1.8	2.6
A. scirpeum	6.5	8.7
A. junceum	3.5	2.1
A. curvifolium	1.9	2.5
A. dasystachyum	3.3	4.6

[a]Determined by the method of Gorham et al. (22).

Table 10.9.   Effects of Salinity on the Glycinebetaine and Proline Contents (mmol kg^{-1} fresh weight) of Some Elymus Species

Species	Glycinebetaine		Proline	
	Low Salt Control	NaCl 200 mol m^{-3}	Low Salt Control	NaCl 200 mol m^{-3}
E. giganteus	4.4	4.4	0.2	1.9
E. dahuricus	4.5	5.6	0.3	6.2
E. augustus	4.3	3.2	0.2	1.0
E. sabulosus	3.9	4.0	0.3	0.2
E. triticoides	7.4	14.2	0.6	1.8

tive nor salt induced levels of these solutes were related to salt tolerance (Table 10.9). Perhaps, the only point to emerge is that in the two most efficient salt excluders found so far (*A. junceum* and *E. sabulosus*) the leaf glycinebetaine content is completely unaffected by the stress. In *E. sabulosus* there is no proline induction either, but this has yet to be tested in the *Agropyrum* species.

4. DISCUSSION

The data presented here are derived from a preliminary survey of some members of the Triticeae and must be supplemented by further screening of other alien accessions, and by more detailed work on the aliens which appear to be of physiological or biochemical interest and of agronomic potential. Even these initial results show clearly that some accessions could provide sources of salt tolerance characters for the bread wheats. However, they have also raised other difficult issues. The characters that confer such tolerance and their possible agronomic consequences are remarkably badly defined. In a number of instances there are indications that tolerance is associated with more efficient Na$^+$ or Cl$^-$ exclusion, for example, in *Aegilops squarrosa*, and certainly this is a major character in *Agropyrum junceum* and *E. sabulosus*. However, other species (*E. dahuricus*, possibly *Agropyrum scirpeum* and, to a limited extent, even *Aegilops searsii*) seem to exhibit tolerance (certainly a considerable tolerance in the case of *E. dahuricus*) with a relatively high leaf salt load.

One major issue that must now be faced is whether these differences are of importance agronomically. For example, the difficulty of understanding, on the basis of this very limited evidence, how osmotic adjustment is achieved in *E. sabulosus* introduces the question of the cost of turgor regulation to the plant. As we have argued previously (20), naive calculations indicate that tolerance by salt accumulation requires less diversion of metabolic energy than the alternative of ion exclusion from the leaves and the accumulation of organic solutes. Nonetheless, there is no firm basis for such speculations and in this context a detailed examination of the relative growth rates and patterns of osmotic and turgor regulation and solute accumulation of the two pairings, *E. sabulosus versus E. dahuricus* and *Agropyrum curvifolium versus A. scirpeum*, could be of great value. Certainly, this question is crucial to any program of genetic manipulation. If tolerance by exclusion is compatible with rapid growth and high yields then its introduction from the aliens described in this chapter is a distinct possibility. Furthermore, characters such as a high leaf K^+/Na^+ ratio and low Cl^- content offer the prospect of simple screening procedures, which could be automated to meet the requirements of plant breeders. If, on the other hand, this is not the case then we are faced with the much more difficult physiological and biochemical problem of identifying the characters that are associated with an ability to tolerate a high leaf salt load, and the problem of clearly distinguishing this type of tolerance from the sudden increase in leaf NaCl so often associated with severe growth inhibition (see Fig. 10.4, Cheyenne). Furthermore, if we compare the data presented in Figure 10.4 for the three bread wheats, Cheyenne, Atou, and Kogo II, we have no objective way of explaining the differences between salt uptake into the leaves of Kogo II, which is associated with inhibition of leaf growth, and that in Atou in which it appears to be tolerated. Differences in solute compartmentation may be involved, but there is insufficient hard evidence to support such speculations. Clearly, there is also an urgent need for more fundamental studies if we are to be able to give meaningful advice to plant breeders.

REFERENCES

1. E. Epstein. Response of plants to saline environments. In D. W. Rains, R. C. Valentine, and A. Hollaender, eds., *Genetic Engineering of Osmoregulation.* Plenum, New York, 1980, 381 pp.

2. E. V. Mass and G. J. Hoffman. Crop salt tolerance—Current assessment. *J. Irrig. Drainage Div. ASCE* **103**, 115–134 (1977).

3. C. Torres-Bernal and F. T. Bingham. Salt tolerance of Mexican wheat. I. Effect of NO_3 and NaCl on mineral nutrition, growth and grain production of 4 wheats. *Soil Sci. Soc. Am. Proc.* **37**, 711–715 (1973).

4. C. T. Bernal, F. T. Bingham, and J. Oertli. Salt tolerance of Mexican wheat. II. Relation to variable sodium chloride and length of growing season. *Soil Sci. Soc. Am. Proc.* **38**, 777–780 (1974).

5. R. H. Qureshi, R. Ahmad, M. Ilyas, and Z. Aslam. Screening of wheat (*Triticum aestivum* L.) for salt tolerance. *Pakistan J. Agric. Sci.* **17**, 19–25 (1980).

6. E. R. Sears. Transfer of alien genetic material to wheat. In E. R. Evans and W. J. Peacock, eds., *Wheat Science—Today and Tomorrow.* Cambridge University Press, Cambridge, 1981, 290 pp.

7. N. W. Simmonds. Plant breeding: The state of the art. In C. P. Meredith, A. Kosuge, and A. Hollaender, eds., *Genetic Engineering of Plants.* Plenum, New York, 1983.

8. M. Feldman and E. R. Sears. The wild gene resources of wheat. *Sci. Am.* **244,** 98–109 (1981).

9. Dr. R. Dewey. Salt tolerance of 25 strains of *Agropyron. Agron. J.* **52,** 631–635 (1960).

10. M. C. Shannon. Testing salt tolerance variability among tall wheatgrass lines. *Agron. J.* **70,** 719–722 (1978).

11. A. Mujeeb-Kazi and R. Rodriguez. Cytogenetics of intergeneric hybrids involving genera within the Triticeae. *Cereal Res. Commun.* **9,** 39–45 (1981).

12. R. C. Valentine. Osmoregulatory (Osm) genes and osmoprotective compounds. In C. P. Meredith, T. Kosuge, and A. Hollaender, eds., *Genetic Engineering of Plants.* Plenum, New York, 1983.

13. R. G. Wyn Jones and J. Gorham. Aspects of salt and drought tolerance in plants. In C. P. Meredith, T. Kosuge, and A. Hollaender, eds., *Genetic Engineering of Plants.* Plenum, New York, 1983.

14. W. J. Peacock, W. L. Gerlach, and E. S. Dennis. Molecular aspects of wheat evolution. Repeated DNA sequences. In L. T. Evans and W. J. Peacock, eds., *Wheat Science—Today and Tomorrow.* University Press, Cambridge, 1981, 290 pp.

15. J. R. Harlan. The early history of wheat: Earliest traces to the sack of Rome. In L. T. Evans and W. J. Peacock, eds., *Wheat Science—Today and Tomorrow.* Cambridge University Press, Cambridge, 1981, 290 pp.

16. W. D. Jeschke and H. Nassery. K^+–Na^+ selectivity in roots of *Triticum, Helianthus* and *Allium. Physiol. Plant.* **52,** 217–224 (1981).

17. W. D. Jeschke and F. Moreth. K^+-dependent net Na^+ extrusion in excised rye roots, effects of Ca on K^+–Na^+ selectivity. *Plant Physiol. Suppl.* **63,** 13 (1979).

18. H. Greenway and R. Munns. Mechanisms of salt tolerance in non-halophytes. *Annu. Rev. Plant Physiol.* **31,** 149–190 (1980).

19. O. E. Elzam and E. Epstein. Salt relations of two grass species differing in salt tolerance. I. Growth and salt content at different salt concentrations. *Agrochimica* **13,** 187–195 (1969).

20. R. G. Wyn Jones. Salt tolerance. In C. B. Johnson, ed., *Physiological Processes Limiting Plant Productivity.* Butterworths, London, 1981, pp. 271–292.

21. D. Shimshi, M. L. Mayoral, and D. Atsmon. Response to water stress in wheat and related wild species. *Crop Sci.* **22,** 123–128 (1982).

22. J. Gorham, E. McDonnell, and R. G. Wyn Jones. Determination of betaines as ultraviolet-absorbing esters. *Anal. Chim. Acta* **138,** 277–283 (1982).

NOTE ADDED IN PROOF

Because of recent changes in the nomenclature of the Triticeae (see Dewey, 1983) and confusion of the proper identity of some of wheat grasses in standard collections (McGuire and Dvorak, 1981; Dewey, personal communication), a table is appended giving the presently accepted classification of the accessions discussed in this paper.

Table 1.   Corrected Classification of Wheat Grass Accessions

Correct Accepted Taxon	Received as	Bangor Accession Number	USDA Accession Number (P.I.)
Elytrigia pontica	*Agropyron elongatum*	20003	142012
Elytrigia pontica	*Agropyron elongatum*	20014	283164
Elytrigia intermedia	*Agropyron elongatum*	20004	204383
Elytrigia intermedia	*Agropyron elongatum*	20007	222958
Elytrigia intermedia	*Agropyron intermedium*	20023	
E. elongata ssp. scirpeum	*Ag. scirpeum*	20024	
E. juncea	*Ag. junceum*	20025	
E. curvifolia	*Ag. curvifolium*	20026	287739
Elymus lauceolatus	*Ag. desystachym*	20027	
Elymus dahauricus	*Elymus dahauricus*		
Leymus giganteus	*Elymus giganteus*	30001	
Leymus angustus	*Elymus angustus*	30006	406461
Leymus sabulosus	*Elymus sabulosus*	30007	
Leymus triticoides	*Elymus triticoides*	30004	

The revised classification of the accessions studied in this chapter also suggest two significant modifications to the comments made in the main body of the chapter.

The results (Table 10.5) on the accessions originally received as *A. elongatum* seemed to indicate a very wide range of tolerance in this taxon. However, only that accession now known to be authentic *Eyltrigia pontica* showed substantial salt tolerance in contrast to accessions now identified as *E. intermedia*. These results are therefore in complete agreement with those of McGuire and Dvorak (1981) and Shannon (1978).

The differences between *Elymus dahauricus* and the other '*Elymus*' accessions were noted earlier. It is possible that these have a genetic basis, as all the latter are now classified as *Leymus* species containing JX genomes and are quite distinct from the S & H genomes of *Elymus* species (Dewey, 1983). These results also confirm the salt tolerance of *E. junceum* as previously reported by McGuire and Dvorak (1981).

REFERENCES

Dewey, D. R. Historical and current taxonomic perspectives of *Agropyron, Elymus* and related genera. *Adv. Agron.* (in press).

McGuire, P. E. and Dvorak, J. High salt tolerance potential in Wheatgrasses. *Crop Sci.* **21,** 702–705 (1981).

11

RESEARCH ON SALINITY TOLERANCE IN TUNISIA

A. Ayadi and M. Hamza

Physiologie Végétale Faculté des Sciences
Campus Universitaire Tunis El Menzah
Tunis, Tunisia

The growth of plants in a salty environment is of increasing interest in Tunisia, as in many other countries. Research in this area has intensified, thanks partly to an increased awareness of advances in understanding the physiology of salinity tolerance, and to greater attention being given by international organizations to the problems of agronomic improvement in regions impacted by saline soils, particularly NaCl. In the early 1960s, centers of research working on physio-chemical characteristics of salty soils and their effect on plant production were established with the aid of UNESCO. The Center for Research on Problems of the Arid Zone and the Center for Research on the Utilization of Saline Water in Irrigation are especially noteworthy.

For an experimental viewpoint, two types of vegetation have attracted the attention of Tunisian investigators owing to their impact on the economy of the country: forage crops, which are characteristic of the steppe regions, and cultivated crops in regions irrigated with saline water.

Various investigations are being conducted and studies on plant nutrition and of the mechanisms employed by plants in adapting to growth in a saline environment constitute the essential research activities on "Stress Physiology." In May, 1979, a conference on this theme was held at Sfax, Tunisia, in conjunction with the Tunisian University and the French Society of Plant Physiology.

The first part of this chapter presents results obtained on growth in relationship to nutritional requirements in a saline environment. In the second part we review the Tunisian contribution to research on tolerance mechanisms both from the metabolic (active processes) and the nonmetabolic (role of the cell wall) points of view.

1.   MAJOR RESULTS OBTAINED BY THE TUNISIAN TEAM

1.1.   Growth and Morphology of Plants

Growth, generally expressed by the production of dry matter, was studied on plants reported to be sensitive, such as bean (1–3), squash (3), or citrus (4–6); resistant such as *Atriplex* (7), *Hedysarum carnosum* (1, 2, 8) and cotton (3); or for the intermediate behavior such as that of the sunflower (3, 9) and the Laurier-rose (10–12). The influence of salt was studied as a function of time at a constant concentration (3, 13), an increasing concentration (4), or in an alternating system of saline and nonsaline media (10).

A salt sensitivity scale was constructed for Tunisian species and varieties. This scale takes into consideration not only increases in weight but also root growth relative to aerial parts, foliar surface variations, and the appearance of more or less rapid lesions (3, 4, 6, 13). As would be expected, salt deposits were more detrimental to sensitive plants such as bean than they were to tolerant varieties such as *Hedysarum carnosum* in which the mesophyll tissue is more resistant to dehydration (1).

Field studies on cultivated plants irrigated with saline water were conducted using parameters such as height, plant vigor, and yield. The four varieties of sunflower studied (Airelle, Issanka, Record, and Hybride 53S) were not affected by the concentration range 0.4–4.6 mmhos (14). With soybean, the variety Flora proved to be more tolerant than the variety Amsoy (14). Finally, for citrus, tests have shown that both tree vigor and yield are dependent upon soil salinity and texture (15).

1.2.   Mineral Nutrition

The presence of NaCl in a medium modifies the nutritional requirements of plants. The specific response is dependent on the ecological state of the plant being considered. Tests undertaken in Tunisia on the nutritional response in diverse plants constitutes an important contribution in this field.

1.3.   The Dynamic Distribution of Ions

The study of the entire plant does not allow for a precise determination of the mechanisms involved in tolerance or sensitivity to NaCl. Experiments need to be conducted with organs (roots, stems, and in particular leaves) to describe the dynamics of ion distribution, their interaction, and the role of each organ in the plant's response and adaptive strategy. Many results have been obtained: we have verified that the selectivity ratio (Na/K) is in favor of K (3, 4, 6, 12) and that Cl$^-$ accumulates more than Na$^+$ in leaves of salt-sensitive plants (3, 4, 6, 13). A weak accumulation of NO_3^- has also been suggested (6). Finally, a saline medium limits the availability of K and Ca (11). This limitation could have been brought about by either a reduction in root mass or by the reduction in the efficacy of roots to absorb K and Ca (3) without there being any relationship to either a plant's sensitivity or tolerance.

We were also able to confirm that with beans there is a direct correlation between salt sensitivity and the ease of phloem conduction. Hence, there exists a stable Na gradient in the stem (3). Such a gradient was not observed in Laurier-rose, which is moderately tolerant (12).

Experimental proof of a stable Na^+ gradient in sensitive plants is provided by the application of KCN at the crown of the root. This causes an accumulation of Na^+ in higher portions of the stem. Finally, upon separating the root system into two parts (Fig. 11.1), each respectively placed in saline and nonsaline medium, Slama (3) revealed that Na is found in the root system placed in the nonsaline medium, resulting from a descending flow of Na, which apparently is an important function of plant sensitivity. These observations suggest that sensitive plants are those whose roots are unable to secrete Na^+ into the xylem for transport to aerial portions. The apoplasmic process would be predominant. This fact supports the hypothesis, for example, that the radial apoplasmic process is more easily utilized for K in sweet orange than with bitter orange and would explain that the selectivity appears more important in the latter even if, all things being equal, the root selectively accumulates

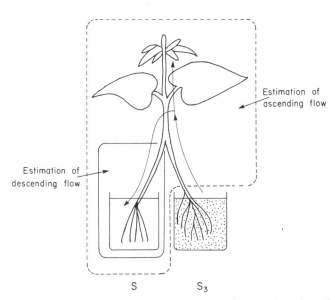

Figure 11.1 Translocation of Na^+ by the phloem. S, nonsaline medium; S_3, saline medium. $(3 \, gl^{-1})$.

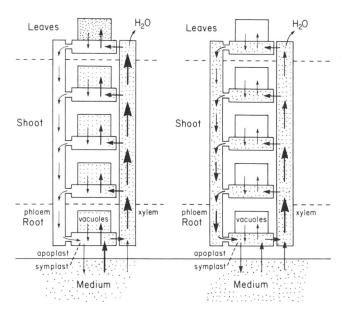

Figure 11.2 Theoretical model for Na transport in a sensitive plant (right) and for a tolerant plant (left). From ref. 3.

K^+ over Na^+ (6). On the contrary, in tolerant plants, the symplasmic process seems to be favored. Salt tolerance is linked to the ability of plants to secrete Na^+ into the xylem and to transport it to aerial portions where it is used as an osmoticum. These facts allow the development of a unitary model (Fig. 11.2) that shows that the apoplasmic or symplasmic process is dominant based on whether the plant is sensitive or resistant.

1.4. The Possible Role of the Wall

The possible contribution of root walls in the adaptive strategy of plants in saline environments is not to be excluded. The study of cell walls was therefore undertaken in collaboration with the team of Dr. Demarty at Rouen University (Prof. Thellier's laboratory).

The cell wall, particularly rich in "polyacids," could play a protective role vis-à-vis salinity. One can study quantitatively the role of the cell wall in the distribution of anions and cations available to the plasmalemma. Figure 11.3 gives an example of the results obtained on Cl^- exclusion by walls equilibrated against diverse concentrations of a KCl solution (16). Curve a is that of an uncharged membrane (no Donnan effect). Curve c was obtained by subscribing to Donnan's law under ideal conditions. Our experimental points are distributed between these two extreme cases (curve b). The exclusion effect is less important than is revealed by the ideal equation. If one examines

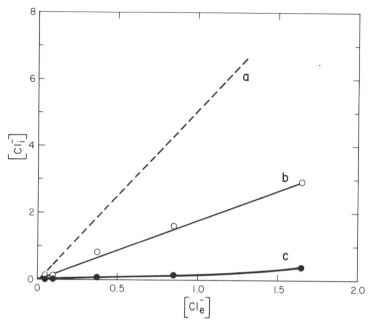

Figure 11.3 Donnan effect on the cell wall. $[Cl_i^-]$ = concentration of chloride in wall phase; $[C_c^-]$ = concentration of chloride in exterior environment. Curve a = non-Donnan effect; curve b = experimental cure; curve c = Donnan's law under ideal conditions.

the case of the Na^+ and Ca^{2+} cations, both of which are plentiful in a saline medium, their reaction can be described by the equation

$$2Na_i + Ca_e \leftrightarrow Ca_i + 2Na_e$$

(i and e refer to the wall phase and the external environment, respectively.) The selectivity constant is

$$K_{Na}^{Ca} = \frac{(Na_e)^2 (Ca_i)}{(Na_i)^2 (Ca_e)} \gg 1$$

The affinity of the cell wall's carboxylic groups for Ca^{+2} corresponds to increasing levels of K.

The presence of a divalent ion in this environment, even at a low concentration, can considerably limit the reaction of a monovalent ion at the cell wall (25). It is possible that the elimination of Ca from a medium where Na is dominant causes the displacement of Na^+ toward the top of the gradient in sensitive plants where the apoplasmic process is dominant (3).

Finally, in a polyelectrolytic phase like the cell wall, the activity coefficients of counter ions (Na^+ or Ca^{2+}) can be less than 1 and are dependent on the density of fixed charges. The polyion is characterized by a structural parameter

$$\xi = \frac{e^2\alpha}{K\epsilon Tb}$$

where e = the electron charge

α = total polyion ionization

ϵ = the dielectric constant of the medium

K = Boltzmann's constant

T = absolute temperature

b = the repetition length along the axis of an average polyion.

Critical ξ values exist, for which the activity of the divalent ions (and possibly also that of the monovalent ions) can become very low, even with high concentrations of these ions in the bathing medium.

Other aspects more or less linked to the action of NaCl have been investigated. By virtue of example, we cite the works carried out on the photosynthesis and assimilation of CO_2 (17), on nitrogen metabolism (18, 19) or glucoside metabolism (20), on oxygen absorption of NaCl treated soybean (21), or on the ultrastructural modification of *Medicago* mitochondria under the effect of saline treatment (22). Furthermore, the study of plant characteristics in a gypsum or calcarious environment, equally abundant in Tunisia, was carried out from an ecological and morphological aspect (23), and from a metabolic aspect (24).

2. CONCLUSION

Studies on the effects of salinity and the reaction of plants to this stress continue to be an object of research both from the fundamental (basic) and from the applied aspects. But what seems encouraging and promising in this area is the recent undertaking of new research strategies based on the complementation of two types of approaches: a reductionist approach at the cellular level, and an approach that envisions the problem at the whole plant level. This scientific step has already proved to be fruitful through the cooperative programs that unite Tunisian laboratories with certain French laboratories.

A good portion of the results obtained recently by Hajji, Slama, and Zid subscribes precisely to the framework of this new strategy within which the contribution of French partners (notably the team of Prof. Grignon, National Superior Agronomic School of Montpellier, on the mechanisms of ionic exchange at the cellular level) has allowed the development of interpretative schemes for different types of plant behavior vis-à-vis saline environments.

The Tunisian investigators wish to diversify scientific contacts and to extend them into countries such as the United States, West Germany, Great Britain, and so on. This would allow the research carried out in Tunisia to be conceived and discussed in relation to that carried out in such countries.

REFERENCES

1. M. Hamza. Influence des conditions climatiques et du régime d'apport du NaCl au milieu sur les limites de tolérance d'une espèce résistante: l' *Hedysarum carnosum* Desf., Soc. Bot. Fr., Actualités Botaniques, no. 3-4, 45-51 (1978).

2. M. Hamza. Influence du régime d'apport du NaCl au milieu sur la régulation du bilan hydrique et de la teneur ionique chez une espèce tolérante, l'*Hedysarum carnosum* Desf., et une espèce sensible. Le Haricot, *Phaseolus vulgaris* L. Soc. Bot. Fr., Actualités Botaniques, no. 3-4, 177-187 (1978).

3. F. Slama, Effet du chlorure de sodium sur la croissance et la nutrition minérale: étude comparative de 6 espèces cultivées. Thèse de Doctorat d'Etat, Tunis, 1982.

4. E. Zid. Croissance et alimentation minérale du jeune bigaradier cultivé en présence de chlorure de sodium. Effet de variations de la concentration du potassium. *Fruits* 30 (6), 403-410 (1975).

5. D. Cherif, E. Zid, A. Ayadi, and M. Thellier. Effet du chlorure de sodium sur la croissance et l'alimentation minérale de *Citrus aurantium* L. (Bigaradier) et de l'hybride *poncirus trifoliata Citrus sinensis* (Citrange troyer). *C. R. Acad. Sci, Paris* **292**, 879-882 (1981).

6. D. Cherif, E. Zid, G. Ghorbal, A. Ayadi, and C. Grignon. Réponse de très jeunes plantes de *Citrus* à NaCl: Étude comparative de deux porte-greffes Citrange Troyer et Bigaradier. *Oecol. Plant.* 3 (17), 79-85 (1982).

7. E. Zid and M. Boukhris. Quelques aspects de la tolérance de l'*Atriplex halimus* L. au chlorure de sodium. Multiplication, croissance, composition minérale. *Oecol. Plant.* **12**(4), 351-362 (1977).

8. M. Hamza. Réponses des végétaux à la salinité. *Physiol. Vég.* **18** (1), 69-81 (1980).

9. F. Slama and A. Bouzaidi. Effet de la salure sur la croissance et la production de quatre variétés de tournesol. (*Helianthus annuus* G.) Informations techniques CETIOM, Paris, 1978.

10. M. Hajji. Comportement éco-physiologique du Laurier-rose (*Nerium oleander*) en milieu salé. *Oecol. Plant.* **13** (1), 59-73 (1978).

11. M. Hajji. Effet du sel sur la croissance et l'alimentation minérale du Laurier-rose. *Physiol. Vég.* **17** (3), 517-524 (1979).

12. M. Hajji. La responsabilité de la racine dans la sensibilité du Laurier-rose au chlorure de sodium. *Physiol. Vég.* **18** (3), 505-515 (1980).

13. M. Hamza. Action de différents régimes d'apport du chlorure de sodium sur la physiologie de deux légumineuses: *Phaseolus vulgaris* (sensible) et *Hedysarum carnosum* (tolérante). Relations hydriques et relations ioniques. Thèse de Doctorat d'Etat, Paris, 1977.

14. F. Slama and E. Bouaziz. Absorption et distribution interne du sodium chez le Soja cultivé en milieu salé. Effet sur la production. *Agrochimica* **22**(2), 128-133 (1978).

15. K. Belkhodja, S. Khalfallah, and M. Lasram. Etude de la qualité des fruits de l'oranger maltaise demi-sanguine en Tunisie en fonction de la nature du sol et de la qualité des eaux d'irrigation. Commission Agro-Technique du CAZF, 1972.

16. A. Ayadi. Etude thermodynamique d'échanges d'ions alcalins par la *Lemna minor*,

L: Interprétation des effets d'ions alcalino-terreux. Thèse d'Etat, Rouen, France, 1975.

17. S. Laouar. Influences comparées de la contrainte hydrique et de la salinité sur l'évolution de la capacité photosynthétique et sur le rythme annuel d'assimilation chez l'Oranger (*Citrus sinensis* L.). *Physiol. Veg.* **18,**(1), 188, (1980).

18. A. Soltani and T. Bernard. Sur le métabolisme azoté d'*Hedysarum carnosum* Desf. cultivé en présence de chlorure de sodium. *C. R. Acad. Sci., Paris (D)* **284,**174–178 (1977).

19. A. Soltani, T. Bernard, Sur le métabolisme azoté d'*Hedysarum coronarium* L. cultivé en présence de chlorure de sodium. *C. R. Acad. Sci. Paris (D)* **287,** 1123–1126. (1978).

20. A. Soltani, M. Briens, and M. Goas. Sur le métabolisme glucidique d'*Hedysarum coronarium* L. cultivé en présence de chlorure de sodium. *C. R. Acad. Sci. Paris (III)* **293,** 297–300 (1981).

21. M. Bejaoui. Effet du NaCl sur l'élongation, la géoaération et l'absorption d'oxygene de segments apicaux de racines de Soja (*Glycine Max.* (L) Merr.). *Physiol. Vég.* **18,**(4), 737–747 (1980).

22. W. Chaibi-Cossentini. Modifications ultrastructurales au niveau des apex radiculaires de *Medicago sativa* L. (cultivar gabès) traités par une solution saline. *Physiol. Vég.* **18,** 184 (1980).

23. M. Boukhris and T. Lossaint. Aspects écologiques de la nutrition minérale des plantes gypsicoles de Tunisie. *Rev. Ecol. Biol. Sol.* **12,**(1), 329–348 (1975).

24. B. Henchi and M. Boukhris. Etude du metabolisme azoté de quelques plantes phosphaticoles de la Tunisie méridionale. *Oecol. Plant.* **14,**(4), 475–482 (1980).

25. A. Ayadi, A. Monnier, M. Demarty and M. Thellier. Echanges ioniques cellulaires: cas des plantes en milieu salé. Rôle particulier des parois cellulaires. *Physiol. Vég.* **18**(1), 89–104 (1980).

12

THE RESPONSES OF HALOPHYTES TO SALINITY: AN ECOLOGICAL PERSPECTIVE

R. L. Jefferies and T. Rudmik

Department of Botany
University of Toronto
Toronto, Ontario, Canada

Species colonizing saline environments are from a variety of plant families, indicating that the evolution of salt tolerance is not restricted to one or two families, and that tolerance has evolved independently in different taxa. Many halophytes, however, are members of three families (namely, the Chenopodiaceae, Gramineae, and Compositae) in which adaptive radiation is pronounced, and to which important crop species belong. However, the extent of the distribution of salt tolerance among members of the plant kingdom suggests there are limits to the process of the evolution of tolerance. Tolerance to seawater has not evolved apparently in a large number of families that contain important crop plants, including the Roseaceae and Rutaceae. This information has bearing on a discussion of the potential for plant improvement for irrigated saline soils. If the necessary genetic information for a complex trait, such as salt tolerance, is absent or only partially present within closely related taxa, the potential for crop improvement using conventional breeding techniques is limited. Most crop plants, with the possible exceptions of barley (1) and sugar beet, apparently fall in this category. Because the prospects for crop improvement using the newer methods of genetic manipulation appear to be long-term, an alternative approach in the short term is to explore the agricultural potential of halophytes (2).

The following questions can be posed: if an increasingly larger number of halophytic species are to be used for different agricultural practices, particularly as a source of forage (see Chapter 16), which characters should be selected from the range of variation present in each species, how will the crop be maintained, and, if applicable, what are the nutritional qualities of the crop?

These questions are not unrelated. An ecological viewpoint has a place in examining and defining those traits or characteristics in halophytes which may be utilized in a breeding program of existing or potential crops. No single mechanism or process governs salt tolerance in plants. Most characters studied in relation to tolerance display continuous phenotypic variation, and are inherited as quantitative traits under the control of multiple genes. Characters of plants such as anatomical dimensions, physiological traits, growth habit, and time of flowering are examples (3).

1. GENETIC VARIATION IN POPULATIONS OF HALOPHYTES

Plant communities within the intertidal zone of coastal ecosystems provide a model to study the growth of plants in irrigated saline soils. Not only are there periodic inundations of tidal water, but salt marshes are nutrient sinks in which there is an abundance of all inorganic nutrients, except nitrogen (4). How much genetic variation occurs in populations of halophytic species at these sites which can be exploited for agricultural purposes? Genetic differentiation between populations of plant species occupying heterogeneous environments is a widely observed phenomenon (cf. 5). Intraspecific variation in salt-marsh species indicates that heritable morphological and physiological differences not only occur frequently between populations in different parts of a marsh, but often genetic differences are evident between groups of plants separated by distances of only a few meters. The origin, maintenance, and selective value of these differences, however, are poorly understood. They appear to be related to the frequency of tidal submergence and/or edaphic conditions prevailing in different parts of a marsh, and probably have little to do with salinity per se. Factors such as mechanical damage from wave action, sediment instability, low oxygen tensions, accumulation of sulfide ions, and competitive ability between component species are of significance in relation to the observed differentiation. Plants from intertidal sites grow and reproduce over a wide range of salinities, a necessary safeguard against environmental uncertainty with respect to fluctuations in salinity. For example, in temperate/subarctic systems salinity may fluctuate between 0.1 and 1.2 M Na^+ in the soil solution at the rooting zone during the growing season (6, 7). However, even in halophytes, high salinities result in poor or slow germination and high seedling mortality (8, 9).

Although the above studies emphasize differentiation between populations, of comparable significance is the amount of genetic variation between individuals of a population. This is important because if individual genotypes with particular traits are to be selected, there is the need to apportion the total variation observed in the phenotype between the genetic and environmental components. There are few studies of this type, but two species within the genera

Puccinellia and *Salicornia* have been examined, both of which are used as a source of forage.

The genus *Puccinellia* comprises a worldwide group of alkali grasses, largely restricted to saline soils. *Puccinellia maritima* is the predominant grass on salt marshes at a number of localities in northwest Europe, and at some sites is heavily grazed by sheep and to a lesser extent cattle (10). In western Australia different species of *Puccinellia* are established in salt flats, and provide valuable grazing on land that would otherwise be nonproductive (11). In *Puccinellia maritima,* a species that is strongly outcrossing (12), populations show a wide variation in biotypes ranging from tall, erect, caespitose, stiff-leaved plants without stolons to small, many tillered, prostrate, small-leaved plants with stolons. Much of this variation has a genetic basis (13). Genotypes originally collected from grazed salt marshes tend to be fast-tillering, small, short-leaved, and prostrate. Although there is a wide variation in the size and shape of different biotypes, the species is a poor competitor in nonsaline soils. Pot and garden cultivation depend on weeding. In competitive experiments with *Festuca rubra* and *Agrostis stolonifera* (14) it outyielded these two species only in water-logged saline soils. These genecological studies provide the basis for the selection and cultivation of suitable biotypes, and the report of Clarke and Malcolm (11) indicates some of the practical difficulties encountered in the large-scale cultivation of *Puccinellia*.

The genus *Salicornia* (annual plants) and closely related genera (perennials) are widespread in coastal and inland saline areas. These genera have also been utilized for food. Glen *et al.* (15) watered *Salicornia europaea* agg. with seawater and obtained a fresh weight yield of 27 metric tonnes per hectare after 4 months; the harvested material was used to feed shrimps. In Australia, samphire, which has become established in saline areas, is used as forage for sheep and to a lesser extent for cattle (16). Although individuals of the genus *Salicornia* show extreme phenotypic plasticity, a recent study has examined genetic variation in two nearly indistinguishable diploid, cleistogamous micro-species (*S. europaea, S. ramosissima*) of this (17). All of the 800 plants examined were homozygous at the 30 loci that were scored, and within each species no electrophoretic variability was observed between individuals. The electrophoretic evidence confirms the reproductive isolation of the species and indicates an apparent lack of genetic variability in both of them. Whether populations throughout the complete geographic range of the two species show a lack of genetic variation remains to be established, but the apparent electrophoretic uniformity of individuals from populations of each of the species examined is consistent with species that enjoy a wide distribution but are restricted to one ecological niche. In the above study the electrophoretic patterns of the enzymes are being used as markers to distinguish the two microspecies.

This approach is of wider application, as genetic markers may be used to identify genotypes within vegetation. The fate of individual genotypes throughout the life cycle can be monitored in different environments in order to

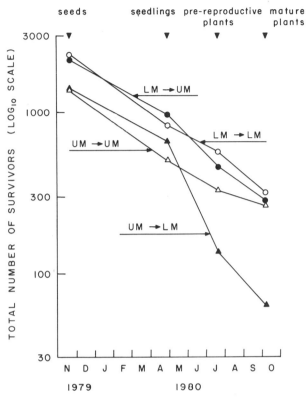

Figure 12.1 Survivorship over whole life cycle of individuals of *Salicornia europaea* (solid symbols) and *Salicornia ramosissima* (open symbols) at Stiffkey salt marsh, United Kingdom. Seeds of each species respectively planted at the sites of collection in the lower marsh (LM) and upper marsh (UM) and transplanted reciprocally between the sites of collection. Unpublished data of Davy and Jefferies.

determine which survive a given perturbation (e.g., drought, salinity, grazing, mowing, harvesting). An example of this approach is shown in Figure 12.1. *Salicornia europaea* and *S. ramosissima,* in both of which there is an apparent absence of genetic variation, are abundant in the lower (seaward) and upper levels of marshes, respectively (18). Known quantities of seeds of the two microspecies were sown in cleared sites in both levels of a marsh, and the fate of individuals followed throughout the life cycle (Davy and Jefferies, unpublished). Overall, approximately 10% of the seeds developed into mature plants 1 year after seeds were sown. However, individuals of *S. ramosissima* at the seedling stage in the lower marsh were particularly susceptible to the adverse effects of waves washing plants from the sediment. Evidence from other studies (Jensen and Jefferies, unpublished) suggested that a low rate of root growth at the seedling stage in this microspecies contributed to the loss of plants. The approach combines genetic, demographic, and physiological studies, in order to determine the fitness of particular genotypes at different stages of their life history.

2. PHYSIOLOGICAL TRAITS

Instead of determining genetic differences within and between populations, and selecting genotypes that produce high yields under brackish or saline conditions, attempts have been made to examine variation in physiological traits in individuals of different species in relation to salt tolerance. The approach serves two purposes. First, it results in the identification and classification of the wide range of responses that plants show in relation to increasing salinity. Secondly, physiological traits that appear to be an adaptation to salinity tolerance can be selected in a breeding program. Often the physiological characters are based on the amounts of the different species of organic and inorganic solutes in leaves. These differences reflect the result of the integration of a large number of processes throughout the plant which not only control ion selectivity and transport, but also the synthesis of organic compounds. The relative contributions and efficiencies of these processes in regulating the internal solute content of cells are not well understood (19; Chapters 3 and 6). In addition, the solute concentrations in the leaves may change during ontogenetic development and when the external salinity fluctuates. These and other complex physiological traits are not controlled by a single gene but involve processes that are probably regulated by a large number of gene loci located on more than one chromosome. If this is correct, the individual genes, irrespective of their number, cannot be identified by their segregations; Mendelian ratios are not displayed, and the methods of Mendelian analysis cannot be applied (3).

Aside from this cautionary note and that of Gorham, Hughes, and Wyn Jones (20), concerning phenotypic plasticity of individuals and genetic variation between populations, studies of physiological traits have revealed fundamental differences in the way plants cope with salinity. One example is the classification of brackish water plants and halophytes, based on the detailed studies of the relative proportions of compatible organic solutes produced by plants (20–25). Two major groups of plants may be recognized: those that produce high levels of soluble carbohydrates and/or polyols, but in which nitrogenous solutes are absent or present at very low concentrations, and plants accumulating nitrogenous compounds under saline conditions (Table 12.1). Can these differences can be correlated with the ecology of the respective groups of plants? A tentative ecological interpretation may be made of an apparent correlation between the distribution of certain terrestrial plants and the amounts of the different organic solutes which accumulate in the tissues, although further studies are required to confirm the conclusion. Those plants (Group 1) that store soluble carbohydrates, such as sucrose and/or polyols as the predominant compatible organic solutes, are either characteristic of freshwater or brackish

Table 12.1

Group 1 Freshwater and Brackish Water Plants That Produce Soluble Carbohydrates in Response to Increasing Salinity	Group 2 Halophytes That Produce Nitrogenous Compounds, in Response to Increasing Salinity[a]
Carex arenaria	*Agropyron pungens*
Carex extensa	*Agrostis stolonifera*[b]
Carex otrubae	*Armeria maritima*
Carex punctata	*Aster tripolium*
Eleocharis uniglumis	*Atriplex hastata*
Glaux maritima	*Atriplex patula*
Iris pseudacorus	*Beta maritima*
Juncus articulatus	*Distichlis spicata*
Juncus effusus var *compactus*	*Festuca rubra*[b]
Juncus gerardii	*Halimione portulacoides*
Juncus inflexus	*Limonium carolinium*
Juncus maritimus	*Limonium vulgare*
Oenanthe lachenalii	*Puccinellia maritima*
Phragmites communis	*Salicornia europaea*
Plantago maritima	*Spartina alterniflora*
Schoenoplectus tabernamontani	*Spartina patens*
Scirpus maritimus	*Spartina* × *townsendii*
	Spergularia media
	Suaeda maritima
	Triglochin maritima

[a] See text for references.
[b] Salt-marsh populations of the species.

conditions, or else their rooting systems grow below a layer of highly saline sediment which frequently lies close to the surface in salt marshes. In contrast, those plants (Group 2) that have a capacity to accumulate nitrogenous substrates rarely occur outside of saline habitats, at least in northwest Europe. It is important to appreciate that under conditions of increasing salinity not only do the amounts of individual solutes change in tissues, but also a number of these organic compounds fulfill other metabolic roles in addition to that associated with osmoregulation (26). The use of nitrogenous compounds as compatible osmotic solutes may be linked with the relative abundance of nitrogen within salt marshes (27) and the low amounts of carbon needed in proline and glycinebetaine compared with sucrose to produce equivalent molal solutions. Carbon my be a limiting resource in halophytes, because under conditions of increasing salinity stomatal closure restricts the flux of carbon dioxide to the leaves (see Chapter 20). Plants growing in brackish sites rarely experience high salinities; hence, in the absence of strong selection pressures soluble carbohydrates serve as osmotic solutes. As pointed out by Yancy et al. (28), phylogenetically diverse organisms, such as bacteria, unicellular algae, vascular plants, invertebrates, and vertebrates, all utilize low molecular weight organic solutes. This striking example of convergent evolution suggests that similar strong selection pressures must have operated on the different groups (28).

Another classification based on the internal ion content of the leaves of halophytes (20, 29–31) indicates the following:

1. Most monocotyledenous halophytic plants have relatively high amounts of K^+ and low amounts of Na^+ in their leaves (i.e., K^+/Na^+ ratio >1.0).

2. Exceptions include *Spartina* spp., *Distichlis* spp., and *Triglochin maritima* in which Na^+ is the predominant ion.

3. Halophytic members of the Chenopodiaceae, and species such as *Aster tripolium,* have the ability under saline conditions to accumulate large quantities of Na^+ and Cl^- in their tissues. This ability is often expressed at relatively low external salt levels.

The amounts and localization of the two ions within the tissues reflect the relative contributions of the processes that control ion selectivity and transport in plants (see Chapters 3 and 6). Because studies on compartmentation of ions in higher plants are few, the contribution these ions make toward π_s is not well documented. It is assumed that most of the sodium is located within the vacuole, or the cell wall, but this assumption and the magnitude of the fluxes of these ions throught the plant require further study.

The ability of a number of halophytes to accumulate Na^+ ions, even at low external salinities, forms the basis of an additional scheme discussed by Storey and Wyn Jones (32), in which halophytes may be described as osmoregulators or osmoconformers. The latter group of plants lower their osmotic potential in accordance with a fall in external water potential, whereas in the former group the osmotic potential is maintained at a low value throughout the season, and is relatively independent of external salinity. The osmoregulators are mainly succulents, such as *Halimione, Suaeda* and *Salicornia,* which accumulate sodium in their cells, even at low external salinities. These plants, particularly the annuals, grow in the surface layers of sediments where the salinity may be expected to be high and subject to fluctuation. In contrast, the deeper rooted perennials exploit a different ecological niche, where extremes of salinity are absent (6). Little attention has been given to the appearance of osmotic solutes in different plant organs during germination and establishment in relation to the prevailing external salinity. This is of interest because in salt marshes seedlings of most perennial plants are infrequent, unlike young plants of the annuals (*Salicornia, Suaeda* spp.) mentioned above.

Perhaps one of the most urgent questions in relation to crop improvement in saline lands is whether genetic or physiological traits may be used as an indicator in screening programmes to measure the responses of large numbers of individuals to salinity. The rooting method (33) developed for heavy-metal tolerance testing has been used to measure salinity tolerance (34–36) and osmotic stress in plants (37). Response to salinity is expressed as an index of tolerance, which is the root length in the salt solution, divided by the root length in the control solution, multiplied by 100. Although there are some difficulties with this method, large numbers of individuals can be assayed at any one time. Venables and Wilkins (35) found a close correlation between salt tolerance (determined by rooting tests) and the salt content of the soil. However, damage may occur directly to the leaves, particularly in localities where there is a heavy salt spray. Recently, Humphreys (38) has developed a test based on the percentage of green leaf remaining after treatment of plants with salt spray. There was a close positive correlation between subsequent survival and the amount of green leaf of *Festuca rubra*.

3.   GROWTH, PRODUCTION AND RESOURCES

As in the case of the evolution of heavy-metal tolerance in plants, halophytes show adaptations to saline soils beyond those manifest at the cellular level. These adaptations involve growth processes that affect developmental patterns, productivity, and yield. In addition, when an agronomist or plant breeder discusses yield he is referring to the property of a community. Values of net primary productivity of higher plants for different coastal communities have been summarized by Keefe (39) and Turner (40). High values have been obtained at some sites, particularly in more southerly marshes in the Northern Hemisphere. Not all values are corrected for internal salt content, or turnover of leaves. For example, the mineral content of *Salicornia europaea, Suaeda maritima, Aster tripolium, Beta maritima,* and *Atriplex hastata* is between 30 and 50% of the dry weight (25). Recently, Hussey (41) working in Long's laboratory has shown that in stands of *Puccinellia maritima* on the Essex coast in England, the turnover rates of leaves may be substantial. The most conservative estimate for annual net above-ground primary productivity taking these factors into account was 650 g ash-free dry wt m^{-2} yr^{-1}, whereas the corresponding estimate based on peak biomass was only 267 g m^{-2} yr^{-1}. Calculations of relative growth rate (R) of above-ground biomass of most salt-marsh plants indicate that low values are to be expected (Table 12.2). These above-ground rates were maintained for only 1 or 2 weeks in the season. The bursts of growth are probably achieved by the utilization of underground reserves in perennating organs, so that much of the annual aboveground growth is the result of internal recycling. It is difficult to be precise about the magnitude of the contribution because of a lack of reliable data, but it may be between

**Table 12.2   Maximum Relative Growth Rates
(R) of shoots of Halophytes in the Upper Marsh
at Stiffkey, Norfolk, United Kingdom**[a]

Species[b]	R(wk^{-1}]
Armeria maritima (P)	0.53
Enteromorpha sp. (alga)	0.29
Festuca rubra (P)	0.62
Limonium vulgare (P)	0.87
Plantago maritima (P)	0.89
Salicornia ramosissima (A)	0.25
Triglochin maritima (P)	0.75

[a]Measurements of standing crop were made at
weekly intervals throughout the season on 10 turves
(17.5 × 17.5 cm)(Byers and Jefferies, unpublished
data).
[b]P = perennial; A = annual.

30 and 50% of the total annual above-ground production in some herbaceous perennials that have a marked seasonal growth pattern. If these perennial plants are cropped, mowed or grazed does this lead to the destruction of the sward? When a *Spartina* marsh was grazed by sheep in England, the grazing favored a significant increase in the spread of *Puccinellia maritima* (42). The length of time for *Puccinellia* marsh to develop at the site was estimated to be the order of 8–10 years. These results suggest that some halophytes with an upright growth habit may not be able to regenerate if cropped frequently, because reserves which are normally recycled are depleted. However, in some salt marshes grazing actually enhanced net primary productivity of vegetation. Grazing by lesser snow geese in a marsh at La Perouse Bay on the Hudson Bay coast, dominated by two prostrate species, *Puccinellia phryganodes* and *Carex subspathacea,* increased net above-ground primary productivity 35 and 80%, respectively, during two consecutive years (43). In addition, the nitrogen content of the leaves of grazed plants was higher than that of ungrazed individuals. This enhancement of production appears to be linked to an increase in the rate of turnover of nitrogen within the system and the development of extra-axillary stolons in the grass.

These examples indicate that although the requirement for salt-tolerant plants is a necessary prerequisite for this type of marginal agriculture, the appropriate management techniques must be developed if the crop is to be maintained. In the examples given above this is closely tied to the growth patterns of the plants (i.e., determinate or indeterminate growth) and the intensity of cropping. A further consideration is the buildup of pest populations, particularly in monospecific stands. For example, in communities of *Salicornia* near Woods Hole, a small beetle and a lepidoteran may reduce seed output by 70% (C. Cogswell, Marine Biological Laboratory, Woods Hole, unpublished data).

The integration and control of the different responses of whole plants to salinity are not well understood. Although under increasing salinity dry matter production of halophytic plants decreases, the allocation of dry matter and resources to individual plant organs is poorly documented. Rates of turnover of leaves and roots and changes in clonal and sexual reproduction are largely unexplored. For example, sexual and asexual reproduction in *Borrichia ritescens* (a composite halophytic shrub) varied significantly, but in opposite directions along a salinity gradient at Sapelo Island, Georgia, with asexual reproduction greater in the high salt environment (44). At high salinities the availability of resources for sexual reproduction may be limited. Coastal marshes contain an abundance of all nutrients, except nitrogen, and experimental studies indicate that inadequate supplies of this element in some marshes restrict plant growth. However, the biogeochemical cycling of nitrogen in different halophytic plant communities is poorly documented. Blue-green algae may fix between 4.0 and 460 kg of N ha^{-1} yr^{-1} [45]. There is also evidence of N-fixing bacteria in the rhizosphere of species such as *Spartina alterniflora* (46). How much of the nitrogen from these two sources is taken up by plants is unknown. Much of it may be exported from the marsh (47). Under limited supplies of this element nitrogen, which otherwise may be used in growth and reproduction, may end up in compatible organic solutes at the expense of growth. Concurrent with the accumulation of organic solutes as the salinity increases are reductions in both the size and numbers of vegetative structures and in the frequency of sexual reproduction. Although these adjustments are consistent with the above hypothesis and reflect the plasticity shown by halophytes, it is not clear whether each of these plastic responses is truly adaptive. Shifts in the percentage allocation of the total nitrogen to different organs or processes may simply be a consequence of overall adjustments of whole plants to salinity. Traditionally, reductions in growth (yield) have been viewed as detrimental. However, reductions in size of plants and changes in the rates of turnover of leaves and roots affect a large number of processes, and provide evidence of an adjustment at the level of the whole plant to salinity. The initial fall in yield may not necessarily reflect a decrease in the efficiency of metabolic processes under increasing salinity, but the ability of the individual to modify its architecture so that resource requirements and levels of respiration are minimal for survival.

Evidence to support this thesis comes from two sources. Results of regular additions of inorganic nitrogen (NH_4 or NO_3) to permanent field plots over a 4- to 5-year period indicated that the perennial plant communities showed a strong degree of constancy toward perturbations. The availability of large amounts of nitrogen did not result in a burst of vegetative growth and the standing crops of each of the perennial species in the different plots were not significantly different from one another (48). Under high salinities prevailing in these plots, there has been selection for small plants with low growth rates (49).

Recently, we have examined the growth and development of plants of *Triglochin maritima* (Juncaginaceae) at different salinities. This circumboreal species is widespread in saline and calcareous habitats. The results show that the plants exhibit considerable phenotypic plasticity in response to salinity. Cloned, 2-year-old plants of *Triglochin* were grown in sand cultures, which were watered with a dilute seawater solution (1:20), supplemented with nitrogen and phosphorus. The birth and death of all leaves was recorded, so that after a complete turnover of leaves the age structure of leaves on each plant was known. At this stage half of the cloned plants received supplemented full strength seawater. Subsequent changes in leaf demography were observed in plants treated with one or the other solution. The results showed that there was a shift in the age structure of the population of leaves in response to increased salinity. Birth rate per module (shoot system) did not alter (like rice, see Chapter 8) but death rate fell (unlike rice which is salt sensitive). This meant that the turnover rate of leaves in the population decreased, a characteristic of stress tolerance (50). There were however, more significant changes. Mean leaf length of all age classes was less than half of that of leaves treated with diluted seawater, the percentage volume occupied by lacunae was much reduced, as was cell size. Osmotic adjustment resulted in a substantial increase in the concentrations of proline and sodium and chloride ions in the leaves. There were indications that the photosynthetic rates of the leaves of plant in full strength seawater were higher, irrespective of the units of expression. If the external salinity was lowered to the original value, the existing leaves were replaced by those characteristic of brackish conditions. Although the detailed ontogeny has not been worked out yet, it appears that the type of leaf produced depends on the environment to which the plant is exposed. Changes in the size and shape of successive leaves occur as external salinity is altered. There is good evidence that leaf shape is controlled by many genes and that their effect can be modified by environmental conditions (51, 52). These types of adjustment represent the capacity of individuals to adjust to a wide range of external salinities, a characteristic shared by individuals of only a few species.

One aspect of this study requires particular comment. The reduction in leaf length and the size of parenchyma cells of plants grown in seawater results in a set of leaf dimensions similar to that of juvenile leaves from the control plants. However, the physiological characteristics of the cells are very different in the two groups of plants. Reduction in size also influences the nature and magnitude of the different components of energy exchange, as well as the rates of gaseous and liquid diffusion. Steudle et al. (53) have discussed the relationship between cell size, turgor pressure, and wall elasticity. They suggest that smaller cells require less turgor to achieve their growth potential than do larger cells, and that much more turgor pressure is required to initiate extension of cell walls in larger cells. This reduction in cell size may also be associated with the maintenance of tension in the cell wall within prescribed limits, irrespective of the salinity. Three variables affect this tension: the hydrostatic (turgor) pressure, the radius of the cell, and the thickness of the cell wall. Shifts in one or more of these affect the magnitude of the tension within the wall. Perhaps, the plethora of reported changes in leaf anatomy in relation to salinity may be related to maintenance of tension within certain limits.

4. CONCLUSIONS

Populations of relatively few species have evolved salt-tolerance. Most of the characters associated with salt-tolerance appear to be under polygenic control. Table 12.3 gives a list of traits discussed in this paper which characterize this group of plants. The traits are not necessarily unique to halophytes but are found in stress-tolerant plants in general (50). However, halophytes exhibit a high level of physiological plasticity coping with environmental uncertainty, and there is evidence at least in some halophytes of morphogentic changes in response to salinity. If the use of salt-marsh plants for agricultural purposes

Table 12.3.   Characteristics of Coastal Halophytes

1. Majority of plants are perennial and long-lived (exceptions some members of Chenopodiaceae).
2. Lower incidence of sexual reproduction under increasing salinity.
3. Different methods of clonal reproduction well developed within halophytic plants as a group. Leaves often linear-lanceolate.
4. Much of the total biomass below ground in perennial plants (70%). Organs of perennation well developed.
5. Relative growth rates low compared with crop plants. Maximum rates in perennial plants ($<0.9 \, wk^{-1}$) achieved early in season as a result of mobilization of reserves in perennating organs.
6. Internal salt content in leaves between 10 and 50% of the dry weight. Concentrations of soluble nitrogenous compounds high.
7. Sparse or sporadic germination at salinities equivalent to that of seawater. Seedling mortality high under these conditions.
8. Considerable genetic variation between populations of perennial halophytes, some of which can be related to environmental conditions (e. g., low R in populations growing under highly saline conditions; prostrate growth forms under influence of grazing).
9. Majority of species are C_3 plants. However, notable exceptions (e.g., *Spartina* spp.)
10. Evidence of morphogenetic responses to increasing salinity (e.g., fall in shoot/root ratio, development of new leaves with different characteristics).
11. Considerable physiological plasticity in response to increasing salinity. Ability to grow and reproduce over a wide range of salinities.
12. Traits associated with salt tolerance appear to be under polygenic control.

is to be extended, much more information on the different traits will be required for their integration at the level of the whole plant. In turn, this will provide a basis on which to attempt modifications of existing crop plants.

ACKNOWLEDGMENTS

I wish to thank Drs. J. Dainty and R. G. Wyn Jones for constructive criticism of the manuscript.

REFERENCES

1. E. Epstein and J. D. Norlyn. Sea-water based crop production: A feasibility study. *Science* **197**, 249 (1977).
2. P. J. Mudie. The potential economic uses of halophytes. In R. J. Reimold and W. H. Queen, eds., *Ecology of Halophytes*. Academic Press, New York, 1974.
3. D. S. Falconer. *Introduction to Quantitative Genetics,* 2nd ed. Longman, New York, 1981.
4. L. R. Pomeroy. The strategy of mineral cycling. *Annu. Rev. Ecol. Syst.* **1**, 171 (1970).
5. A. J. Gray, R. J. Parsell, and R. Scott. The genetic structure of plant populations in relation to the development of salt marshes. In R. L. Jefferies and A. J. Davy, eds., *Ecological Processes in Coastal Environments*. Blackwell, Oxford, 1979.
6. R. L. Jefferies, A. J. Davy, and T. Rudmik. The growth strategies of coastal halophytes. In R. L. Jefferies and A. J. Davy, eds., *Ecological Processes in Coastal Environments*. Blackwell, Oxford, 1979.
7. R. L. Jefferies, A. Jensen, and D. Bazely. The biology of the annual, *Salicornia europaea* agg., at the limits of its range in Hudson Bay. *Can. J. Bot.* **61,** 762 (1983).
8. Y. Waisel. *Biology of Halophytes*. Academic Press, New York, 1972.
9. I. A. Ungar. Halophyte seed germination. *Bot. Rev.* **44**, 233 (1978).
10. D. S. Ranwell. *Ecology of Salt Marshes and Sand Dunes*. Chapman and Hall, London, 1972.
11. A. J. Clarke and C. V. Malcolm. *Puccinellia*—Grass for salt land. Western Australian Department of Agriculture Farmnote 29/76, 1976.
12. A. J. Gray and R. Scott. Biological Flora of the British Isles: *Puccinellia maritima* (Huds.) Parl. *J. Ecol.* **65**, 699 (1977).
13. A. J. Gray and R. Scott. A genecological study of *Puccinellia maritima* Huds. (Parl.) 1. Variation estimated from single-plant samples from British populations. *New Phytol.* **85**, 89 (1980).
14. A. J. Gray and R. Scott. The ecology of Morecambe Bay. VII. The distribution of *Puccinellia maritima* and *Festuca rubra*. *J. app. Ecol.* **14,** 299 (1977).
15. E. Glen, M. Fontes, S. Katzen, and B. Colvin. Nutritional value of halophytes irrigated with hypersaline water in the Sonoran Desert. In A. San Pietro, ed., *Biosaline Newsletter 3*. Indiana University Press, 1981.

16. C. V. Malcolm. Wheat belt salinity—Review of the salt land problem in South Western Australia. Western Australian Department of Agriculture Technical Bulletin. No. 52, 1982.

17. R. L. Jefferies and L. D. Gottlieb. Genetic differentiation of the microspecies *Salicornia europaea L. (sensu stricto)* and *S. ramosissima*. J. Woods. *New Phytol.* **92,** 123 (1982).

18. R. L. Jefferies, A. J. Davy, and T. Rudmik. Population biology of the salt marsh annual *Salicornia europaea* agg. *J. Ecol.* **69**, 17 (1981).

19. H. Greenway and R. Munns. Mechanisms of salt-tolernce in non halophytes. *Annu. Rev. Plant Physiol.* **31**, 149 (1980).

20. J. Gorham, L. L. Hughes, and R. G. Wyn Jones. Chemical composition of salt-marsh plants from Ynys Mon (Anglesey): The concept of physiotypes *Plant Cell Environ.* **3**, 309 (1980).

21. R. Storey and R. G. Wyn Jones. Responses of *Atriplex spongiosa* and *Suaeda monoica* to salinity. *Plant Physiol.* **65**, 156 (1979).

22. G. R. Stewart, F. Larher, I. Ahmad, and J. A. Lee. Nitrogen metabolism and salt tolerance in halophytes. In R. L. Jefferies and A. J. Davy, eds., *Ecological Processes in Coastal Environments.* Blackwell, Oxford, 1979.

23. J. Rozema. Population dynamics and ecophysiological adaptations of some coastal members of the *Juncaceae* and *Gramineae*. In R. L. Jefferies and A. J. Davy, eds., *Ecological Processes in Coastal Environments.* Blackwell, Oxford, 1979.

24. A. J. Cavalieri and A. H. C. Huang. Evolution of proline accumulation in the adaptation of diverse species of marsh halophytes to the saline environment. *Am. J. Bot.* **66**, 307 (1979).

25. M. Briens and F. Larher. Osmoregulation in halophytic higher plants: A comparative study of soluble carbohydrates, polyols, betaines and free proline. *Plant Cell Environ.* **5**, 257 (1982).

26. J. S. Pate. Transport and partitioning of nitrogenous solutes. *Annu. Rev. Plant Physiol.* **31**, 313 (1980).

27. G. R. Stewart, J. A. Lee, and T. O. Orebamjo. Nitrogen metabolism of halophytes. 1. Nitrate reductase activity in *Suaeda maritima. New Phytol.* **71**, 263 (1972).

28. P. H. Yancey, M. E. Clark, S. C. Hand, R. D. Bowlus, and G. N. Somero. Living with water stress: Evolution of osmolyte system *Science* **217**, 1214 (1982).

29. R. Albert and H. Kinzel. Underscheidung von Physiotypen bei Halophyten des Neusiedlerseegebietes (Osterreich) *Z. Planzenphysiol.* **70**, 138 (1973).

30. R. Albert. Salt regulation in halophytes. *Oecologia* **21**, 57 (1975).

31. R. Albert and M. Popp. Chemical composition of halophytes from the Neusiedler Lake region in Austria. *Oecologia* **27**, 157 (1977).

32. R. Storey and R. G. Wyn Jones. Taxonomic and ecological aspects of the distribution of glycinebetaine and related compounds in plants. *Oecologia* **27**, 319 (1977).

33. D. A. Wilkins. A technique for the measurement of lead tolerance in plants. *Nature* **189**, 73 (1957).

34. N. J. Hannon and A. D. Bradshaw. Evolution of salt tolerance in two coexisting species of grass. *Nature* **220**, 1342 (1968).

35. A. V. Venables and D. A. Wilkins. Salt tolerance in pasture grasses. *New Phytol.* **80**, 613 (1978).

36. L. Wu. The potential for evolution of salinity tolerance in *Agrostis stolonifera* L. and *Agrostis tenuis* Sibth. *New Phytol.* **89**, 471 (1981).

37. I. Ahmad and S. J. Wainwright. Tolerance to salt, partial anaerobiosis and osmotic stress in *Agrostis stolonifera*. *New Phytol.* **79**, 605 (1977).

38. M. O. Humphreys. The genetic basis of tolerance to salt spray in populations of *Festuca rubra* L. *New Phytol.* **91**, 287 (1982).

39. C. Keefe. Marsh production: a summary of the literature. *Contrib. Mar. Sci.* **16**, 163 (1972).

40. R. E. Turner. Geographic variations in salt marsh macrophyte production: A review. *Contrib. Mar. Sci.* **20**, 47 (1976).

41. A. Hussey. *The net primary production of an Essex salt marsh, with particular reference to Puccinellia maritima.* Ph.D. thesis, University of Essex, England, 1980.

42. D. S. Ranwell. Spartina salt marshes in Southern England. I. The effects of sheep grazing at the upper limits of *Spartina* marsh at Bridgewater Bay. *J. Ecol.* **49**, 325 (1961).

43. S. M. Cargill. (1981). *The effects of grazing by lesser snow geese on the vegetation of an arctic salt marsh.* M. Sc. thesis, University of Toronto.

44. A. E. Antlfinger. The genetic basis of microdifferentiation in natural and experimental populations of *Borrichia frutescens* in relation to salinity. *Evolution* **35**, 1056 (1981).

45. K. Jones. Nitrogen fixation in a salt water marsh. *J. Ecol.* **62**, 583 (1974).

46. D. G. Patriquin and C. R. McClung. Nitrogen accretion, and the nature and possible significance of N_2 fixation (acetylene reduction) in a Nova Scotian *Spartina alterniflora* stand. *Mar. Biol.* **47**, 227 (1978).

47. I. Valiela and J. M. Teal. The nitrogen budget of a salt marsh ecosystem. *Nature* **280**, 652 (1979).

48. R. L. Jefferies and N. Perkins. The effects on the vegetation of the additions of inorganic nutrients to salt marsh soils at Stiffkey, Norfolk. *J. Ecol.* **65**, 867 (1977).

49. R. L. Jefferies. Growth responses of coastal halophytes to inorganic nitrogen. *J. Ecol.* **65**, 847 (1977).

50. J. P. Grime. *Plant Strategies and Vegetation Process.* Wiley, New York, 1979.

51. E. Ashby. Studies in the morphogenesis of leaves. 1. An essay on leaf shape. *New Phytol.* **47**, 153 (1948).

52. A. D. Bradshaw. Evolutionary significance of phenotypic plasticity in plants *Adv. Genet.* **13**, 115 (1965).

53. E. Steudle, U. Zimmerman, and U. Lüttge. Effects of turgor pressure and cell size on wall elasticity of plant cells. *Plant Physiol.* **59**, 285 (1977).

PART II

CROP SELECTION AND IMPROVEMENT

13

BREEDING, SELECTION, AND THE GENETICS OF SALT TOLERANCE

M. C. Shannon

U. S. Salinity Laboratory
U. S. Department of Agriculture
Riverside, California

Salinity of soil and waters is caused by the presence of soluble salts originating from deteriorating and dissolving rock and concentrated by evaporation and plant transpiration. At low concentrations salt suppresses plant growth but higher concentrations can cause death. In semiarid regions especially, the scarcity of water and the hot, dry climates frequently cause salinity concentrations that limit or prevent crop production. Poor drainage in agricultural areas may also cause saline soils by raising water tables too near the surface.

Saline soils cover a substantial portion of our earth's surface. Estimates vary from 400 to 950 $\times$ 10^6 ha (1). Reclamation, drainage, and water control can minimize the extent and spread of saline soils; however, engineering and management costs are high. Increasing costs for water and energy accent the need for new strategies.

One example of a new strategy is breeding crops for increased salt tolerance. It is a promising, energy-efficient approach that can be meshed with water and land management alternatives. The adaptation of crops to salinity is a formidable challenge for plant breeders and geneticists. This chapter will cover our present state of knowledge and progress in meeting this challenge.

Throughout this chapter, comparisons will be made between halophytes or salt-tolerant plants and glycophytes. Halophytes are plants that have been naturally selected to grow in saline environments. This does not infer that salinity is obligatory for the growth of halophytes, but just that they have a survival or competitive advantage in this environment over nonhalophytes. Extreme examples of halophytes are plants that live and thrive in the sea, and plants growing in coastal environments. Glycophytes are plants that do not grow well in the presence of salinity, even salinity concentrations much more dilute than seawater. Almost all crop species are glycophytes. Glycophytes have a selective advantage in nonsaline soils over halophytes because, as a general rule, their rate of growth is faster. Natural selection favors plants that survive and compete well, but these are not the same selective criteria used by plant breeders. Plant breeders favor high yields and quality improvements in the development of new varieties. The acquisition of the character of salt tolerance would increase yields of plants grown for food or income by increasing the acreage of land that could be dedicated to growing these economically important plants.

The concept of breeding and selecting for salt tolerance has recently been reviewed (2–5). Basic breeding principles require the availability of genetic diversity in salt tolerance among many species. Unfortunately, efforts to breed for salt tolerance have resulted in the release of only a single variety that has high salt tolerance as a discernible character (6). Slow progress is due to a combination of many factors: (i) incomplete knowledge of the effects of salinity on plants, (ii) inadequate means of detecting and measuring salinity, (iii) ineffective selection methods, (iv) a poor understanding of the interactions of salinity and environment as it affects the plant, (v) the vague or nonspecific effects other than growth of moderate salt stress, (vi) the interacting nature of the ionic and osmotic properties of salts on plants, and (vii) changes in salt tolerance with plant development. In this chapter some aspects of salt tolerance are presented that will be useful to the breeder interested in developing this character. Specific physiological mechanisms believed involved in plants that express tolerance will be examined in the context in which they relate to the practical aspects of screening and selection.

1. PLANT RESPONSES TO SALINITY

1.1.   General Salinity Effects

Bernstein (7, 8) proposed that salt has both ionic and osmotic effects on plants and most of the known responses of plants to salinity are intimately linked to these effects. Ionic effects may be further subdivided into nonspecific and specific ion effects. At low salinities, different ionic salts have similar osmotic effects on plants. Only in special instances does salinity cause deficiencies or toxicities of nutrient ions (e.g., Ca^{2+} or PO_4^{2+}) (9–11). At high salinities, most harmful effects of salts are still nonspecific, but some specific symptoms of plant damage may be recognized, such as leaf tip burn due to Na^+ or Cl^- ions. High ionic concentrations may disturb membrane integrity and function, interfere with internal solute balance, or cause a shift in solute concentrations. Na^+ and Cl^-, usually the most prevalent ions in saline soils or water, account for most of the deleterious effects that can be related to specific ion toxicities. Osmotic effects of salts on plants are a result of a lowering of the soil water potential due to increasing solute concentration in the root zone. At very low soil water potentials, this condition interferes with the plant's ability to extract water from the soil and maintain turgor. Thus, in some aspects salt stress may resemble drought stress. However, at low or moderate salt concentrations (higher soil water potentials), plants adjust osmotically (accumulate internal solutes) and maintain a potential gradient for the influx of water. Under these conditions, growth may be moderated, but unlike drought stress, the plant is not water deficient.

Salinity slows growth. Beyond this one symptom, there are few morphological effects that are universally associated with nonlethal salt stress. Other symptoms that are observed in some plants and not others are a darkening of leaf color and an increase in leaf succulence. Fundamental physiological and biochemical processes of cells are unaltered at nonlethal salinities. Specific enzymes, total protein, and nucleic acids are reduced in quantity because the plants are smaller; however, specific activities and concentrations remain constant, in balance with a slower growth rate (12). Nieman and Maas (13) have suggested that the supply of metabolic energy may be the basic limitation to plant growth under saline stress. Plants use a certain proportion of their energy supply for cellular maintenance. The remainder is available for normal growth processes. Salinity increases the amount of osmotic and ionic work necessary for normal cellular maintenance; as a consequence, there is less energy for growth requirements.

Development in salt-stressed plants follows the normal sequence of ontogeny; thus, expression of the genetic code is unaltered and control mechanisms continue operating in balance to one another. Only under severe salinity stress do symptoms such as leaf injury, morphological changes, and death occur.

1.2. Environmental Interactions

Plant phenotype is a product of genotype and environment. Salinity and other environmental effects interact in several ways that may obscure inheritance studies of salt tolerance (3, 12). Nonheritable variance may reflect differences in plant microenvironments, cultural practices, plant development, or chance or unexplainable occurrences. Low humidity, high temperature, and light intensity potentiate the effects of salinity on plants (14–17). Air pollution does not affect salt-treated plants as much as untreated plants (18). Cultivation practices, soil fertility, and irrigation management can contribute significantly to an apparent increase in salt tolerance of plants growing in the field (11, 19–22). It is extremely important that researchers have a basic understanding of salinity–environment interactions so that accurate assessments of salt tolerance can be made and reported.

2. BREEDING AND GENETICS

2.1. Past Efforts in Breeding

The beginnings of salinity studies concerned with the breeding and genetics of salt tolerance is obscure. Certainly one of the earliest accounts of variability and its inheritance was made by Lyon (23) in a study conducted at the United States Salinity Laboratory. He noted that *Lycopersicon pimpinellifolium* was less sensitive than *L. esculentum* to $NaSO_4$ salinity as measured by root dry weight, average fruit weight, and mean weight per fruit. The interspecific F_1 hybrid derived from these two species exhibited the characteristics of the sensitive parent. Later, Strogonov (24) found that Lamarckian genetics did not apply to salt-stressed cotton. Strogonov reported that seeds from plants grown under saline conditions did not become more salt tolerant or acquire a greater resistance to soil salinity. Strogonov (25) also suggested that increased tolerance could be obtained by crossing selections for salt tolerance, taken from a saline field, with vigorous plants, taken from nonsaline locations, and then screening the progeny of subsequent generations in a salinized field. He reported a 7–43% increase in cotton yield and an improvement in fiber quality using this procedure. However, there is no evidence that a new variety has ever been developed from this material or that this technique has been applied to other crops. Bernstein (personal communication) found that a similar procedure resulted only in increased vigor in hybrids and not in salt tolerance. Vigor was lost in subsequent generations.

Around 1952, intraspecific differences were noted in the germination and yield responses of certain barley and wheat varieties at high salinities (26, 27). Subsequent studies with lettuce, onion, carrot, and green beans indicated only minor intraspecific differences among these horticultural species (28–31). This led Bernstein (32) to conclude, "For most crops, the relative uniformity in salt tolerance among varieties suggests little likelihood of improving salt tolerance by any (genetic) combination of the commonly available varieties." He suggested the use of "wild" germ plasm resources to introduce the necessary variability into such species.

Dewey (33) noted a wide range of responses in salt tolerance among species and collections of *Agropyron*, and outlined the following breeding plan, which was to follow initial seed procurement and plant establishment (34).

1. *First year.* (a) Collection of seed and screening for salt tolerance during germination in the laboratory (10% survival level). (b) Establishment of selected seedlings into nonsalinized field plots. (c) Salinization of field plots to differential response level. (d) Selection of salt tolerant plants and removal to greenhouse for cloning.

2. *Second year.* (a) Production of at least 8 clones from each selection. (b) Replication of plantings into field plots.

3. *Third year.*   (a) Salinization of field plots. (b) Transplanting of selections to isolated plots for controlled cross pollination.

4. *Fourth year.*   Harvest seed from the plots and begin a new selection cycle.

It is interesting to note that Dewey suggested screening at germination initially and then for yield. To date there is no evidence that these selection procedures have been attempted but they do represent a logical procedure for the selection of salt tolerant grasses.

In a study on salt-induced sterility in rice, Akbar and Yabuno (35–37), found that rice varieties subjected to salinity responded quite differently from one another during flowering. They classified three types of salt-induced sterility. F_1 crosses between the relatively salt-tolerant 'Jonah 349' and the salt-sensitive 'Magnolia' varieties were highly tolerant to salinity. The F_2 populations included some salt-tolerant individuals and salt-tolerant progeny were selected from the F_3 and F_4 populations. Inheritance of the salinity-induced delayed type panicle sterility was investigated in the F_2 of the backcross populations derived from 'Magnolia' $\times$ 'Jonah 345' crosses. Resistance to this type of sterility was found to be a dominant character and presumably was controlled by at least three pairs of genes.

2.2.   Past Efforts in Genetics and Physiology

The search for intraspecific variation in salt tolerance has progressed in parallel with studies on physiology of ion transport. Burstrom (38) found that certain sugar beet strains translocated Na^+ from the root to the leaves at a greater rate than other strains. Low Na^+ levels in the root were found to be a genetically dominant characteristic. K^+ levels in roots likewise were under genetic control (39). In 1963, the evidence for varietal differences in the selective uptake of nutrient ions was reviewed (40) and Epstein and Jefferies (41) summarized the work that linked selective ion transport to specific genes. Greenway (42a) showed that salt-sensitive barley varieties had higher rates of Cl^- uptake in the roots and higher leaf Cl^- contents than tolerant varieties. In 1964, Abel and MacKenzie found distintive differences in shoot Cl^- accumulation between two soybean (*Glycine max*) varieties grown under NaCl stress (43). Later, a single Cl^- exclusion gene (*Ncl*) which made the variety 'Lee' more tolerant was found to be dominant over its recessive allele (*ncl*) in 'Jackson' (44). Varietal differences in Cl^- uptake and tolerance had been noted previously in *Citrus* (45, 46), and Furr and Ream (47) reported on its usefulness as a screening marker for improved salt tolerance in this species. They found that F_1 hybrids showed continuous variation and concluded that salt tolerance was a quantitative character. In the same year Bernstein et al. (48) discovered that salt-tolerant grape varieties also were better Cl^- excluders. Differential transport of Cl^- between root and shoot was responsible for the difference between tolerant and sensitive varieties. Later, Downton (49) showed that grape vines on rootstocks of Cl^- excluding varieties contained less Cl^- than own-rooted vines.

In many studies it is not possible to determine whether the toxic effects observed are due to Cl^-, Na^+, or a combination of the two, since the concentrations of both anion and cation increase concurrently in root solutions and plant tissue. More recently, studies have indicated that Cl^- is the damaging ion species in woody species (citrus and grapes). However, beans and corn can better tolerate low salinity levels by excluding Na^+ from their leaves (50, 51). Na^+ is excluded by reabsorption from the xylem sap into surrounding tissues of the proximal region of the root.

Studies on species other than crops have contributed useful information concerning possible genetic and physiological mechanisms on the accumulation of Na^+ and Cl^- within plants. Differences in salt tolerance have been observed in naturally occurring populations of grasses. Hannon and Barber (52) found that salt tolerant clones of *Festuca rubra* and *Agrostis stolonifera* restricted accumulation of both Na^+ and Cl^- in the shoots. Populations of *A. stolonifera* collected from maritime habitats had lower shoot Na^+ concentrations and higher K^+/Na^+ ratios than inland populations when both were grown over a range of salinities. Likewise, tolerant accessions of *Agropyron elongatum* were more effective excluders of Na^+ and Cl^- than sensitive accessions (53, 54).

Ion exclusion is one of the most frequently reported differences between salt-sensitive and salt-tolerant cultivars. At the cellular level, ion selectivity may occur at the plasmalemmas, or tonoplasts. In higher plants, bidirectional pumps have been detected at both the plasmalemma and the tonoplast. Pitman and Saddler (55) located inward K^+ and Cl^- pumps and an outward Na^+ pump at the plasmalemma of barley roots. It has been suggested that dual mechanisms exist for ion transport across plant membranes into cells. One mechanism may operate at concentrations below 1 mM and another at 1–50 mM (56–58). Membrane transport systems can be highly selective for particular ion species or the degree and specificity of selectivity may differ with ion concentration and among tissues. There is good evidence that genetically determined controls regulate the structure and function of membrane transport systems and their orientation in the membranes of which they are a part. Kuiper (59, 60) noted differences in the extent of double bonding in the fatty acid portions of root membrane lipids among tolerant and sensitive grape varieties subjected to salt. The integrity of the membrane matrices that maintain the function and orientation of membrane ion transport systems may be impaired by salinity at some specific threshold concentration.

Although it would seem that there is a growing body of evidence linking ion exclusion to salt tolerance, there is also evidence contradicting this thesis. Certain halophytic species take up substantially high concentrations of ions (accumulation) which is believed to contribute to positive osmotic relations and high salt tolerance (37, 61–63). Rush and Epstein (64, 65) have noted that salt tolerance differences between the wild halophytic *Lycopersicon cheesmanii* and the salt-sensitive cultivated species (*L. esculentum*) are associated with the former's ability to accumulate Na^+ and selectively absorb K^+. Similar associations have been found between *L. esculentum* and the wild species of *L. peruvianum* and *Solanum pennellii* (66, 67). Halophytes tend to accumulate salt even at low salinities (67a). At higher external Na^+ concentrations additional NaCl is accumulated within the plant at the expense of a loss in K^+/Na^+ selectivity (68). The accumulation of salt reduces the requirements for increased wall extensibility, leaf thickness, and water permeability that might otherwise be required to maintain positive growth and turgor at low soil water potentials.

Other than solute accumulations just described, basic biochemical differences between halophytes and glycophytes have not been detected. For instance, most enzymes isolated from halophytes and glycophytes are inhibited in the same manner and to the same degree by NaCl in *in vitro* systems (61, 69). This may indicate that NaCl is not distributed uniformly throughout the cell but is contained in compartments separated from cellular sensitive sites. The consensus of researchers in this field is that NaCl is located primarily in the vacuoles, but concentrations of salt in vacuoles may differ extensively depending on plant species, specific tissues, and tissue maturity (37, 61, 62, 70, 71). Thus, even in halophytes, salt tolerance involves a combination of either salt exclusion or accumulation, strategies based on the plant's abilities to accumulate salt in vacuoles, and the ability to decrease the internal osmotic potentials to maintain water influx and thus maintain a positive turgor needed for cell enlargement.

In conclusion, the effects of salinity on plant growth have been studied for nearly a century, but the primary site of salinity stress on plant metabolism has not been identified and it is unlikely that a simple one gene relationship with total salt tolerance will be found. A comprehensive bibliography on the effects of salinity on morphology, nutrition, yield, and other features of agronomic interest of plants has been compiled by Francois and Maas (72) together with a pending supplement; it consists of over 3300 references. Maas and Hoffman (18) have developed an overall interpretation of much of this work as it relates to yield responses in 76 crops. Even so, biochemical and physiological aspects of salt effects on plants or the effects of salinity at the cellular and subcellular level are not included in these summaries, but have been covered elsewhere in several reviews (9, 61, 73–77).

3. MEASUREMENT OF TOLERANCE

3.1. Definitions

Salt tolerance may be defined generally as sustained growth of plants in an environment of NaCl or combinations of mixed salts. Levitt (75) has associated salt tolerance with the absence of negative effects on growth in plants that accumulate salt within their tissues. In so doing, he has distinguished it from salt avoidance mechanisms and has used the term salt resistance to refer to a combination of tolerance and avoidance strategies. Examples of salt-avoidance mechanisms include delayed germination or maturity until favorable conditions prevail; the exclusion of salt at the rootzone or preferential root growth into nonsaline areas; compartmentation of salt into and secretion from specialized organelles such as salt glands and salt hairs; or storage in older leaves. In practice, the terms salt tolerance or salt resistance have been used interchangeably to define true cytoplasmic resistance to salinity or, in conjunction with salt avoidance, to describe all mechanisms that may give the plant a selective advantage during saline stress. Literally, tolerance means "to endure, sustain, or put up with"; whereas resistance implies the exertion of opposite effort. Plants generally exert considerable effort to sustain themselves in the presence of salt. Thus, reasonable arguments can be made for either definition. Levitt's classification leaves much to be desired. There is a growing body of evidence to indicate that all plants are essentially salt avoiders. None excludes salt in an absolute sense.

3.2. Criteria

Salt tolerance can be measured by a number of criteria. Survival at high salt concentrations has recently been the fundamental selection criterion for barley, wheat, and tomato (64, 65, 78). However, if we consider the mechanisms that plants use for survival, they may not be the same ones that are necessary for the maintenance of high growth rate at moderate salinities. For example, many halophytes withstand high salinities by such strategies as temporary dormancy, increased succulence, or shortening the growing season (75). Dormancy is not compatible with high yields, and increasing succulence contributes nothing to dry weight yield. A shortened growing season, could be beneficial in some agronomic situations. Moreover, dormancy may be important in contributing to survival during temporary periods of high osmotic stress due to low soil water potentials.

Another method of measuring salt tolerance is the determination of growth or yield response under saline conditions. This criterion, as noted previously (3, 76) may be more a reflection of vigor than actual tolerance. For example, the absolute yields of three muskmelon (*Cucumis melo*) cultivars may differ substantially at different salinities (Fig. 13.1). 'Top Mark' produced higher yields than 'PMR 45' or 'Hale's Best' under nonsaline and low salinity conditions, but yields from 'PMR 45' were better than other cultivars at high salinities. At moderate salinities the cultivar yields are not significantly different (79). From the standpoint of the breeder, the 'Top Mark' cultivar has high yield and vigor but may not be especially salt tolerant with respect to specific heritable physiological characters; however, at moderate salinities a grower may benefit by using 'Top Mark' rather than 'PMR 45'.

Salt tolerance can be measured as a relative reduction in yield as a function of increasing soil salinity (Fig. 13.2) (76). Soil salinity can be monitored by several

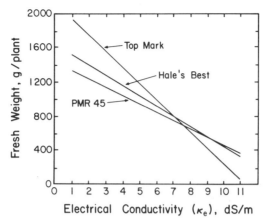

Figure 13.1   Fresh weight yield of three muskmelon (*Cucumis melo*) cultivars in response to salinity. κ_c is the electrical conductivity of a saturation extract.

methods that are reviewed, explained and defined elsewhere (80). In Figure 13.2 salinity is measured as electrical conductivity of the soil saturation extract (81). Each curve in Figure 13.2, for each species, may be described by its threshold and slope value. Slope values are a reflection of relative yield reductions per unit salinity increase. They are useful in predicting yield over a range of salinities. Threshold values (which are intercepts of the two linear portions of the response curve) give an estimate of the amount of salinity that a plant can withstand with no significant reduction in yield. The threshold value may be one of the more important indicators of true salt tolerance, but threshold values are difficult to measure accurately.

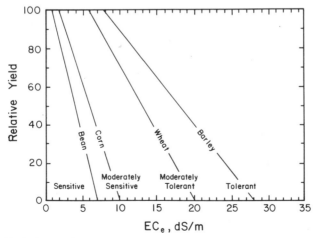

Figure 13.2.   Typical crop yield response curve for plant species differing in salt tolerance (76). EC_e is the electrical conductivity of a saturation extract.

Relative salt tolerance is the fraction of growth (yield) under saline conditions as compared with growth under nonsaline conditions. Judged by such a criterion, a slow growing plant may rate well if salinity does not substantially reduce its growth further. Absolute salt tolerance is directly related to the maximum growth or yield potential of a given plant under salinity regardless of its growth achievement under nonsaline conditions. Under salinity stress, a vigorously growing variety that is drastically affected by salinity would have low relative salt tolerance by this criterion. However, if it still yields more than a slow growing variety whose growth is not as severely affected by salinity, then it would be the preferred choice of the grower. Although absolute yield under high salinity is of great importance to the grower, relative yield must be considered by the breeder and geneticist. If a variety is relatively tolerant but grows slowly or has poor quality characteristics, its high relative tolerance should not be ignored. That characteristic may be transferable to high-yielding cultivars. Relative tolerance may also contribute to the overall fitness of the species to produce well in a range of saline and nonsaline environments, a character referred to as environmental plasticity and measured as mean productivity. Strogonov (25) advocated crossing tolerant lines with those showing high vigor to obtain progeny of high mean productivity. More recently, it has been indicated that selecting for high stress tolerance may be incompatible with high yield under nonstress conditions (82). Certainly, more work needs to be done on the relationships among salt tolerance, genetic plasticity, and various other markers being used as selection criteria.

3.3. Partitioning

The breeder is ultimately concerned with the effects of salinity on yield, and yield is usually partitioned into various components. For forages and leafy vegetables yield is closely correlated with vegetative growth; however, with other crops, salinity sometimes affects marketable yield differently than vegetative growth. For example, fruit (grain, bolls, etc.) yields are generally not as sensitive to salinity as vegetative growth, probably because the plant effectively repartitions its available resources for survival as a species. Often, moderate salinity slightly increases yields or maturation. As another example, salinity stimulates root growth and thus increases root to shoot ratios perhaps in response to the osmotic effect that salt has in reducing water potential. This affects yields of root, tuber, or bulb crops (14, 30, 31, 83, 84). Finally, salinity may alter quality and thus affect marketable yield. In some instances, salt increases soluble solids and improves the quality of fruit crops such as tomato or melon (64, 65, 79, 85), whereas in other cases, salinity reduces fruit size or shelf life or inhibits quality characteristics like head formation in lettuce (86).

Although salt tolerance may be a complex polygenic trait, it could be more simply studied if plant responses were partitioned in some manner. Thus, the number of genes that contribute to each specific response, and the number of interacting factors for each response, would be fewer. For instance, salt tolerance responses are frequently studied at particular growth stages that vary according to the whims of the investigator or with respect to a certain physiological criterion. But, although partitioning of salt or growth responses can be a useful way to simplify the study of salt effects, its usefulness can be over simplified. Tolerance is known to differ with ontogeny (12, 26, 73). Thus, the tolerance of a particular species may be different at each of its physiological stages of development, such as germination, seedling growth, vegetative growth, flowering, and fruiting. Barley, sugar beet, and cotton, for instance, are among the most salt-tolerant agricultural crops but each is relatively sensitive during either germination or early seedling growth. Rice may be sensitive during seedling and flowering stages. During specific growth stages the influences of salinity may be related to a few specific gene loci; consequently, selection during these particular stages may be easier. However, the level of tolerance at one growth stage cannot be assumed to be the same level of tolerance at any other.

4.   SCREENING AND SELECTION

General conclusions from the above studies for the plant breeder are that the physiological effects of salinity are not fully known, the measurement of salt tolerance is difficult, and the concept is poorly defined. Almost nothing is known about the genes that affect salt tolerance. It is definitely known that salinity imposes an environmental restraint on plant growth. Since growth and yield reductions are quantitative parameters, they can be measured using the relatively new disciplines of biometric and quantitative genetics. However, growth reduction is a symptom of the influence of other factors besides salinity. Also, since yield and growth reductions are the only measurements of salinity stress, genetic relationships will be complex. The longer the chain of events between the initial readings of information in the genes and its expression in the phenotype, the greater the complexities that may arise from a single alteration.

These difficulties raise problems in measuring the development of salt tolerance in plants. However, such difficulties are not insurmountable. Disease and pest resistant plants have been developed without complete knowledge of vectors, toxins, or alleopathic substances; varieties tolerant to heat, cold, drought, and numerable other environmental and biological stresses have been bred without full knowledge of the basic physiological links between the genes and site of injury. Breeding for tolerance to salinity has proceeded and, for a time, must continue to proceed in this manner. The efforts of breeders may eventually provide plant materials that will aid the physiologists in unraveling the basic mechanisms of salt damage.

4.1. Field and Plot Work

Selection of salt-tolerant plants from saline fields or plots would seem a logical step for most plant breeders; however, this procedure has not produced good results in the past. The most common problem is that soil salinities vary substantially with time, location, and depth. Selection techniques in fields would be improved if proper precautions were taken to uniformly presalinize the soil profile and to maintain salinities by applying saline water at uniform rates. This often requires big investments for control and monitoring devices for irrigation and salinity. At high salt concentrations certain ions may exert specific toxic effects on plants and the physical nature of soils may be changed by salt–soil chemical interactions. Saline solutions for screening and selection studies should be composed of a realistic mixture of salts, not just NaCl. Salt imbalance in the nutrient solution is the most frequent deficiency in screening studies and cultivar assessments. In soils, Na^+ and Ca^{2+} are usually present in 2:1 to 5:1 molar ratios, whereas in seawater, ratios exceed 40:1. Exceedingly high Na:Ca ratios in clay-containing soils cause loss of permeability and soil structure. Sodium absorption ratios (SAR) must be considered in soils that have substantial clay contents. SAR is calculated from $SAR = Na/\sqrt{(Ca+Mg)/2}$, where Na, Ca, and Mg are expressed in units of ionic equivalents (81).

The effects of salts on plants in soils with high SAR (sodic soils) and permeability problems are substantially different from those on plants in nonsodic saline soils. Each condition should be dealt with separately by the breeder. The same is true for soils that have a particular mineral toxicity or deficiency. Other problems with high soil Na:Ca ratios include partial loss of plant membrane permeability (deficient Ca^{2+}) or possible Na^+ induced Ca^{2+} deficiency symptoms in plants with low Na:Ca selectivity (28). The importance of maintaining proper Na:Ca ratios has been reviewed by Greenway and Munns (74). The 2:1 molar ratio used extensively in United States Salinity Laboratory studies can be used in both field and pot (soil) culture studies and does not result in a large change in the SAR.

In soils, tolerant plants may actually have higher root zone salinities than surrounding plants if their additional growth and water use concentrates the salts through exclusion processes. Thus, soils may actually buffer the growth responses of salt-tolerant plants, whereas plants that happen to be located in slightly less saline areas may appear more tolerant, but in reality are not. Screening for salt tolerance in fields may be improved if selections are based on plant response as associated with soil salinity and water content measurements. A device such as the four-probe instrument developed by Rhoades would be useful for rapid salinity measurements (87).

Although irrigated agriculture has aggravated salinity problems, it is quite possible that improved management can aid breeding efforts to develop salt-tolerant cultivars. High-frequency irrigation, minimum leaching technology, and the recycling of agriculture drainage waters have been advanced in a unified concept as an alternative or adjunct to desalinization strategies (88–90). These proposals advocate more refined control of water application and the use of saline waters, two principal requirements necessary for large-scale breeding approaches to develop more salt-tolerant plants. Drip, bubbler, and sprinkler systems are easily modified and can be applied to techniques for obtaining breeding objectives. Water application can be closely controlled and monitored with these systems. High-frequency irrigation decreases the variability in soil moisture content and soil water salinity. Thus, soil salinity would be more directly a function of applied water. Additionally, agricultural drainage waters can be used as a source of saline water for the plant breeder. The combination of these technologies would allow breeders to uniformly distribute saline water to large field plots. However, as stated previously, the cost of these technologies is high.

4.2. Greenhouse and Laboratory Methods

Problems with environmental interaction can be alleviated by screening under controlled conditions in the greenhouse or laboratory. Such methods are popular because screening can be accomplished in smaller spaces and shorter times; however, most methods are distinctly different from soil systems and may not be useful in selecting for traits associated with root–soil interactions. Generally, a wide variety of media and criteria can be used for such screening to include simple devices like germination dishes or more advanced techniques such as tissue culture and recombinant DNA.

Fryxell (91) advocated selection for germination under high osmotic stress as a method for screening for salt tolerance. However, it has been shown repeatedly that no relationship exists between tolerance at germination and later growth stages (35, 43, 92). Species that have high yield tolerances may be relatively tolerant (barley) or sensitive (sugar beet) at germination (Table 13.1). Conversely, sensitive species may be either sensitive (onion) or tolerant (rice) during germination and emergence. Seedling tolerance will also differ.

Initial plant establishment, seed collection, and subsequent screening at germination has been recommended as a method to begin genetic improvement of salt tolerance in wheatgrass (34) (see Section 2.1). Initial selection during germination could eliminate valuable gene resources for tolerance at later growth stages. A safer method might be to select for salt tolerance during germination and emergence independently from seedling or later growth stages. It is likely that different genes will become involved in various mechanisms of salt tolerance as the plant develops. Some investigators, however, have reported rather remarkable success in the improvement of salt tolerance using constant salt stresses from seed to seed (2, 78). To date, no salt-tolerant varieties or breeding lines have been released from this material.

An important aspect of Dewey's (34) work was the rejuvenation of his seed

Table 13.1. Relative Salt Tolerance of Various Crops at Emergence and during Growth to Maturity[a]

Crop		Electrical Conductivity of Soil Saturation Extract (dS/m)		
Common Name	Botonical Name	50% Yield	50% Emergence	Reference
Barley	*Hordeum vulgare*	18	16–24	92–95
Cotton	*Gossypium hirsutum*	17	15	95,96
Sugar beet	*Beta vulgaris*	15	6–12	92, 93
Sorghum	*Sorghum bicolor*	15	13	93, 97
Safflower	*Carthamus tinctorius*	14	12	98
Wheat	*Triticum aestivum*	13	14–16	93, 94, 99
Beet, red	*Beta vulgaris*	9.6	13.8	94
Cowpea	*Vigna unguiculata*	9.1	16	100
Alfalfa	*Medicago sativa*	8.9	8–13	92–94
Tomato	*Lycopersicon Lycopersicum*	7.6	7.6	94
Cabbage	*Brassica oleracea,* Capitata group	7.0	13	96
Corn	*Zea mays*	5.9	21–24	92, 93, 95, 99
Lettuce	*Lactuca sativa*	5.2	11	94
Onion	*Allium cepa*	4.3	5.6–7.5	93, 101
Rice	*Oryza sativa*	3.6	18	95, 102
Bean	*Phaseolus vulgaris*	3.6	8.0	92

[a]Emergence of saline treatment determined when nonsaline control attained maximum germination. Values for 50% yield from ref. 18.

supply prior to screening. Poor seed vigor is known to reduce germination rate, viability, and subsequent seedling vigor. The additional effect of stress would intensify these effects and obscure relative salt damage among varieties or seed lots. It is important that seed used for screening and selection be of good viability and vigor. Other problems associated with germination and emergence tests include the choice of a screening medium. Germination plates have been criticized because moisture may condense on the undersurface of the top of dishes and drip back onto the filter paper, creating nonsaline micro-environments; also, this technique does not distinguish between seeds that may have only successful radicle protrusion and those that can emerge through a soil covering. The emergence of seedlings from sand cultures has been reported as a useful technique for determining emergence potential under salt stress (103). More recent advancements include the use of salinized agar gels or physical measurements of the force exerted by radicles under salt stress (103a).

Familiarity with agricultural practices for specific crops will enable the breeder to evaluate the importance of selecting for tolerance at particular growth stages. For instance, in most of the world, rice is transplanted from nurseries into the field. The practice of establishing rice seedlings in nurseries is for reasons other than preventing salinity damage; thus, it may not be of benefit to growers for breeders to screen for salt tolerance at germination when management practices can overcome the problem so easily. In the United States, however, rice is usually seeded from the air. Here, breeding for salt tolerance at germination may be beneficial. Screening techniques for Al^{3+} toxicities have used root growth as an index for tolerance (104, 105). Root growth is not a reliable indicator of salt tolerance because in some species there is a stimulation of growth. Usually, relative root growth at low salinities is far less sensitive to salinity than vegetative growth (3). Other screening techniques have successfully used resistance to leaf chlorosis as a selection criterion for salt tolerance (54, 106). The absence of chlorosis is generally indicative of varieties that are better salt excluders. The use of emergence and vegetative growth of seedlings has been reported as a reliable indicator of absolute tolerance in lettuce and has been used as a criterion to rank over 80 cultivars for tolerance; however, no distinctions were made for cultivars that have good relative salt tolerance (107).

The development of screening techniques for salt tolerance that are based on physiological or morphological characteristics has recently been reviewed and will not be examined here (4). In general, these techniques have not been proved to be reliable. Tissue culture techniques have also yielded only limited success (108), and have not produced the rapid progress that was originally anticipated (109–111).

4.3. Biochemical and Genetic Techniques

It would be convenient if it were possible to choose which morphological, developmental, and physiological characteristics are markers for salt-tolerant plants, and to engineer a plant specifically for the saline environment. Unfortunately, salt glands and hairs are the only morphological characters that have been unequivocally identified with increased salt tolerance. The feasibility of using biochemical markers, such as proline, glycinebetaine, or sugar accumulation, or Na/K ratios, as selection criteria has been reviewed (4) and does not show any current promise. This is probably because salt tolerance is the combined characteristic of many plant features, both morphological and physiological. Proper evaluation of such characters would have to be made against a common genetic background. Isolines, lines that are nearly identical in all features except for that character in question, would have to be developed. This would be a valuable tool for the study of mechanisms of salt tolerance. Another genetic method that can be used to improve measurement of fitness to salinity is progeny testing. This method, like isoline development, is laborious and time consuming. Progeny tests can be used to circumvent the need for refined salinity and environmental control.

If the effects of salinity cannot be distinguished by a specific physiological character, we must measure salinity effects upon growth. Growth is a continuous variable. Continuous variations in growth are due to genes which we cannot expect to be easily recognizable in segregating populations. We can, in some instances, overcome the intrinsic difficulties by separating growth into components (somatic analysis) and partitioning the available energy resources more efficiently. Obviously, there are still imaginative ways in which genetic techniques can be used to answer some of the fundamental questions concerning salinity effects on plants.

5.   SALT-TOLERANT IDEOTYPES

In 1968, C. M. Donald stated that eventually most plant breeding will be based on ideotypes (112). Ideotypes are conceptualized models that are perceived as ideal plants for a defined environment. Our theoretical capability to design such a model is based on the existence of sufficient knowledge, adequate genetic diversity, and suitable techniques of breeding or genetic engineering. Although we do not yet have much understanding of the physiology of salt tolerance, nature has provided models for us to follow. Current genetic and breeding approaches to increasing salt tolerance in crops include variety selection and recombination, the use of wide crosses between wild salt-tolerant species and their crop relatives, and the development of agronomic potential in undomesticated salt-tolerant species. In each case, arbitrary decisions will be necessary since we still do not have enough information to choose which characters to incorporate into the new salt-tolerant variety and which to discard. There are still many unanswered questions that must be solved before it becomes possible to define the several ideotypes needed for the different agronomic situations and environments.

5.1.   Accumulation versus Exclusion

Halophytes accumulate certain inorganic ions in high concentrations and use them to maintain an osmotic potential within their tissues lower than the external potential (61, 113). In so doing, they can tolerate salinities higher than sea water. The adaptation of ion accumulation in halophytes is undoubtedly linked to an efficient partitioning of solutes between vacuole and cytoplasm, successful osmotic adjustment, and the plant's ability to extract water from a highly saline medium. Reports of high increases in yield in halophytes grown at high salinities should be examined closely. Maximum fresh weights of *Atriplex spongiosa* and *Suaeda monoica*, for instance, have been reported in solution cultures containing 100 and 150 mol m^{-3} of NaCl, respectively. A closer examination indicates that fresh weight increases in *A. spongiosa* are due primarily to increases in succulence, whereas dry weight does not significantly increase (63). Furthermore, in spite of an almost 300% increase in fresh weight of *S. monoica* at 500 mol m^{-3} NaCl, it was found that dry weights increased substantially (~30%) only up to 150 mol m^{-3} NaCl; and at least one-half of the dry weight increase could be attributed directly to the accumulation of Na$^+$ salts. Glycophytes have evolved in a divergent manner and are adapted to areas with frequent rains and salt-free environments. Under such situations, salt concentration and water availability vary inversely and certain salt avoidance mechanisms have important advantages (61). Ion exclusion protects many glycophytes from the toxic effects of salinity, but it also contributes to osmotic imbalances. Osmotic adjustment in glycophytes may include the energy-dependent synthesis of organic molecules. Proline, sugars, and glycinebetaine are among the leading organic osmoregulators. Glycophytes that exclude salt must rely more extensively on the synthesis of osmotica to facilitate the uptake of water into the plant. But at high salt concentration, ion exclusion somehow fails. Above a critical threshold, salts rush into sensitive tissues and death soon results. The tolerance of many salt excluders has been related to differences in this threshold level.

Osmotic adjustment in plants ensures the availability of water. Water potential in soils decreases as a result of evaporation, water use by plants, and salt accumulation. Decrease in plant water potential must immediately be offset by a decrease in the osmotic potential through an increase in solute content for turgor potential to be maintained. Salt uptake on a limited basis is beneficial for rapid osmotic adjustment. On the other hand, high ionic concentrations have deleterious effects on membranes and metabolic reactions. Thus, plants that survive the highest salt concentrations are those that use the salts beneficially. The major selective pressure in nature is survival rather than growth. Plants have evolved finely tuned mechanisms to regulate salt influx, efflux, and its separation between tissues and organelles. Photosynthetic and meristematic regions seem to be the most protected portions of the plant. Meristematic tissues may have some natural protection due to small cell size and high solute concentrations (higher cytoplasm/vacuole ratio). They may also be more protected from water loss by enveloping tissues and poorly developed xylem elements.

Growth potential in salt excluders and accumulators is ultimately dependent on a release from salt stress followed by recovery to "normal" growth or the ability of a plant to continue to grow at a specific salt concentration. There is little good evidence to indicate the conditions under which salt accumulators or excluders are more efficient with respect to these strategies.

5.2.   The Energy-Efficient Plant

As indicated in Section 1.1, it has been hypothesized that the production and use of energy by the salt-affected plant may be closely correlated with salt tolerance. In saline environments the cost of cellular maintenance in terms of energy utilization is higher to the plant than it is in nonsaline environments. Respiration and protein synthesis rates do not decrease under salt stress and the former sometimes increases (16, 114). The slight stimulation of growth at low salinities in halophytes may be an expression of this increased respiration rate. In other words, the allocation of metabolic energy to ion transport and osmotic adjustment reduces the availability of energy for growth processes.

In glycophytes, the exclusion of Na^+ and/or Cl^- from shoots may possibly involve ion compartmentation within root xylem parenchyma into transfer cells, the exclusion of ions from the root, and the functioning of all the associated energy-dependent ion transporting processes (115). Another sink for energy is in the synthesis of organic solutes such as proline, glycinebetaine, or sugars necessary for osmotic adjustment.

In some instances, synthesized osmotic regulators may have a second role as carbon, nitrogen, and energy reserves should salinity stresses be alleviated. Proline and sugars are subject to enzymatic breakdown and may be rapidly converted to useful metabolic intermediates for growth and energy such as other amino acids or tricarboxylic acid intermediates. On the other hand, glycinebetaine, which has four N groups per molecule, is not readily catabolized upon cessation of salinity stress (116). In either case, organic compounds, although not being directly associated with the normal growth processes, may accumulate in stressed plants in such large amounts as to significantly contribute to dry weight increase.

True halophytes are known for their abilities to accumulate ions as osmotica (74). The usefulness or advantage of this strategy has not been proved conclusively relative to maintenance of high growth rates. Since ions are probably selectively sequestered within tonoplasts, the metabolic costs of maintaining these ion gradients may be substantial. But, estimates of energy necessary for ion pumping in algae have been made and amount to only about one-eighth of a percent of the energy being produced by photosynthesis and respiration (117, 118). As pointed out by Dainty (119), the halophyte must maintain a higher ion gradient between the vacuole and cytoplasm. The costs of this additional pumping is unknown. Additionally, a change in external salinity environment may require additional energy (62).

It may be hypothesized that the strategy of ion pumping may require more energy than do exclusion processes, but offers the selective advantage of faster osmotic adjustment and the necessity of having membranes that are physically capable of withstanding high salt concentrations. Conversely, the exclusion habit of glycophytes may give them a better, competitive ability under low salinity (see Section 2.2) but at the same time predispose them to lower survivability at high salt concentrations. Much of the available literature supports such a theory.

High yielding genotypes generally show a sharper relative decrease in yield than low yielding genotypes (Fig. 13.1). This may be a basic phenomenon of stress (82) (see Section 3.3), but from a physiological viewpoint it may be because the nonstressed high yielding genotypes are already consuming almost all their energy output. Slow growing genotypes may have a greater energy reserve or potential. It may be further speculated that this phenomenon may be linked to, and thus measured by, the threshold value of the salt tolerance curve (see Section 3.2). This idea needs development and testing.

Finally, it should be reiterated that plants may partition their resources differently in response to salt stress. Some of the effects of salt stress may be overcome by better energy efficiency within the salt-stressed plant.

5.3. Agronomic Fitness

Because salt tolerance is affected by environment and since many agromanagement techniques can be used to partially alleviate some of the salt stress, salt-tolerant ideotypes can be tailored to different agricultural management needs. Fitness to salinity would include different combinations of avoidance, exclusion, and other tolerance mechanisms depending on the specific agronomic and cultural practices that would be followed. The possibilities of selecting for increased tolerance during a particular growth stage has been discussed previously. The use of drip and bubbler systems have been advocated for more efficient water use in semiarid and arid regions (89). Special salt-tolerant plants may be selected for use under such conditions. In these ideotypes, characteristics for the salinity-related, drought tolerance may be minimized while other mechanisms, such as Na^+ occlusion at the root, may be optimized. The need to develop only shallow root systems would allow more growth potential to be diverted to yield components.

Some expensive cultural processes may be eliminated by the development of more tolerant ideotypes. Lettuce is often sprinkled-irrigated for 48–72 hr after planting to prevent drying and salinity buildup. Increased tolerance at germination would offer great potential for reducing the costs of plant establishment in lettuce production. Small plant size resulting from high salinities may be offset by increasing population densities or producing cultivars capable of partitioning a larger percentage of their reduced growth into marketable products. Since salinity stimulates higher sugar production in some vegetable crops, yield reductions may be tolerable if improved market quality is the result. Additional research should be directed toward the development of undomesticated plant species for crop use. These species may be productive forage or biomass crops capable of growing at high salinities. The development and use of such species on marginal lands and saline waters will free more favorable land for other agronomical uses.

6. CONCLUSIONS

Generally terrestial plants grow better without salt and plant growth in saline environment requires the expenditure of energy. It is probably this energy expenditure which ultimately decreases growth. In the last two decades new knowledge concerning ion absorption and transport and related areas of plant physiology contributed to a more thorough understanding of how plant cells and tissues react to salt stresses. It is currently inconceivable that plant yields in extremely saline environments can rival those in nonsaline areas. However, salt-tolerant plants may soon be developed that can withstand moderate increases in salt without significant yield reductions. Other varieties may be developed that will produce some yield at salinities which they may not presently survive. These objectives are within our grasp.

Out immediate goals should be to achieve a better understanding of the metabolic energy costs of the different mechanisms that we now know contribute to salt avoidance or tolerance, and then to engineer appropriate characteristics into new varieties that are designed, not only to survive and avoid salt under saline stress, but also to produce maximum yields under those conditions.

REFERENCES

1. F. I. Massoud. FAO/UNEP Expert Consultation on Soil Degradation, FAO, Rome (1974).

2. E. Epstein. Genetic potentials for solving problems of soil mineral stress: Adaptation of crops to salinity. In M. J. Wright, ed., *Plant Adaptation to Mineral Stress in Problem Soils.* Cornell University Press, Ithaca, New York, 1976, p. 73.

3. R. H. Nieman and M. C. Shannon. Screening plants for salinity tolerance. In M. J. Wright, ed., *Plant Adaptation to Mineral Stress in Problem Soils.* Cornell University Press, Ithaca, New York, 1976, p. 359.

4. M. C. Shannon. *Hort. Sci.* **14,** 587 (1979).

5. M. C. Shannon. Genetics of salt tolerance: New challenges. In A. San Pietro, ed., *Biosaline Research. A Look to the Future (Environmental Science Research*, Vol. 23). Plenum, New York, 1982, p. 271.

6. W. E. Cordukes. *Can. J. Plant Sci.* **61,** 761 (1981).

7. L. Bernstein. *Am. J. Bot.* **48,** 909 (1961).

8. L. Bernstein. *Am. J. Bot.* **50,** 360 (1963).

9. L. Bernstein. *Annu. Rev. Phytopathol.* **13,** 295 (1975).

10. H. G. Gauch and C. H. Wadleigh. *Soil Sci.* **59,** 139 (1945).

11. R. H. Nieman and R. A. Clark. *Plant Physiol.* **57,** 157 (1976).

12. E. V. Maas and R. H. Nieman. Physiology of plant tolerance to salinity. In G. A. Jung, ed., *Crop Tolerance to Suboptimal Land Conditions (ASA Special Publication Number 32).* ASA, CSSA, SSSA, Madison, Wisconsin, 1978, p. 277.

13. R. H. Nieman and E. V. Maas. *6th Intl. Biophys. Congr. Abstr.* p. 121 (1978).

14. G. J. Hoffman and S. L. Rawlins. *Agron J.* **63,** 877 (1971).

15. A. Meiri, G. J. Hoffman, M. C. Shannon, and J. A. Poss. *J. Am. Soc. Hort. Sci.* **107,** 1168 (1982).

16. R. H. Nieman and L. L. Poulsen. *Bot. Gaz.* **128,** 69 (1967).

17. R. H. Nieman and L. L. Poulsen. *Bot. Gaz.* **132,** 14 (1971).

18. E. V. Maas, G. J. Hoffman, S. L. Rawlins, and G. Ogata. *J. Environ. Qual.* **2,** 400 (1973).

19. L. Bernstein. *USDA Inform. Bull.* **205** (1959).

20. L. Bernstein and L. E. Francois. *Soil Sci.* **115,** 73 (1973).

21. L. Bernstein and L. E. Francois. *Agron. J.* **67,** 185 (1975).

22. L. Bernstein and L. E. Francois, and R. A. Clark. *Agron. J.* **66,** 412 (1974).

23. C. B. Lyon. *Bot. Gaz.* **103,** 107 (1941).

24. B. P. Strogonov. *C. R. Acad. Sci. URSS.* **54,** 645 (1946).

25. B. P. Strogonov. *Z. Obshch. Biol.* **6,** 460 (1954).

26. A. D. Ayers. *Agron. J.* **45,** 68 (1953).

27. A. D. Ayers, J. W. Brown, and C. H. Wadleigh, *Agron. J.* **44,** 307 (1952).

28. A. D. Ayers, C. H. Wadleigh, and L. Bernstein. *Proc. Am. Soc. Hort. Sci.* **57,** 237 (1951).

29. L. Bernstein and A. D. Ayers. *Am. Soc. Hort. Sci. Proc.* **57,** 243 (1951).

30. L. Bernstein and A. D. Ayers. *Am. Soc. Hort. Sci. Proc.* **61,** 360 (1953).

31. L. Bernstein, and A. D. Ayers. *Am. Soc. Hort. Sci. Proc.* **62,** 367 (1953).

32. L. Bernstein. *Proc. Am. Soc. Civil Eng.* **87,** 1 (1961).

33. D. R. Dewey. *Agron. J.* **52,** 631 (1960).

34. D. R. Dewey. *Crop Sci.* **2,** 403 (1962).

35. M. Akbar and T. Yabuno. *Jpn. J. Breed.* **24,** 176 (1974).

36. M. Akbar and T. Yabuno. *Jpn. J. Breed.* **25,** 215 (1975).

37. M. Akbar and T. Yabuno. *Jpn. J. Breed.* **27,** 237 (1977).

38. H. Burstrom. *Ann. Agric. Col. Sweden* **5,** 89 (1938).

39. J. W. Dudley and L. Powers. *J. Am. Soc. Sugar Beet Technol.* **11,** 97 (1960).

40. P. B. Vose. *Herbage Abstr.* **33,** 1 (1963).

41. E. Epstein and R. L. Jefferies. *Annu. Rev. Plant Physiol.* **15,** 169 (1964).

42. H. Greenway. *Aust. J. Biol. Sci.* **18,** 763 (1965).

42a H. Greenway. *Aust. J. Biol. Sci.* **15,** 16 (1962).

43. G. H. Abel and A. J. McKenzie. *Crop Sci.* **4,** 157 (1964).

44. G. H. Abel. *Crop Sci.* **9,** 697 (1969).

45. W. C. Cooper and B. S. Gorton. *Proc. Am. Soc. Hort. Sci.* **59,** 143 (1952).

46. W. C. Cooper, B. S. Gorton, and C. Edwards. *Proc. Rio Grande Valley Hort. Inst. Proc.* **5,** 46 (1951).

47. J. R. Furr and C. L. Ream. *Proc. 1st Int. Citrus Symp.* **1,** 373 (1969).

48. L. Bernstein, C. F. Ehlig, and R. A. Clark. *J. Am. Soc. Hort. Sci.* **94,** 584 (1969).

49. W. S. Downton. *Aust. J. Plant Physiol.* **4,** 183 (1977).

50. B. Jacoby and A. Ratner. Mechanism of sodium exclusion in bean and corn plants—A reevaluation. In X. Wehrmann, ed., *J. Plant Analysis and Fertilizer Problems.* German Society of Plant Nutrition, Hannover, 1974, p. 175.

51. C. L. Richter and H. Marschner. *Z. Pflanzenphysiol.* **70,** 211 (1973).

52. N. J. Hannon and H. N. Barber. *Search* **3,** 259 (1972).

53. P. E. McGuire and J. Dvorak. *Crop Sci.* **21,** 702 (1981).

54. M. C. Shannon. *Agron J.* **70,** 719 (1978).

55. M. G. Pitman and H. D. W. Saddler. *Proc. Natl. Acad. Sci. USA* **57,** 44 (1967).

56. E. Epstein and C. E. Hagen. *Plant Physiol.* **27,** 457 (1952).

57. E. Epstein, D. W. Rains, and O. E. Elzam. *Proc. Natl. Acad. Sci. USA* **49,** 684 (1963).

58. D. W. Rains and E. Epstein. *Plant Physiol.* **42,** 314 (1967).

59. P. J. C. Kuiper. *Plant Physiol.* **43,** 1367 (1968).

60. P. J. C. Kuiper. *Plant Physiol.* **43,** 1372 (1968).

61. T. J. Flowers, P. F. Troke, and A. R. Yeo. *Annu. Rev. Plant Physiol.* **28,** 89 (1977).

62. R. L. Jefferies. *BioScience* **31,** 42 (1981).

63. R. Storey and R. G. Wyn Jones. *Plant Physiol.* **63,** 156 (1979).

64. D. W. Rush and E. Epstein. *Plant Physiol.* **57,** 162 (1976).

65. D. W. Rush and E. Epstein. *J. Am. Soc. Hort. Sci.* **106,** 699 (1981).

66. K. Dehan and M. Tal. *Irrig. Sci.* **1,** 71 (1978).

67. M. Tal. *Aust. J. Agric. Res.* **22,** 631 (1971).

67a. R. Albert. *Oecologia (Berlin)* **21,** 57 (1975).

68. M. G. Pitman. *Aust. J. Biol. Sci.* **19,** 257 (1966).

69. H. Greenway and C. B. Osmond. *Plant Physiol.* **49,** 256 (1972).

70. J. L. Hall and T. J. Flowers. *J. Exp. Bot.* **27,** 658 (1976).

71. D. J. Weber, H. P. Rasmussen, and W. M. Hess. *Can. J. Bot.* **55,** 1516 (1977).

72. L. E. Francois and E. V. Maas. Plant Responses to Salinity: Indexed Bibliography. USDA Agric. Rev. and Manual Ser., 1978.

73. L. Bernstein and H. E. Hayward. *Annu. Rev. Plant Physiol.* **9,** 25 (1958).

74. H. Greenway and R. Munns. *Annu. Rev. Plant Physiol.* **31,** 149 (1980).

75. J. Levitt. In T. T. Kuzlowsky, ed., *Responses of Plants to Environmental Stresses.* Academic Press, New York, 1972.

76. E. V. Maas and G. J. Hoffman. *J. Irrig. Drainage Div. ASCE* **102,** 115 (1977).

77. B. P. Strogonov. Physiological basis of salt tolerance of plants. *Jerusalem Isr. Prog. for Sci. Trnasl.,* 1962.

78. E. Epstein and J. D. Noryln. *Science* **197,** 249 (1977).

79. M. C. Shannon and L. E. Francois. *J. Am. Soc. Hort. Sci.* **103,** 127 (1978).

80. J. D. Rhoades. *Proc. Am. Soc. Chem. Eng.,* Irrigation and Drainage Division, Spec. Conf., Reno, Nev., July 1977, p. 85.

81. United States Salinity Laboratory Staff, Diagnosis and improvement of salinity and alkali soils. *U.S. Dept. Agric. Handb.* **60** (1954).

82. A. A. Rosielle and J. Hamblin. *Crop Sci.* **21,** 943 (1981).

83. L. Bernstein, A. D. Ayers, and C. H. Wadleigh. *Am. Soc. Hort. Sci. Proc.* **57,** 231 (1951).

84. F. M. Eaton. *Plant Physiol.* **16,** 545 (1941).

85. Y. Mizrahi. *Plant Physiol.* **69,** 966 (1982).

86. M. C. Shannon. *J. Am. Soc. Hort. Sci.* **105,** 944 (1980).

87. J. D. Rhoades and J. van Schilfgaarde. *Soil Sci. Soc. Am. J.* **40,** 647 (1976).

88. G. J. Hoffman, C. Dirksen, R. D. Ingvalson, E. V. Maas, J. D. Oster, S. L. Rawlins, J. D. Rhoades, and J. van Schilfgaarde. *Agric. Water Manage.* **1**, 233 (1978).

89. S. L. Rawlins. *Ann. N. Y. Acad. Sci.* **300**, 121 (1977).

90. J. D. Rhoades. Monitoring soil salinity: A review of methods. *Establishment of Water Quality Monitoring Programs*, American Water Resources Association, 1978, p. 150.

91. P. A. Fryxell. *Agron. J.* **46**, 433 (1954).

92. A. D. Ayers and H. E. Hayward. *Soil Sci. Soc. Am. Proc.* **13**, 224 (1948).

93. F. S. Harris and D. W. Pittman. *Utah Agric. Exp. Stn. Bull.* **168**, 23 (1919).

94. G. Lopez. Germination capacity of seeds in saline soil. In H. Boyko, ed., *Saline Irrigation for Agriculture and Forestry*. W. Junk, The Hague, 1968, pp. 11–23.

95. A. Wahhab. *Proc. Teheran Symp. Arid Zone Res.* **14**, 185 (1961).

96. B. V. Mehta and R. S. Desai. *J. Soil Water Conserv. India* **6**, 168 (1958).

97. L. Lyles and C. D. Fanning. *Agron. J.* **56**, 518 (1964).

98. L. E. Francois and L. Bernstein. *Agron. J.* **56**, 38 (1964).

99. C. L. Mehrotra and B. R. Gangwar. *J. Indian Soc. Soil Sci.* **12**, 75 (1964).

100. D. W. West and L. E. Francois. *Irrig. Sci.* **3**, 169 (1982).

101. M. A. Aziz, A. Fattah, A. S. A. Salam, I. A. Elmofty, and M. M. Abdel-Gawwad. *Desert Inst. Bull. A. R. E.* **22**, 157 (1972).

102. G. A. Pearson, A. D. Ayers, and D. L. Eberhard. *Soil Sci.* **102**, 151 (1966).

103. K. D. Beatty and C. F. Ehlig. *J. Am. Soc. Sugar Beet Technol.* **17**, 295 (1973).

103a. P. D. Sexton and C. J. Gerard. *Agron. J.* **74**, 699 (1982).

104. C. D. Foy. General principles involved in screening plants for aluminum and magnesium tolerance. In M. J. Wright, ed., *Plant Adaptation to Mineral Stress in Problem Soils*, Cornell University Press, Ithaca, New York, 1976, p. 255.

105. D. A. Reid. Screening barley for aluminum tolerance. In M. J. Wright, ed., *Plant Adaptation to Mineral Stress in Problem Soils*, Cornell University Press, Ithaca, New York, 1976, p. 269.

106. G. L. Ream and J. R. Furr. *J. Am. Soc. Hort. Sci.* **101**, 265 (1976).

107. M. C. Shannon, J. H. Draper, and J. D. McCreight. *J. Am. Soc. Hort. Sci.* **108**, 225 (1983).

108. M. W. Nabors, S. E. Gibbs, C. S. Bernstein, and M. E. Meis. *Z. Pflanzenphysiol. Bd.* **97**, 13 (1980).

109. P. J. Dix and H. E. Street. *Plant Sci. Lett.* **5**, 231 (1975).

110. M. W. Nabors, A. Daniels, L. Nadolny, and C. Brown. *Plant Sci. Lett.* **4**, 155 (1975).

111. D. W. Rains, T. P. Croughan, and S. J. Stavarek. Selection of salt tolerant plants using tissue culture. In D. W. Rains, R. C. Valentine, and A. Hollaender, eds., *Genetic Engineering of Osmoregulation: Impact on Plant Productivity for Food, Chemicals, and Energy*. Plenum, New York, 1980, p. 279.

112. C. M. Donald. *Euphytica* **17**, 385 (1968).

113. Y. Waisel. *Biology of Halophytes*. Academic Press, New York, 1972.

114. R. H. Nieman. *Bot. Gaz.* **123**, 279 (1962).

115. W. D. Jeschke and H. Nassery. *Physiol. Plant* **52,** 217 (1981).

116. R. G. Wyn Jones and R. Storey. Betaines. In *Physiology and Biochemistry of Drought Resistance in Plants.*Academic Press, New York, 1981, p. 171.

117. A. B. Hope and N. A. Walker. *Physiology of Giant Algal Cells.* Cambridge University Press, Cambridge, 1975.

118. U. Lüttge and M. G. Pitman, Transport and energy. In U. Lüttge and M. G. Pitman, eds., *Encyclopedia of Plant Physiology* (New Series, Vol. 2, Part A). Springer-Verlag, New York, 1976, p. 251.

119. J. Dainty. The ionic and water relations of plants which adjust to a fluctuating saline environment. In R. L. Jefferies and A. J. Davy, eds., *Ecological Processes in Coastal Environments.* Blackwell, Oxford, 1979, p. 201.

14

ROLE OF CULTIVAR TOLERANCE IN INCREASING RICE PRODUCTION ON SALINE LANDS

F. N. Ponnamperuma

The International Rice Research Institute
Los Baños, Laguna, Philippines

1. RECENT DEVELOPMENTS

Six recent developments have stimulated interest in the potential of saline lands for crop production:

1. Rapid population growth.
2. Scarcity of arable land.
3. Land degradation.
4. High energy costs.
5. Limitations of modern crop technology.
6. Success in breeding salt-tolerant crop plants.

1.1. Rapid Population Growth

If present estimated trends persist, by the year 2000 the world population will likely be 6.3–6.7 billion and the food need 3 billion t (1). But food production growth rates are barely sufficient to keep pace with population growth especially in the poor populous countries (2), and a cereals deficit of 85 million t in 1985 is projected for developing countries (3).

1.2. Scarcity of Arable Land

Until about 1950, the increase in world food demand was met by an increase in the cultivated area. But from 1950 to 1975 world population increased faster than the cropped land area leading to a drop in per capita area from 0.241 to 0.184 ha (4). The pressure on the land is most severe in densely populated Asia "where population density is already extremely high, population growth is increasing and undeveloped land resources are limited" (5). Overcrowding, food shortages, and land scarcity are compelling developing countries to bring under crops, lands lying idle because of salinity and other soil stresses.

1.3. Land Degradation

Erosion, salinization, and denudation cause land degradation. About one-third of the world's crop land is severely eroded, and one-tenth of the 210 million ha of irrigated land has deteriorated because of salization (6). Soil denudation is moving the Sahara desert southward on a 3000-mile front.

1.4. High Cost of Energy

The increasing cost of fertilizers, chemicals, and power is a serious obstacle to increasing crop yields by intensifying agriculture, especially in developing countries.

1.5. Limitations of Modern Crop Technology

Use of modern technology enabled farmers in affluent countries to achieve impressive yield increases from 1961 to 1971. The introduction of new wheat and rice varieties helped even the Third World countries to benefit by the new technology. But yields started to plateau or decline in affluent countries, while the green revolution in rice bypassed most farmers of South and Southeast Asia. Thus, the new technology alone is not likely to prevent the huge food deficits anticipated a decade from now.

1.6. Advent of Salt-Tolerant Cultivars

Although cultivar differences in salt tolerance have been known for over 50 years, it is only during the past decade that organized efforts have been made to select and breed salt-tolerant crop plants. The results have been encouraging (7). Cultivar tolerance for salinity is a substitute for amendments on moderately saline soils and a supplement to amendments in strongly saline ones.

2. SALINE SOILS

2.1. Extent and Distribution

There are 344 million ha of saline soils on the earth's land surface. Of these, 230 million ha are not strongly saline (Table 14.1) and have crop production possibilities. About 56 million ha are in densely populated South and Southeast Asia where both food and arable land are scarce (Fig. 14.1). Of the latter, 27 million ha are lands in the humid tropics climatically, physiographically, and hydrologically suited to rice but lying uncultivated largely because of salinity.

2.2. Definition

Saline soils contain sufficient salt in the root zone to impair the growth of crop plants. But because salt injury depends on species, variety, growth stage, environmental factors, and nature of the salts, it is difficult to define saline soils precisely. Current definitions are based on salt content alone or in conjunction with texture, morphology, or hydrology (9–13). The most widely accepted definition of a saline soil is one that gives an electrical conductivity in the saturation extract (EC_e) exceeding 4 mmho cm^{-1} (4 dS m^{-1}) at 25° C. FAO/UNESCO (11) mapped soils with EC_e's exceeding 15 mmho cm^{-1} as strongly saline soils or solonchaks.

2.3. Formation

A saline soil forms when the influx of salt is greater than the efflux. The balance depends on climate, geomorphology, relief, and hydrology.

Table 14.1.   Distribution of Saline Lands[a]

Region	Area (million ha)		
	Strongly Saline	Moderately Saline	Total
Africa	16.5	37.0	53.5
Australasia	16.6	0.79	17.4
Mexico and Central America	0.24	1.72	1.96
North America	0	6.2	6.2
South America	10.5	58.9	69.4
North and Central Asia	22.5	69.2	91.7
South Asia	47.2	36.1	83.3
Southeast Asia	0	20.0	20.0
Total	113.5	230.0	343.5

[a]From ref. 8.

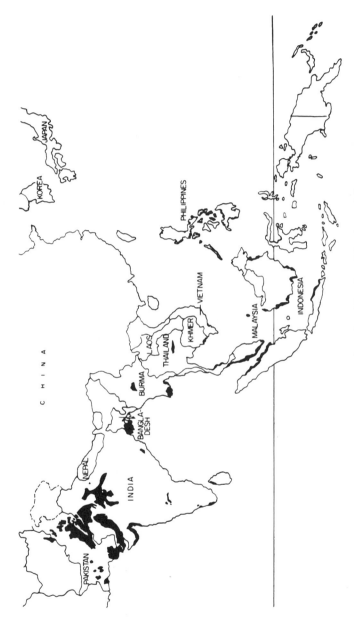

Figure 14.1 Saline soils of South and Southeast Asia.

The bulk of the world's saline soils occurs in arid and semiarid regions (8), where evapotranspiration exceeds precipitation. Evaporation of salt solutions brought into poorly drained depressions by surface runoff, groundwater rise, artesian activity, or interflow leads to salt accumulation. Anthropic salinization occurs in arid and semiarid areas due to waterlogging brought about by improper irrigation.

Another type of salinity occurs in coastal areas subject to tides in both arid and humid regions. The main cause is intrusion of surface or underground seawater.

2.4. Characteristics

Saline soils vary widely in their chemical and physical characteristics, salt dynamics, and hydrology. The variables include salt source; nature and content of salts; lateral, vertical, and seasonal distribution of salt; soil pH; nature and content of clay; organic matter content; nutrient status; water regime; relief; temperature; and soil toxicities. Some of these differences are shown in Tables 14.2 and 14.3. These differences have implications for the management of saline soils and breeding varieties for salt tolerance.

2.5. Reclaiming Saline Lands

Saline lands can be converted to crop lands by preventing the influx of salt water, leaching the salts out of the root zone, and correcting soil toxicities and nutrient deficiencies. Rice is the crop best suited to saline lands because the standing water that is necessary for leaching the salts causes chemical changes in the soil that benefit rice (14, 15). The reclamation costs can be reduced by growing salt-tolerant rice cultivars.

Table 14.2.   Some Characteristics of Saline Soils

Characteristic	Range
Texture	Sandy to clayey
pH	2.5 to 11
EC_e (dS m^{-1})	4 to > 50
Salt (%)	0.1 to 5
Organic C (%)	< 1 to > 30
Fertility	Very low to moderately high
Clay mineralogy	2:1 types to hydrous oxides

Table 14.3.   Kinds of Saline Soils and Associated Stresses[a]

Kind of Saline Soils	Accessory Growth-Limiting Factors
Arid saline	High pH; deficiencies of zinc, nitrogen, and phosphorus
Neutral and alkaline coastal saline	Zinc deficiency, boron toxicity, deep water
Acid saline	Iron toxicity, phosphorus deficiency
Coastal acid sulfate	Iron and aluminum toxicities, phosphorus deficiency, deep water
Coastal organic	Deficiencies of nitrogen, phosphorus, zinc, copper; toxicities of iron, hydrogen sulfide, and organic substances; deep water

[a]From ref. 16.

3. SALINITY AND PLANT GROWTH

3.1. Salt Injury in Plants

Strongly saline soils are barren. Less strongly saline soils in arid regions are characterized by patchy growth of halophytic grasses and shrubs (10), whereas coastal saline soils of the humid tropics and subtropics are characterized by the presence of mangrove species (17).

Most plants suffer salt injury at EC values exceeding 4 dS m^{-1}, but Boyko (18) states that many crop plants can stand much higher concentrations than generally supposed if the solution consists of physiologically balanced salts as in seawater. Epstein and Norlyn (19) grew barley in sand dunes irrigated with seawater and obtained about one-half the national yield per ha.

According to Maas and Hoffman (20) crop yield decreases markedly with increase in salt concentration, but the threshold concentration and rate of yield decrease vary with the species. They describe the effect of salt concentration on yield by the equation

$$Y - 100 - B(EC_e - A)$$

where Y is relative yield, EC_e is electrical conductivity of the saturation extract, A is the salinity threshold value in dS m^{-1}, and B is yield decrease per unit of salinity increase. A was higher and B lower for two salt-tolerant rice cultivars than for two sensitive ones (21).

There are marked interspecific differences in crop tolerance for salinity and within a species ecotypes exist that can tolerate much higher salt concentrations than normal populations (22).

3.2. Salt Injury in Rice

Rice is moderately susceptible to salinity (12, 20). The degree of injury, however, depends on the nature and concentration of salts, soil pH, water regime, method of planting, age of seedling, growth stage of the plant, duration of exposure to salt, and temperature (23–25).

Most rice cultivars are severely injured in submerged soil cultures at an EC_e of 8–10 dS m^{-1} at 25°C; sensitive ones are hurt even at 2 dS m^{-1}.

At comparable EC_e's injury was less in seawater than in solutions of common salt, in neutral and alkaline soils than in acid soils, at 20°C than at 35°C, and in 2-week old seedlings than in 1-week old plants (23).

Rice is tolerant during germination, becomes very sensitive during the early seedling stage, gains tolerance during vegetative growth, again becomes sensitive during pollination and fertilization, and then becomes increasingly more tolerant at maturity. Salinity at the reproductive stage depresses grain yield much more than salinity at the vegetative growth stage (25).

The symptoms of salt injury in rice are stunted growth, rolling of leaves, white leaf tips, white blotches in the laminae, drying of the older leaves, and poor root growth (26).

4. SALT TOLERANCE IN RICE

4.1. Selecting and Breeding for Salt Tolerance

Rice breeders have used genetic variability to produce cultivars that have high yield potential and that resist disease and insect damage and that tolerate cold, drought, and even floods. But apart from some sporadic work in Sri Lanka (27) and India (28, 29), little was done until recently to identify and breed cultivars adapted to adverse soil conditions such as salinity (30, 31).

For centuries, farmers have grown salt-tolerant cultivars on the saline soils of India, Burma, Thailand, Indonesia, and the Philippines. But because of lodging and susceptibility to disease and insect damage, yields are about 1 t/ha. Traditional cultivars selected and bred for salt tolerance during the past 40 years have not done much better.

Recognition of the potential of saline lands for rice production in the densely populated countries of South and Southeast Asia prompted the inclusion of salt tolerance as a component of the Genetic Evaluation and Utilization (GEU) program of the International Rice Research Institute (IRRI).

The GEU program (23) is an interdisciplinary effort on the part of plant breeders and problem area scientists to build into the germplasm tolerance for or resistance to the adverse conditions that small farmers (who constitute the bulk of rice producers in developing countries) encounter in their environments and which they are often unable to correct. These adverse factors include pest damage, drought, floods, and adverse soil conditions. Of the adverse soil conditions, salinity received the most attention because of its widespread occurrence in current and potential rice lands.

Breeding rices for salt tolerance included the following steps:

1. Developing screening techniques.
2. Identifying salt-tolerant germplasm.
3. Combining salt tolerance with good agronomic characteristics and pest resistance.
4. Testing thousands of rices generated by the hybridization program for salt tolerance.
5. Selecting in the field salt-tolerant breeding lines possessing other desirable traits.
6. Multilocational international testing.
7. Conducting yield trials under controlled field conditions at IRRI.
8. Testing in farmers' fields.

4.2. Screening Techniques

A study of the factors affecting salt injury indicated that:

1. The discriminating salinity level is 8–10 dS m^{-1} at 25°C.
2. A nearly neutral submerged clay soil treated with 0.5% common salt is a suitable medium.
3. Transplanting a 2-week-old seedling raised in culture solution is better than direct seeding.
4. The percentage of dead leaves is a good measure of salt injury (24).

The greenhouse screening technique IRRI uses is based on those observations. About 4000 rices can be screened in 12 months on 100 m^2 of bench space.

Salt-tolerant cultivars identified in the greenhouse are used by IRRI's Plant Breeding Department as parents in the hybridization program. Because plant breeders like to field test breeding lines under insect, disease, and other stresses, progeny from the breeding program are screened for salt tolerance at IRRI on a 0.1 ha block of a nearly neutral clay treated with common salt to give an EC$_e$ of 8–10 dS m^{-1}. In a year 1200 rices can be screened.

Under natural conditions a rice crop is exposed to many hazards other than salinity. Thus a line selected at IRRI may fare poorly in a different environment. To sample diverse environments, IRRI has the International Rice Salt and Alkali Tolerance Observational Nursery (IRSATON) within its International Rice Testing Program (IRTP). In the IRSATON program, promising cultivars, selections, as well as material from the breeding program of IRRI and its collaborators are grown in replicated tests and scored visually for salt tolerance in the field in experiment stations in different countries.

4.3. Results of Screening

We began our search for salt tolerance in the world collection of germplasm with 48 cultivars that were either collected from salt-affected areas or were reported to be salt tolerant. Six were found to have an acceptable degree of salt tolerance and were used as donors in the salt tolerance hybridization program. Then emphasis shifted to the progeny from IRRI's salt tolerance breeding program.

During the period 1972–1982, 60,261 rices, the bulk of which came from IRRI's breeding program, were screened, and 10,369 were found to be tolerant. The vast majority of the tolerant rices came from the salt tolerance hybridization program, but there were in each test some elite lines from IRRI's general breeding program that had an acceptable degree of salt tolerance.

The salt-tolerant rices that were identified included the traditional varieties, Kalarata, Pokkali, Nona Bokra, Getu, SR 26 B, CSSR 1, and CSSR 2; the lines IR4432-20-2, IR4595-4-1, IR4630-22-2, IR4763-73-1, IR9852-19-2, IR9884-54-3, IR9975-5-1, IR10206-29-2, and IR13646-2 from the salt tolerance hybridization program; the elite lines IR2153-26-3, IR2863-35-3, IR4422-6-2, and IR4563-52-1 from the general breeding program; and the modern cultivars IR42, IR46, IR50, and IR52.

Observations in the international tests during the period 1978–1981 revealed that the traditional varieties Pokkali, Getu, Nona Bokra, and the following lines derived from them are salt-tolerant: IR4432-28-5, IR4595-4-1, IR4630-22-2, IR4763-73-1, IR5657-33-2, IR9884-54-4, IR10198-66-2. Among IR cultivars that have salt tolerance are IR42, IR43, IR46, IR52, and IR54. In 1981, Dokri, Pakistan, where several rices perished, IR4595-4-1 and IR5657-33-2 showed no salt injury (32).

5. YIELD TRIALS AND THEIR PRACTICAL SIGNIFICANCE

Promising breeding lines are tested for yield potential in replicated field experiments at IRRI and in farmers' fields.

5.1. Yield Trials under Controlled Conditions

The first field test is done at IRRI on a nearly neutral, irrigated, artificial saline soil, with adequate fertilization and pest control. Results indicate that even among salt-tolerant breeding lines the capacity to produce grain varies markedly.

In the 1980 dry season the grain yields of 28 rices ranged from 0.9 to 2.6 t/ha at an EC_e of 8.4 dS m^{-1}. In the wet season the yields ranged from 1.4 t/ha for IR48 to 3.1 t/ha for IR52 at an EC_e of 7.3 dS m^{-1}. For 24 tolerant breeding lines tested in the 1981 dry season, the yields were 1.7 to 3.1 t/ha at an initial EC_e of 9.7 dS m^{-1}, compared with 2.8 to 5.0 t/ha on a nonsaline soil (Calimon, M. C. Q. E., IRRI, unpublished).

The low yield of some rices identified visually as salt tolerant may be due to their poor agronomic characteristics or the poor correlation between salt tolerance at the vegetative stage (when scoring is done) and that in the reproductive phase.

5.2. Experiments in Farmers' Fields

During the period 1977–1980, IRRI evaluated a total of 64 modern salt-tolerant insect- and disease-resistant rices in a total of 25 tests at 15 sites in coastal saline tracts in the Philippines.

The rice plants in the experimental plots experienced the rigors and environmental vagaries that farmers' crops undergo in their fields: excess salt, flooding, drought, and damage by buffaloes, rats, and birds. Table 14.4 illustrates some of the hazards. The plots received a moderate amount (50 kg N and 25 kg P per ha) of fertilizers. In 1981 even that was discontinued without adverse effects on yield (33).

Because of severe salt, flood, or drought injury in some experiments the rices produced no grain. But scores confirmed the salt tolerance of IR42, IR4432-28-5, IR4595-4-1, IR4630-22-2, and IR9884-54-3. The mean minimum yield was 1.5 t/ha and the mean maximum 3.6 t/ha averaged for all 25 experiments (34). More than half the yields exceeded 2 t/ha (Fig. 14.2). Salt tolerance in modern rices conferred a comparative yield advantage of 2 t/ha.

Table 14.5 compares the performance, in some recent tests, of three lines from the salt tolerance breeding program with that of six IR cultivars. Four of the latter have moderate salt tolerance although no planned effort was made to achieve it.

In the 1981 tests of 27 rices in 8 saline fields with EC_e's ranging from 1 to 18 dS m^{-1}, yields ranged from <1 t/ha to nearly 5 t/ha. The mean yields for the rices for which data were available from at least five experiments were 2.3–3.0 t/ha (33).

Table 14.4. Summary of Tests of 12 Varieties and Lines on Coastal Saline Soils in Farmers' Fields in the Philippines, 1980 Dry and Wet Seasons[a]

Site	Salt Source	EC (dS m⁻¹)	Soil pH	Soil Organic Matter Content (%)	Other Soil Problems	Yield Range (t/ha)	Best Yielders	Remarks
					Dry Season			
Lubao, Pampanga	Creek	14	6.9	1.7	—	—	—	Intrusion of strongly saline water killed the plants
Mexico Pampanga	River	11.7	7.0	0.7	Zinc deficiency	1.0–4.3	IR10168-3	Plants were harvested before high salinity
Samal, Bataan	Creek	4.3	6.5	1.5	—	0.9–4.4	IR50 IR9884-54 IR10206-29	Salinity was low
Taal, Batangas	River	4.6	7.1	1.4	Zinc deficiency Boron toxicity	—	—	Frequent deep flooding
Sinacaban, Misamis Occidental	Seepage	7.6	4.1	4.0	Iron toxicity	1.6–3.9	IR4630-22 IR5657-33 IR4432-28-5	Rains depressed salinity but tungro disease reduced yield
					Wet Season			
Cebu, Cebu	Creek		7.4	3.9	Zinc deficiency Boron toxicity	4.5–6.6	IR9884-54 IR42	Low salinity
Minalin, Pampanga	Creek	9.2	5.8	1.6	Zinc deficiency	0.8–3.6	IR9884-54	Rains depressed salinity
Bani, Pangasinan	Creek	8.0	6.1	0.8	—	1.1–2.6	IR5657-33 IR52, IR42, IR9884-54	Drought depressed growth

[a]From ref. 21.

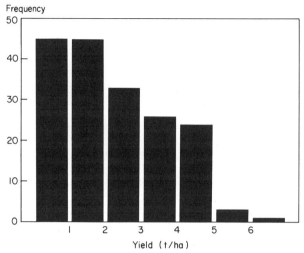

Figure 14.2. Frequency distribution of 177 mean rice yields in experiments in coastal saline soils in the Philippines.

Because coastal saline soils in the humid tropics are usually well supplied with nitrogen and have adequate supplies of phosphorus and potassium, 10 salt-tolerant rices-were grown in saline fields with EC_e values of 2–10 dS m^{-1} at planting at two locations in the Philippines with and without 50 kg N/ha but no P or K. There was no significant response to nitrogen fertilizer (33) at a yield level of 3–4 t/ha. If these findings apply widely, and if modern salt-tolerant varieties are used, coastal saline soils may produce 3 t/ha of rice without land reclamation or fertilizer application.

Table 14.5.   Grains Yields of Nine Rices on Coastal Saline Soils in the Philippines, 1980–1982[a]

Cultivar/Line	Times Tested	Yield (t/ha) Range	Yield (t/ha) Mean	Bred for Salt Tolerance
IR5657-33-2-1	5	2.6–4.6	3.4	Yes
IR10198-66-2	6	2.6–4.1	3.4	Yes
IR9884-54-3	5	1.9–3.8	2.9	Yes
IR52	6	1.2–3.7	2.8	No
IR42	7	0.4–4.8	2.7	No
IR50	7	1.1–4.0	2.6	No
IR54	7	1.3–3.8	2.6	No
IR36	5	1.4–3.6	2.2	No
IR28	2	1.2–2.2	1.7	No

[a]Adapted from ref. 35.

In reclaimed lands where rainfall is adequate and well distributed, salt-tolerant modern varieties have yielded over 4 t/ha per season at EC_e levels of 3.5 to 4.5 dS m^{-1}. At one such site a total yield of over 8 t/ha was obtained on land that was an idle coastal swamp only 1 year before (36).

5.3. Two Crops of Rice Where None Grew Before

Most coastal saline lands in the Philippines are virtually uncultivated because of strong salinity in the dry season and deep flooding in the wet. Experiments in such fields at three locations revealed that by planting IR 50 (an early maturing, salt-tolerant, improved variety) at the end of the wet season and again early in the wet season, salt injury in March–April and flood damage in July–August can be overcome. Average yield from the three sites was 6 t/ha per year without costly inputs (33).

Two rice crops are possible on vast tracts in South and Southeast Asia with similar salt and water regimes if early maturing, salt-tolerant cultivars are grown.

6. PROBLEMS IN USING CULTIVAR TOLERANCE

The problems in using cultivar tolerance include

1. Paucity of information on the mechanism of salt tolerance.
2. Inadequacy of the level of salt tolerance found in rice.
3. Unsuitability of the modern, short-statured, photoperiod-insensitive plant type for flood-prone coastal areas.
4. Soil stresses.
5. Biotic stresses.

6.1. Mechanism of Salt Tolerance in Rice

Information on the physiological basis of salt tolerance in rice is meager. That is an obstacle to the development of rapid and reliable methods for screening thousands of progeny from a breeding program.

Janardhan and Murty (37) found a higher water content in the shoots of two salt-tolerant cultivars than in those of two susceptible ones. Work done at IRRI (36, 38) indicates that salt injury in rice involves water stress; tolerance is associated with a high electrolyte content in the roots and a low content in the shoot; the ability to accumulate potassium in the shoot correlated well with salt tolerance; salt-tolerant Pokkali accumulated 13 times more proline in the shoot than when grown in normal soil; chloride is not toxic; and salt tolerance is associated with tolerance for both high salt concentration and a high sodium adsorption ratio.

6.2. Level of Salt Tolerance in Rice

Even Pokkali, one of the most salt-tolerant rices, does not accumulate enough salts to give it the degree of salt tolerance that barley and wheat have. This is an obstacle to breeding rices suited to strongly saline soils or soils that are periodically inundated by saline water. To enhance salt tolerance, the following strategies are being explored:

1. Use of population breeding methods aimed at gene accumulation from diverse sources.
2. Combining different sources of salt tolerance.
3. Use of cell and tissue culture.
4. Use of exotic germplasm.
5. Chemical mutagenesis.

6.3. Unsuitability of Some Modern Rices

The modern photoperiod-insensitive, short-statured, salt-tolerant rices may not be suited to areas where deep seasonal flooding is a problem. For such areas, photoperiod-sensitive cultivars of intermediate stature are required.

6.4. Soil Toxicities

Soil stresses other than salinity, such as alkalinity, strong acidity, iron toxicity, and excess organic matter may lower the yield potential of the modern salt-tolerant rices. To obtain yield stability over wide soil conditions, multiple stress tolerance is desirable (34, 39). IR42 has multiple stress tolerance.

6.5. Biotic Stresses

Salt tolerance, once genetically incorporated, will not change appreciably, but resistance to disease and insect pests changes. So a salt tolerant variety may succumb to disease or insects and may lose its yield potential in about 5 years.

7. CONCLUSIONS

1. The modern, disease-and insect-resistant, salt-tolerant rice cultivars have a comparative yield advantage of about 2 t/ha, over nontolerant ones.
2. In the tropics, on slightly saline lands with water control use of such cultivars will enable farmers to obtain 8 t/ha per year without reclamation measures.
3. On lands that are strongly saline in the dry season and flooded deeply in the wet season, use of early-maturing, salt-tolerant cultivars can overcome both obstacles and deliver 6 t/ha per year.
4. Where salinity is severe and reclamation measures are necessary, costs can be reduced by using salt-tolerant cultivars.
5. Because most coastal saline soils in the tropics are fairly well supplied with plant nutrients, yields of about 3 t/ha per season are possible without fertilizers.
6. The advent of the modern, salt-tolerant rices offers the hope of increasing rice yields on current saline lands and bringing into rice production millions of ha of idle lands without costly inputs.
7. Use of salt-tolerant rices may be a solution to soil salinity in arid and semiarid regions caused by waterlogging or use of saline irrigation water.

REFERENCES

1. W. D. Hopper, Recent trends in world food and population. In R. G. Woods, ed., *Future Dimensions of World Food and Population.* Westview Press, Boulder, Colorado, 1981, pp. 35–55.
2. T. N. Barr. The world food situation and global prospects. *Science* **214,** 1087–1095 (1981)
3. S. C. Hsieh, J. C. Flinn, and N. Amerasinghe. The role of rice in meeting future needs. In *Rice Research Strategies for the Future.* International Rice Research Institute, Los Baños, Philippines, 1982, pp. 27–49.
4. L. R. Brown. Worldwide loss of cropland. In R. G. Woods, ed., *Future Dimensions of World Food and Population.* Westview Press, Boulder, Colorado, 1981, pp. 57–96.
5. J. W. Willet. *The World Food Situation: Problems and Prospects to 1985.* Oceana Publications, New York, 1976, 1136 pp.
6. L. R. Brown. World population growth, soil erosion, and food security. *Science* **214,** 995–1002 (1981).
7. F. N. Ponnamperuma. Breeding crop plants to tolerate soil stresses. In I. K. Vasil, W. R. Scowcroft, and K. J. Frey, eds., *Plant Improvement and Somatic Cell Genetics.* Academic Press, New York, 1982, pp. 73–97.

8. F. I. Massoud. Salinity and alkalinity as soil degradation hazards. FAO/UNDP Expert Consultation on Soil Degradation. June 10-14, 1974. FAO, Rome, 1974, 21 pp.

9. K. H. Northcote, and J. K. M. Skene. Australian soils with saline and sodic properties. *Commonwealth Scientific and Industrial Research Organization* (CSIRO). *Soil Publ.* **27,** 62 (1972).

10. FAO-UNESCO (Food and Agriculture Organization-United Nations Educational, Scientific, and Cultural Organization). *Irrigation, Drainage, and Salinity. An International Source Book.* Hutchinson, London, 1973, 510 pp.

11. FAO-UNESCO (Food and Agriculture Organization-United Nations Educational, Scientific, and Cultural Organization). Soil map of the world 1:500,000. Legend. UNESCO, Paris, 1974.

12. United States Salinity Laboratory Staff. Diagnosis and improvement of saline and alkali soils. *Agric. Handb.* **60.** USDA, Washington, D.C., 1954, 160 pp.

13. Soil Science Society of America. *Glossary of Soil Science Terms.* Madison, Wisconsin, 1978, 36 pp.

14. F. N. Ponnamperuma. Electrochemical changes in submerged soils and the growth of rice. In *Soils and Rice* International Rice Research Institute, Los Banos, Philippines, 1978, pp. 421-444.

15. F. N. Ponnamperuma. Some aspects of the physical chemistry of paddy soils. In *Proceedings of the Symposium on Paddy Soils.* Science Press, Beijing, People's Republic of China, 1981, pp. 59-94.

16. F. N. Ponnamperuma. Saline soils of South and Southeast Asia for rice production. In International Commission on Irrigation and Drainage Second Regional Afro-Asian Conference, Manila, Philippines, 1978, pp. 126-135.

17. V. J. Chapman. *Mangrove Vegetation.* J. Cramer, 1977, 447 pp.

18. H. Boyko. Basic ecological principles of plant growing by irrigation with highly saline or seawater. In H. Boyko, ed., *Salinity and Acidity.* Junk, The Hague, 1966, pp. 131-200.

19. E. Epstein and J. D. Norlyn. Seawater based crop production: A feasibility study. *Science* **197,** 249-251 (1977).

20. E. V. Maas and G. K. Hoffman. Crop salt tolerance-current assessment. *J. Irrig. Drainage Div. ASCE* **193** (IR 2). Proc. Pap. 12993. 1977.

21. IRRI (International Rice Research Institute). *Annual Report for 1980.* Los Baños, Philippines, 1981, 467 pp.

22. A. Lauchli. Genotypic variation in transport. In U. Luttage and M. G. Pitman, eds., *Encyclopedia of Plant Physiology, New Series,* Vol. 2, Part B. Springer-Verlag, Berlin, 1976, pp. 372-393.

23. IRRI (International Rice Research Institute). *Annual Report for 1974.* Los Baños, Philippines, 1975, 384 pp.

24. F. N. Ponnamperuma. Screening rice for tolerance to mineral stresses. *IRRI Res. Paper Ser.* **6,** (1977).

25. M. Akbar and F. N. Ponnamperuma. Saline soils of South and Southeast Asia as potential rice lands. In *Rice Research Strategies for the Future.* Los Baños, Philippines, 1982, pp. 265-281.

26. F. N. Ponnamperuma and A. K. Bandyopadhya. Soil salinity as a constraint on food production in the humid tropics. In *Priorities for Alleviating Soil-related Constraints to Food Production in the Tropics*. IRRI, Los Baños, Phillipines, 1980, pp. 203–216.

27. L. H. Fernando. The performance of salt resistant paddy, Pokkali in Ceylon. *Trop. Agric.* **105**, 124–126 (1949).

28. R. K. Bhattacharya. Promising rice selections suited to coastal saline soils. *J. Soc. Exp. Agric.* **1**, 21–24 (1976).

29. S. Chandra. Genetics and plant breeding. In *A Decade of Research*. Central Salinity Research Institute, Karnal, India, 1979, pp. 80–98.

30. F. N. Ponnamperuma and R. U. Castro. Varietal differences in resistance to adverse soil conditions. In *Rice Breeding*. International Rice Research Institute, Los Baños, Philippines, 1972, pp. 677–684.

31. H. Ikehashi and F. N. Ponnamperuma. Varietal tolerance of rice for adverse soils. In *Soils and Rice*. International Rice Research Institute, Los Baños, Philippines, 1978, pp. 801–823.

32. IRRI (International Rice Research Institute). *Preliminary Report of 1981 IRTP nurseries*. International Rice Research Institute, Los Baños, Philippines, 1982.

33. IRRI (International Rice Research Institute). *Annual Report for 1981*. Los Baños, Philippines, in press (1982).

34. F. N. Ponnamperuma. Genotypic adaptability as a substitute for amendments on toxic and nutrient-deficient soils. In *Plant Nutrition* Vol. 1. Proceedings of the Ninth International Plant Nutrition Colloquium, Warwick University, England, Aug. 22–27, 1982, pp. 467–473.

35. R. Y. Reyes, N. B. Uy, M. C. Q. E. Calimon, and F. N. Ponnamperuma. Role of varietal tolerance in rice production on coastal saline soils in the Philippines. Proc. 13th Crop Sci. Soc. Phil. Sci. Mtg. 28-30 April, 1982, Cebu City, Philippines.

36. IRRI (International Rice Research Institute). *Annual Report for 1978*. Los Baños, Philippines. 1979, 478 pp.

37. K. V. Janardhan. and K. S. Murty. Effect of sodium chloride on leaf injury and chloride uptake by young rice seedlings. *Indian J. Plant Physiol.* **13**(2), 225–232 (1970).

38. IRRI (International Rice Research Institute). *Annual Report for 1977*. Los Baños, Philippines, 1978, 548 pp.

39. M. M. Mahadevappa, H. Ikehashi, and F. N. Ponnamperuma. The contribution of varietal tolerance for problem soils to yield stability in rice. *IRRI Res. Paper Ser.* **42**, (1979).

15

SCREENING WHEAT AND BARLEY GERMPLASM FOR SALT TOLERANCE

J. P. Srivastava

Cereal Improvement Program
International Center for Agricultural Research in Dry Areas
Aleppo, Syria

S. Jana

Department of Crop Science and Plant Ecology
University of Saskatchewan
Saskatoon, Canada

In order to identify genotypes capable of exploiting limited resources, Boyer (1) suggested that selections should be made under the adverse conditions likely to be encountered rather than solely in favorable environments. Considerable success can be expected if plant improvement includes selection under conditions that are often unfavorable for growth (1, 2).

The accumulation of salts in agricultural soil is a problem that has been with man since he started cultivation. Salinity of the arable land is an increasing problem of many irrigated, arid, and semiarid areas of the world and it is a significant factor in reducing crop productivity from salt-affected lands.

Area affected by soil salinity is rapidly increasing in the Middle East and North Africa. Large areas in Pakistan, Iran, Iraq, and Egypt have been rendered agriculturally unproductive. The climate of the region, with its alternating pattern of winter rainfall and summer drought, results in a concurrent pattern of moisture accumulation in the soil in the winter, followed by progressive reduction of soil moisture in the summer. In many areas, the rainfall is inadequate to remove accumulated salts from the top soil by leaching. Thus, vast areas under rain-fed conditions are affected by salt buildup in the top soil.

The level of salinization in such areas determines the cropping pattern. Whereas soils with an electrical conductivity (*EC*) higher than 4 mmhos/cm are usually considered marginal (3), barley and sometimes durum wheat is grown by farmers at salinity concentrations of more than 10 mmhos/cm. Though plant death may occur under extreme conditions, a reduced vegetative growth and subsequent reduction in grain yield is the most common symptom encountered with salinity (4). In the arid areas, the combined effects of salinity, drought, and high temperature further reduce the crop productivity.

Systematic work on exploiting genetic variability to develop increased salt tolerance in barley and wheat cultivars is still in its infancy. Efforts to improve salt tolerance in cultivated varieties have been hampered due to lack of understanding of the mechanisms of salt tolerance and how it is influenced by environmental factors (5). Tolerance varies with the stage of plant growth and is affected by soil moisture, temperature, relative humidity, light, and nutrient availability (6). Now nonavailability of reliable and efficient field and laboratory screening tests for salt tolerance at the practical plant breeding level is another reason for little progress in improving salt tolerance in the cultivated varieties.

Genetic differences for salinity tolerance have been established by several workers (2, 7–12). Although selection for salt tolerance has been sporadic, there is sufficient genotypic variation within the existing germplasm to provide an opportunity for improved yields under saline conditions (1).

1. FIELD SCREENING OF GERMPLASM

The field screening of wheat, barley, and triticale germplasm was initiated in 1979 at a site on the shore of the Jabboul salt lake near Hegla village. The site is located in an arid area 50 km southeast of the city of Aleppo in Northern Syria. Soil salinity levels ranged from 20 to 100 mmhos at the site and varied with the availability of soil water. The crop growing period is typified by winter precipitation, which is often low (in the past 3 years, precipitation has ranged from less than 200 mm to around 300 mm annually) and erratic in distribution, and by high temperature, and cessation of rainfall at the time of anthesis and grain development. Soil moisture and salinity stress are further aggravated by hot, dry winds during the grain filling period. All field screening is conducted in rain-fed conditions.

In the first year (1979–1980), entries used for field screening consisted of ICARDA's advance breeding lines and part of the world germplasm collection. They comprised 4779 barley, 2839 durum wheat, 506 bread wheat, and 87 triticale lines. Each accession was planted in two rows 1.5 m long and 30 cm apart. A standard variety grown in every twentieth row served as the check. Observations were recorded on germination in the field, seedling vigor at 5-6 leaf stage, adult plant vigor, and ability to produce fully developed kernels. The seedling and plant vigor were visually evaluated on a 1 to 9 scale. One was assigned to most vigorous plants, and 9 represented very little growth and development.

Table 15.1   Results of the Field Screening Test for Salt–Drought Tolerance in 1980–1981

Crop	Number of Accessions Tested	Number of Promising Lines Identified	Percent Promising Lines
Triticale	87	9	10.3
Barley	779	69	8.8
Durum wheat	948	64	6.8
Bread wheat	506	7	1.4

Selected lines from 1979–1980 field screening and some new accessions were subjected to field screening near the old site during 1980–1981. About 30 seeds of each accession were sown in rows 1.5 m long and 30 cm apart. Each accession was replicated thrice along the salinity gradient. Entries within each replicate were independently randomized. A standard variety grown in every twelfth row served as the check. The visual scores (1–9 scale) were given independently by three persons and were found to be consistent. The mean score of an accession over three replicates was used as a measure of its response to salinity stress.

The field screenings led to the identification of several lines that scored well for germination, plant growth, and producing fertile spikes. Whereas our prime

Table 15.2   A List of Most Promising Salt-Tolerant Lines Based on Field Screening Tests

Barley	*Durum Wheat*
Hembar/Beecher 9L	Qfn/Rabi's' L0505-4L-1AP-0AP
H272	P66/270//T.dur.Ram/G11's'//
MD/Atl//CM IB-4-2-B-B	F3TUN/3/Cr's'/4/Plc's'/5/G58128
WI 2137-2	CD15575-1S-3AP-0AP
ER/Apm	Dumaroc Battandier PI 174623
KY-1324	Fg's'/Sincape 9
	CD 15708-3S-2AP-0AP
	Bit's'/Alder's'//Mex's'/3/Gta's'/Fg's'
	CD 16677-A-7M-1Y-3M-0Y
	Waha's'
	Rabi's'/Fg's'
	CM 10162-76M-0Y-1B
	Cr's'/Stk's'
	L92-6AP-2AP-0AP
Bread Wheat	*Triticale*
Lerma Rojo	Cin-Pi × Pro/Bgl
	× 16350-0AP
	M2A-1GA × IA-KLA
	x 11286-C-4M-4Y-0Y

interest in screening at this site was to find salinity tolerance, other stresses, especially drought, were undoubtedly present in both years. Thus, the selected lines were those that performed best under a combined stress which was typical of such areas in the region. Table 15.1 gives the numbers of promising lines of each crop identified in 1980–1981. A list of the most promising lines for their ability to produce vegetative growth and well filled kernels in the 2-year field screening tests is presented in Table 15.2.

2. SEEDLING GERMINATION TESTS

It has been demonstrated that reaction to salt stress varies with the stage of plant development and that a given cultivar may be tolerant at one stage and sensitive at another (7, 11, 13). Significant and nonsignificant associations between tolerance at the germination stage and adult plant growth and development have been indicated (7, 11, 13, 14). Although relative salt tolerance for seed germination and that of the adult plant may differ, salt tolerance during germination is known to reflect the tolerance of the adult barley plant (15, 16). Abo-Elenin et al. (8) have identified salt-tolerant barley lines, using field screening, lysimeters with saline irrigation water, and controlled sand culture techniques.

On the basis of their performance in the field screening tests for salt tolerance, 80 lines of barley and 80 lines of durum wheat were selected for seedling germination screening at controlled salt concentration levels in growth chambers. The seedling germination tests were conducted at the Crop Science Department, University of Saskatchewan, Saskatoon, Canada.

Sterile vermiculite, 6.5 cm deep, was placed in the bottom of rectangular ($28 \times 25 \times 14 \text{ cm}^3$) plastic boxes. This was covered with vertically placed styrene seedling tubes, which were filled to three-quarters of their height with sterile vermiculite. Each box contained 195 seedling tubes. EC values of 20.0, 24.0 and 30.0 mmhos/cm of nutrient solution was produced by adding NaCl to a half-strength Hoagland's solution. The solution was poured into each germination box in equal amounts, enough to saturate the vermiculite. Four hours later one seed was sown in each tube and covered with 5 mm of sterile vermiculite. Moisture needed for germination was maintained by periodically adding the original solution. Twelve seeds from each accession were sown in a row of twelve seedling tubes. Each germination box contained 15 lines and one check variety. After sowing, the germination boxes were placed in growth chambers at a constant temperature of 20°C with 12 hr of daylight. Barley lines were screened at EC = 30 mmhos/cm and Bonanza was used as the check variety. Durum lines were screened at EC = 20 and 24 mmhos/cm and the widely grown durum variety Wascana was used as check. Electrical conductivity of ocean water is around 52 mmhos/cm. Percent germination and seedling vigor were recorded 3 weeks after planting.

Based on their germination in saline solution (EC of 30 mmhos/cm), 14 barley lines exhibited significantly superior salt tolerance as compared to the check (5% level). Five lines appeared very sensitive and their germination ability was much lower than the check variety Bonanza. The remaining varieties fall in between these two groups. The germination abilities of these two groups of lines are presented in Table 15.3.

Table 15.3   Salt Tolerance of Barley Lines Expressed as Seedling Survival at Salt Concentration of EC 30 mmhos/cm

Line/Cross	Seedling Survival (% of control)
Tolerant Lines	
Api/CM 67	156.3
CMB72-60-500-502Y-503Y-502B-0Y	
ER/Apm	140.7
Bussell/Aurori/Esp = Ketade	125.0
1L-5AP-0AP	
Assala's'	125.0
ICB76.21-4L-3AP-0AP	
CN42/CI7772//FUN/3/FUN/TCH/4/FUN/KI	125.0
II 11694-2B-BULK7N-4N	
R.T. Ramage selection	113.6
79-B-62216	
Baladi Local 46/Dier Alla 106	113.6
R.T. Ramage selection	112.6
1976. 77-0AP	
4977/Benton//Bigo	109.4
4L-1AP-0AP	
H. Odesskji 17/DL 71	109.4
CMB 75A-817-5S-4AP-0AP	
Tanekase 2/Baitori//Aths	109.4
Harbing.//Avt/Aths	109.4
CYB 18-2A-3A.2AP-0AP	
Moracaine 079	104.1
KY-1324	103.6
Sensitive Lines	
Cross 366-16-2	10.4
2762/Beecher-6L	11.3
Union/CI376//Coho	11.3
ICB 76.216-12L-2AP-0AP	
C63	34.1
Mean of checks (Bonanza)	69.9
LDS 5%	33.4

Line, Early Russian, crossed with Apam (ER/Apm) and KY-1324 exhibited salt tolerance in the field screening tests as well as in germination tests indicating apparent association between field response and seedling survival. This is in agreement with earlier findings that certain cultivars, such as California Martiout, appear to be tolerant during all growth stages (10, 12, 13). However, a number of lines that appeared promising on the basis of field response exhibited only on intermediate level of salt tolerance in the germination test. Similarly a number of lines that appeared to tolerate salinity during germination and seedling emergence were not picked up during field screening tests. These preliminary results confirm the importance of utilizing both seedling as well as field screening tests.

Germination tests for durum wheat cultivars were conducted at two salt concentrations, EC values of 20 and 24 mmhos/cm. Based on germination and seedling survival at the two salt concentrations, the 80 durum lines can be placed in three groups: tolerant, moderately tolerant, and sensitive. A list of tolerant and sensitive lines is presented in Table 15.4.

A number of durum lines that appeared to exhibit salt tolerance in the seedling survival test had also been identified as promising lines in the field screening tests. This is in conformity with the findings of Schaller et al. (12), who reported an association between field response and germination tests in 11 barley varieties. Maddur (14) suggested that an association between tolerance at germination and seedling vigor could contribute to more vigorous seedling establishment and subsequent growth. However, the majority of the lines did not show this association indicating differences in genetic systems controlling tolerance at the different stages as suggested by Al-Shamma (7), Maddur (14), Norlyn (11), and Rana (17).

Response of durum lines varied at the two levels of concentrations. Whereas the levels of concentration did not affect germinability of cultivar, Barygon Yaqui, germination of Boohai was reduced at higher salt concentration. In Table 15.4 a number of lines show higher seedling survival at EC = 24 mmhos/cm. This is primarily due to low germination of the check variety, Wascana, at this salinity concentration. It is interesting to note that lines Waha, Bit's'/Alders's'//Mexi's'/3/Gta's'/Fg's', Cr's'/Stk's', and a few others that performed well in the field as well as in the germination tests are high yielding, widely adapted advance lines in the international nurseries. Boohai and Saba 22738 are local cultivars from Ethiopia and Barygon Yaqui is an old Mexican line used by CIMMYT in the crossing program.

About 2200 individual plant samples of *Hordeum spontaneum* Koch. were collected from a wide range of habitats (18). These accessions, mainly from Syria, were evaluated for salt tolerance at germination at an electrical conductivity of 24 mmhos/cm in the manner described earlier. Some populations of wild barley showed significantly larger variability for salt tolerance than others, but the range of salt tolerance in the *H. spontaneum* collection was not larger than that found in a highly variable composite cross population

Table 15.4   Salt Tolerance of Durum Wheat Lines Expressed as Seedling Survival at Levels of Salt Concentrations

Line/Cross	Seedling Survival (% of control)	
	20 mmhos/cm	24 mmhos/cm
Tolerant Lines		
Bit's'/Alder's'//Mexi's'/3/ Gts's'/Fg's'		
CD 16677-A-7M-1Y-3M-0Y	114.2	120.2
Mexi's'/Chap//21563/3/Snipe's'		
CD 14759-2S-1AP-0AP	100.0	140.0
P66/270//T. dur. Ram/G11's'//		
F3 TUN/3/Cr's'/4/Plc's'/5/G58128		
DC 19497-C-1Y-1AP-0AP	114.2	140.0
Boohai	166.6	100.0
Barygon Yaqui	183.2	183.2
Saba 22738		
PI 263421	116.6	100.0
Waha's'	150.0	100.0
Cr's'/Stk's'		
L92-6AP-1AP-0AP	100.0	200.0
Fg's'/Sincape 9		
CD 15708-3S-2AP-0AP	112.6	266.4
Rabi's'/Fg's'		
CM 10162-76M-0Y-1B	112.6	166.4
M2A-lGAxlA-KLA*		
X11286-C-4M-4Y-0Y	112.6	200.0
Sensitive Lines		
Chichicuilote's'		
CD 1314-A-1Y-2Y	37.5	39.9
Stork's'	55.5	50.0
D. Dwarf S-15/Cr's'		
D 33312-7Y-4M-1Y-0M	62.5	66.4
Mean of the check (Wascana)	62.3	31.9
L.S.D. 5%	35.4	52.2

*Triticale line

(CCXXI) of cultivated barley. However, for electrophoretically identifiable isozyme loci, these wild barley populations were much more polymorphic than CCXXI (19). Further studies are in progress involving additional populations of *H. spontaneum* and composite cross populations of cultivated barley.

The goals of our field and controlled environment evaluations are to identify germ plasm lines with high levels of salt–drought tolerance, and using these lines, achieve what is commonly known as germplasm enhancement. It is hoped that this germplasm enhancement program will prove useful to breeders and geneticists in the region for reducing genetic vulnerability of winter cereals to two of the most important production constraints of the region, salt and drought stress.

3. SOIL AMELIORATION

The importance of manipulating the soil environment, so that it becomes more suitable for the plant to grow and produce viable seed, cannot be overemphasized. It is obvious that the development of agronomic paractices that are appropriate for saline conditions is complementary to the genetic manipulation for increasing crop tolerance of salinity. Preliminary experiments were conducted to determine the role of agronomic treatments in salinity amelioration (20).

The first experiment was conducted in 1980–1981 at the experimental site at Hegla. Arabi Abiyad, an old local landrace variety of barley, was sown in a set of

Table 15.5 Effect of Soil Amelioration on Yield and Yield Components of Barley Variety Arabi Abiyad on Salt-Affected Soil

Treatments	Grain Yield (g/m²)	Total Yield (g/m²)	Plants per m²	Seeds per Spike	1000 Kernel Weight
Ridge sown before rains					
+ P₂O₅	22.9	59.3	101	5.2	31.0
Ridge sown before rains	18.6	51.2	73	4.8	31.1
Ridge sown after rains					
+ P₂O₅	30.4	74.1	112	5.1	32.9
Ridge sown after rains	26.6	67.1	117	4.3	30.2
Furrow sown before rains					
+ P₂O₅	29.7	75.2	119	5.3	32.4
Furrow sown before rains	14.3	43.8	100	3.9	28.1
Furrow sown after rains					
+ P₂O₅	16.2	55.5	121	3.3	31.9
Furrow sown after rains	12.1	38.5	115	3.4	30.1
Mulched with barley straw	12.6	36.9	108	3.1	29.6
Mulched with wheat straw	12.6	40.3	114	2.4	27.9
Gypsum (2 tons/ha)	12.6	37.9	101	3.6	28.5
Control	11.9	34.9	86	3.7	27.8
LDS (5%)	3.6	7.5	8.0	0.5	N.S.

combinations of ridges and furrows with sowing times, phosphate fertilizer, and soil treatments. Soil salinity levels varied from 20 to 100 mmhos/cm of EC. Significant differences among treatments were found for plant density, total biological yield, grain yield, and number of seeds per spike (Table 15.5).

Plant establishment was not consistently related to grain yield. However, total biological yield was related to grain yield, which in turn was associated with number of seeds per spike. This preliminary experiment suggested that sowing barley on the ridge or furrow was better than sowing on undisturbed level ground. The addition of straw mulch or gypsum did not affect this result. Increased moisture availability in the ridge or furrow treatments is a possible reason for the increased yield under soil moisture and salt stress that prevailed at the experimental site. In general, Arabi Abiyad responded favorably to the addition of phosphate by producing higher grain and total biological yield.

4. CONCLUSIONS

The results obtained so far are encouraging, but more suggestive than conclusive. They suggest that both genetic and agronomic manipulations of plants for increasing salt tolerance and improving crop yield are possible. Combined agronomic and genetic efforts and a better understanding of different salt tolerance mechanisms are essential for further progress. The limited progress made so far is based on empirical evaluation and selection at a site where the combined salinity and water stress problems exist. Development of more reliable and time-saving evaluation and selection techniques will greatly accelerate the progress. The lack of reliable large-scale field screening techniques still seems to be the biggest problem in genetic improvement of salt and drought tolerance in crop plants.

ACKNOWLEDGMENTS

Our thanks are due to Dr. W. K. Anderson for the use of data on the soil amelioration experiment and valuable suggestions in the preparation of the manuscript. We also thank Dr. Karl Harmsen for soil conductivity analysis, Dr. Samir Ahmed for assistance in planting the 1980–1981 field screening tests and Mr. Bahij Kawas for his overall assistance in the project. Financial support of the Natural Sciences and Engineering Research Council of Canada for the work of S. Jana is gratefully acknowledged.

REFERENCES

1. J. S. Boyer. Plant productivity and environment. *Science* **218**, 443–448 (1982).
2. E. Epstein, J. D. Norlyn, D. W. Rush, R. W. Kingsbury, D. B. Kelley, G. A. Cunningham, and A. F. Wrona, Saline Culture of Crops: A genetic approach. *Science* **210**, 399–404 (1980).
3. K. Mengel and E. A Kirby. *Principles of Plant Nutrition.* International Potash Institute, Berne, 1978.
4. H. C. Weltzien, and J. P. Srivastava. Stress factors and barley productivity and their implications in breeding strategies. Proc. of fourth International Barley Genetics Symposium. Edinburgh, 1982.
5. R. H. Nieman, and M. E. Shanon. Screening for salinity tolerance. In M. J. Wright, ed., *Plant Adaptation to Mineral Stress in Problem Soils.* Proceedings of Workshop, Beltsville, Maryland, 1976.
6. M. C. Shannon and M. Akbar. Breeding plants for salt tolerance. Proc. Biophysics of Cell Membrane Development of Plants for Salt Tolerance. Faisalabad, Pakistan, 1978.
7. A. M. Al-Shamma. A genetic study of salt tolerance in barley. M. S. thesis, Michigan State University, 1979.
8. R. A. Abo-Elenin, M. S. Heakal, A. S. Gomaa, and J. G. Moseman. Studies on salt tolerance in barley and wheat: II. Sources of tolerance in barley germplasm. Proc. of Fourth International Barley Genetics Symposium, Edinburgh, 1982.
9. A. D. Ayers. Germination and emergence of several varieties of barley in salinized soil cultures. *Agron. J.* **45**, 68–71 (1953).
10. T. J. Donovan. and A. R. Day. Some effects of high salinity on germination and emergence of barley. *Agron. J.* **61**, 236–238. (1969).
11. J. D. Norlyn. Breeding salt tolerant plants. In *Genetic Engineering of osmoregulation: Impact on Plant Productivity for Food, Chemicals and Energy.* Plenum, New York, 1980.
12. C. W. Schaller, J. A. Berdegue, C. W. Dennett, R. A. Richards, and M. D. Winslow. Screening the world barley collection for salt tolerance. Barley Genetics 4: 389–393. Edinburgh University Press. 1981.
13. A. D. Ayers, J. W. Brown, and C. H. Wadleigh. Salt tolerance of barley and wheat in soil plots receiving several salinization regimes. *Agron. J.* **44**, 307–310 (1952).
14. N. Maddur. The inheritance of salt tolerance in barley. Ph.D. thesis, Michigan State University, 1976.
15. S. R. Chapman, and L. Hart. Developing salt tolerant crop varieties. In D. Warren, ed., *NOW,* 13, 10–11. Montana State University, 1977.
16. E. Epstein. Genetic potential for solving problems of soil mineral stress: Adaptation of crops to salinity. In M. J. Wright, Proc. of Workshop, Cornell University Exp. Stn. Special Bull., New York, 1977.
17. R. S. Rana. Plant Adaptation to Soil Salinity and Alkalinity. Proc. Indo. Hungarian Seminar, Mangement of Salt Affected Soils. Soil Salinity Research Institute, Karnal, India, 1977.

18. S. Jana. Canada-ICARDA collaboration for cereal germplasm conservation. *Plant Genet. Res. Newslett.* **49,** 5–10 (1982).

19. M. J. Morris, S. Jana, A. Hakim-Elahi, A. L. Kahler, and R. W. Allard. Genetic variability for isozyme characters in Syrian wild barley populations (in preparation).

20. W. K. Anderson. Effect of soil amelioration in salt affected soil. Annual Report. International Centre for Agricultural Research in Dry Areas, Aleppo, Syria, 1981.

16

THE ROLE OF HALOPHYTES
IN IRRIGATED AGRICULTURE

James W. O'Leary

Environmental Research Laboratory
The University of Arizona
Tucson, Arizona

We depend on plants for a substantial amount of our daily resources. All the food we eat ultimately is provided by plants, and much of our fuel, building materials, clothing, pharmaceuticals, industrial chemicals, and so on, comes from plants. Furthermore, most of these things come from domesticated plants. Almost all of the plants selected for domestication were plants that in their native environments were provided with essentially fresh water (1, 2). This was not necessarily intentional, but rather was the consequence of the geographic patterns of the development of human civilization. However, since population growth has greatly increased the demands for plant resources, it has become clear that it is indeed unfortunate that our domesticated plants have come primarily from ancestors that were dependent on fresh water. Fresh water is becoming increasingly scarce in general, and for agricultural use in particular (3).

A brief consideration of the two major aspects of the domestication process will indicate the nature of the present dilemma involving crops and fresh water. The first aspect utilized selection and breeding to increase the desirability of the plants for man's purposes. This involved in most cases, increasing sink size (e.g., increased size and/or number of seeds). The second aspect provided an "energy subsidy" to the plants to enable them to fill the increased sink capacity (4). Depending on the plants' characteristics and the environment in which they are growing, there is some finite amount of energy available to them, resulting from photosynthesis. In their native environments, plants spend a considerable amount of that energy in coping with the environment, such as resisting insects and diseases and competing with other plants for light. One of the most costly aspects involves competing with other plants and environmental losses for water and minerals. This is accomplished by building extensive root systems, which in many plants requires an allocation of over one-half of the plant's available energy (5, 6). However, when those plants are planted in rows to maximize light interception, when other competitors (weeds) are eliminated, and when water and minerals are delivered right to the base of the plants in near optimal amounts, the plants do not have to invest as much of their own energy in extensive root systems, or in stems to raise the leaves up higher than competitors. That energy not used is thereby potentially available for filling the increased sinks (e.g., seeds) on the plants. This is the "energy subsidy" we provide to domesticated plants. Unfortunately, a large part of that energy subsidy often involves providing large amounts of fresh water to the plants. Early in our history, this probably did not seem like a harmful or costly practice, and it is not surprising that agricultural crop production has evolved in this way.

Now that the availability and cost of fresh water has become a major constraint for irrigated agriculture, however, it is clear that the course of events just outlined has, in fact, created a serious problem. That problem is exacerbated even more by the continual loss of good farmland (7). With the loss of farmland to urbanization and salinization, the demand for high productivity per unit of land area is increased on the remaining land. Since productivity and water consumption are tightly coupled, even in wild plants, high productivity can only be achieved at the expense of high water consumption, and since our domesticated plants require fresh water, that high water consumption is high fresh water consumption.

There are two valid approaches to solving the problem. One approach is to increase the salt tolerance of our currently used crop plants. This has been, is now, and will continue to be an important area of agricultural research. This will permit use of water having lower quality and thereby reduce some of the demand on the decreasingly available, higher quality water. It also will extend the time before a given area becomes unusable due to salinity buildup, and it may even open up some marginal, presently unused, agricultural land. This will reduce the impact of the loss of farmland to urbanization and salinization. However, in order to greatly extend the salinity range over which crop growth will be possible, and to open up vast areas of new agricultural land, the second approach becomes necessary.

The second approach is to develop new crops to meet the needs of mankind, new crops that have high productivity when provided with highly saline water, even seawater. That is, rather than increasing the salt tolerance of relatively salt-sensitive, contemporary crop plants, we should improve the crop characteristics, through selection and breeding, of highly salt-tolerant plants. There is no reason to believe, or even suspect, that among those wild plants whose water sources are highly saline, representatives would not also be found with desirable qualities, such as high food value, high potential for fiber or fuel use, significant amounts of valuable chemicals, and so on, and would be good candidates for domestication. Thus, those highly salt-tolerant plants, or halophytes, represent a large, valuable, unexploited resource.

There is no lack of information concerning halophytes in the literature. They have been the subjects of numerous ecological and physiological studies, and evidence for the wealth of available information is given by the number of reviews within the past decade (8–14). Similarly, there has not been a complete lack of attention given to the promotion of halophytes as potential crop plants (15–22). Several times the suggestion has been made that halophytes should be considered good candidates for use as pasture or range forages (15–17) or even as cultivated crop plants (18–22). Much of that promotion has been largely speculation based on extrapolations and correlations. For example, it long has been known that various species of *Atriplex* have been grazed by sheep and other animals (15, 16). Since *Atriplex* will grow in many places where other range plants will not, and since it has a reasonably good nutritive value, it has successfully been used as a supplemental forage (23). There even are intensive efforts under way to select better species and/or varieties of *Atriplex* and other plants to use for seeding rangelands (24). However, those efforts are directed toward noncultivated and nonirrigated uses, and in most cases, to what could best be called mildly saline conditions. As for cultivated, irrigated crops, most of the promotional work has been based on citations of high productivities for halophytes in marshes, estuaries, and so on, that can be found in the literature (25–28). Unfortunately, there have been few cases where test plantings of halophytes simulating cultivated and irrigated conditions were made and productivities measured under those conditions (20, 29, 30).

Since much of that literature has been recently reviewed (20, 30), I will concentrate on the issues that have not been adequately covered in any of those earlier optimistic promotions of halophytes as future crop plants. I will discuss the roles halophytes might occupy in irrigated agriculture, the kinds of crops that might be developed, how the process of halophyte domestication should be accelerated, and what major problems will have to be overcome before halophytes really will become useful, economically important, irrigated crop plants.

1.   POTENTIAL ROLES OF HALOPHYTES

1.1.   Use to Reclaim Salinized Soil

The use of halophytes to harvest salt from a salinized soil at first sounds like an attractive, and even reasonable, scenario, which probably explains why this suggestion has been made (16). However, even if water that is considered fresh water is used to grow a plant that accumulates salt, the amount of salt contributed by the irrigation water may be more than the plant can accumulate, and there is no net removal of salt from the soil by the plant. In fact, the salinity of the soil probably would increase in most cases. Consider the following example. Assume a transpiration ratio of 500 g water per g dry weight of harvested plant material, a representative value. If the irrigation water were 1 ppt, or 1 g salt per 1000 g water, then 0.5 g salt would be delivered to the soil with every 500 g irrigation water. If the plant could accumulate salt to the point where it represented 50% of the total dry weight, then the amount of salt removed by the plant would be exactly equal to the amount added to the soil by the irrigation water during growth of the crop. Most of the time, the salinity level of the irrigation water in such situations probably would be higher, and the tissue salt content probably would be lower, so the net effect would be increased salinization of the soil. Thus, it is unlikely that halophytes will play any role in reclamation of salinized soil in irrigated fields. There are some scenarios in which halophytes could be envisioned in the role of harvesting salt from the soil, but they all are under nonirrigated conditions.

1.2.   Multiple Use of Irrigation Water

There is increasing interest in multiple use of irrigation water as a way of increasing efficiency of water use. This involves irrigating a crop, then using the drain water, or irrigation return water, to irrigate a second crop, and possibly more, before discarding the water. The intent is to minimize the amount of wastewater from irrigation use, and the ultimate goal would be to return none to the source. To fully realize these objectives requires having a series of crops with increasing salinity tolerance. A general model for such a system is presented in Figure 16.1. The Type A crop would be a conventional crop typically irrigated

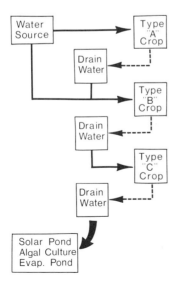

Figure 16.1.   Model for a multiple water use system in irrigated agriculture.

with fresh water from a surface or underground source. The drain water from that crop could be blended with the fresh water and used to irrigate a crop with higher salt tolerance (Type B crop). Some crops (Type C crop) could be irrigated with undiluted drain water of increasing salinity. The total number of crops involved and the number of each type would depend on several factors, but the use of halophytes would greatly extend the range of salinity over which this multiple use could extend. The end of the line could be a pond or ponds in which highly salt-tolerant organisms, such as *Dunaliella*, could be raised, or used as solar ponds to collect energy, before the water evaporates and the salt is harvested.

1.3. Replacement of Present Crops in Existing Irrigated Areas

As the salinity of the soil and/or irrigation water increases to the point where reclamation no longer is feasible, the only prospect now is to retire the involved fields from cultivation. If suitable halophytic crops were available to replace the relatively salt-intolerant crops in such situations, this would extend the time, maybe indefinitely, for continued use of that land. As long as the soil conditions were such to permit continued use of the brackish water, this could be a profitable alternative, since the halophytic crops should have high productivity due to the energy subsidy received by those plants. The salinity levels that make land nonusable for contemporary crops are far below the salinity levels tolerated by halophytes, and when halophytes are grown with water less saline than in their native environments, the productivity increases substantially (18, 31, 32). For example, salinity levels of 5–10 ppt that prevent growth of almost all contemporary crops result in maximum productivity of many halophytes because of the large amount of "subsidizable" energy that is available. Halophytes spend a considerable amount of their energy coping with the high salt content in their native habitats. If the salt content of their environment is reduced to 5–10 ppt (e.g., compared to 30–40 ppt in their native environment), this is a considerable energy subsidy to the plants, which is reflected by increased growth (Fig. 16.2).

1.4. Bringing into Use New Land and Water Sources for Irrigated Agriculture

There are thousands of miles of sandy desert seacoast, largely unused, that are adjacent to an unlimited supply of water, albeit highly saline. For countless generations, people have probably stood on such coastlines wistfully looking out to sea and dreamed of using that water to grow crops in those areas. That dream is within our reach, and always has been. Halophytes could be grown as irrigated crops in many such areas. As the pressures of increasing population

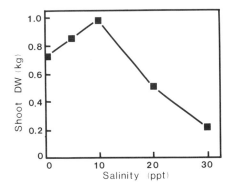

Figure 16.2. Dry weight of above ground biomass from *Atriplex barclayana* grown at five salinity levels. Salinity of irrigation water is given in parts per thousand (ppt) of total dissolved salts. Plants were harvested 11 weeks after seedlings were transplanted into the treatment plots. Each point is the average of nine plants.

and decreasing farmland availability become more severe, it is imperative that we select and develop promising halophytes for crop use in those areas. The questions no longer are can, and should, we do it. The question is how we can most effectively and most quickly do it.

2.   THE MOST LIKELY KINDS OF CROPS

Now that it has been established that there are some feasible potential roles for halophytes in irrigated agriculture, the pertinent question becomes: What kinds of crops might most likely be derived from wild halophytes?

2.1.   Forage and/or Fodder Crops

Since *Atriplex* and other halophytes have been used as food sources by animals already, as described above, it is logical to assume that the most likely initial irrigated halophyte crops would be those used as animal forage or fodder. In fact, these are the types of halophytes primarily chosen by the research groups at the University of Arizona, Ben Gurion University of the Negev, and the University of Delaware for intensive investigation in their seawater irrigation test plots (21, 29, 30). However, careful consideration of the situation indicates that this may not be as optimistic a scenario as first envisaged. All the previous instances in which *Atriplex* and other halophyte foliage were used as animal feed involved plants grown in situations where the salinity was considerably less than that of seawater, or even of reasonably brackish irrigation water. Thus, the ash content of foliage in those cases was probably considerably lower than what it is in plants irrigated with highly saline water. Even in those previous cases, though, the animals concerned usually ate the halophytes only when there was no alternative. If we assume that any halophytes grown as forage or fodder crops will be irrigated with water of reasonably high salinity, then the high salt content of the foliage will be a serious constraint to their use. There are two alternatives for the use of such foliage. One possibility is to wash or leach much of the salt out of the tissue. This seems like an unrealistic approach. The amount of water and energy required would make it a costly crop, too costly to be a feasible forage or fodder crop.

The alternative approach is to incorporate the halophyte crop into a mixed feed. In this way, the plants could be utilized in their unaltered form, and the percentage of the total feed represented by the halophytes would be set by the amount that would fall under the threshold salt level for the animals' diet. For example, in a prepared feed like chicken feed, halophyte tissue with a salt content as high as 40% of the dry weight could be used at levels of 10-20% of the total feed. This would satisfy the entire salt requirement in the feed and could replace entirely the green plant tissue component, such as alfalfa, often included in most chicken feeds in several countries (33, 34). In like fashion, halophyte tissue could be chopped and mixed with other plants to make hay for ruminant animals.

In summary, I think halophytes could be successfully used as fodder type crops in limited ways, but I do not think they could be used as straight forages where that was the primary feed source. On the other hand, there may be some halophytes that will be found to have relatively low tissue salt content when grown with water having high salinity. Screening for this characteristic should be maintained.

2.2. Grain or Seed Crops

Even though the vegetative tissues of a plant may contain extremely high salt levels, the seeds of those same plants normally have low salt levels. In spite of the range of salt levels found among plant leaves, almost all seeds we have examined have salt contents below 10%. Thus, the seeds may be the most likely part of a halophyte to find immediate use. The reason for seed crops not having been considered as likely initial successes as forage/fodder crops by most people probably is due to the relatively small seeds normally found among halophytes. However, if the total number of seeds produced per plant is high enough, even with very small seeds, the seed yield per unit of land could be reasonably high. Another consideration is the amount of variability in seed size that might be found within a species. If there is a wide variability, then selection and breeding could be used to obtain greater seed size. Such a situation exists for *Distichlis*, for example (35). Furthermore, the nutritional qualities, baking characteristics and taste of the flour made from *Distichlis* seeds compares very favorably with wheat flour (N. Yensen, personal communication).

I think increased attention should be given to looking for potential seed or grain crops, and it should not be limited to grasses or cereal-type plants. Oilseed crops may be good possibilities. The advantage of having a crop with two valuable products (vegetable oil and protein-rich meal) increases the chance of it becoming a successful, economically feasible crop.

2.3. Fuel Crops

The possibility of using halophytes, such as mangroves, for fuel crops has been suggested (36) and, in fact, mangroves have been used as a source of charcoal for over a century. There are other woody halophytes that might also be considered and that might be more feasible for use under a wider range of environmental and soil conditions. However, since the concept of fuel crops per se still is in the early developmental phase (37, 38), it does not seem likely that this will be among the first successful use of halophytes. Furthermore, the high ash content of halophytic wood may pose a serious constraint to its use as fuel. Nevertheless, the vicious circle of increasing desertification due to removal of plants for use as firewood and the increasing removal of native plants from more extensive areas due to lowered productivity resulting from desertification in many areas may be a strong enough pressure to force cultivation of halophytic fuelwood crops in certain areas.

2.4. Landscape Plants

Many halophytes are extremely attractive and could be used as landscape plants immediately. With increasing pressure to reduce use of fresh water for nonessential uses, the market for landscape plants that could be grown with lower quality water may increase. If it does, cultivation of halophytes for use as ornamental plants would begin immediately. Desirable plant characteristics already are present, and all that is needed is the market for the product.

Similarly, some of the plants like *Limonium* already have counterparts in the florist industry, and they could almost immediately be cultivated with salt water irrigation and used for cut flowers.

2.5. Others

There is no reason not to believe that among the halophytes there could be found good sources of pharmaceuticals, chemical feedstocks for industry, perfumes, or related products. Screening for such compounds should be increased, since these are all relatively high value products.

3. THE PROCESS OF ACCELERATED DOMESTICATION

In order to exploit this valuable resource, a systematic process of accelerated domestication should be followed. That is, a coordinated plan of attack should be developed and used as a guide to research in this field, and international cooperation should be an integral part of it. The following outline is suggested as such a plan.

1. Assemble from a wide range of sources a germplasm bank of halophytes.
2. Screen and catalog them for their tolerance to various levels of salinity up to seawater and even beyond.
3. Screen them for beneficial properties, such as food value, chemical content, and so on.
4. Determine the productivity of the promising candidates at various salinity levels.
5. Produce and accumulate a large supply of seeds of the most promising candidates.
6. Distribute the seeds and grow the potential crop plants under actual agricultural conditions in representative prospective locations.
7. Use appropriate genetic techniques to improve the desirable crop characteristics.
8. Conduct economic analyses of crop/location scenarios.

Now each of these will be explained in more detail, and some of the progress already being made will be reviewed.

3.1. Germplasm Connection

In order to survey the halophyte population of the world in a systematic manner, it is necessary to assemble, from a wide range of sources, a collection of materials for analysis. These materials should be available for use by all investigators in much the same way that the materials from the germplasm collections of many of our cultivated crops are available. It is not necessary that all of the materials be kept in one location. In fact, it may even be preferable to have many of the items replicated in several locations. A good example of a halophytic germ plasm collection that is designed for specific needs in a specific area is the one started by Malcolm (24) several years ago. Collections have been made from several locations around the world to find suitable forage shrubs for revegetating salt-affected lands in Australia. In contrast, the collections being assembled at the Applied Research Institute of the Ben Gurion University of the Negev and at the Environmental Research Laboratory of the University of Arizona are much broader in scope.

3.2. Screening for Salt Tolerance

One of the problems of making quick collecting trips to unknown areas is the difficulty of knowing how salt tolerant the collected plants or seeds may really be. The location of a plant right next to a seacoast, for example, does not ensure that it is even a halophyte, particularly in the case of annuals. There may be freshwater lenses being tapped by the plant in question, or it may be able to complete its life cycle on fresh water resulting from a rainfall, and so on. Thus, we routinely screen all accessions for their salt tolerance. We have tried to find a rapid screening procedure that allows us to catalog these accessions into one of two categories: those that would be suitable for consideration as potential crop plants using seawater or equivalent, and those that would only be suitable for consideration as potential crop plants in brackish water situations.

Since most halophytes have poor germination at high salinities (14, 39), and since the germination response to salinity level usually has poor correlation with behavior at later stages of growth (39), we have excluded germination response as part of this initial, rapid screening procedure. Seeds are planted in sand, germinated under fresh water, and then transferred to flood tables where they are irrigated with water having a salt content of 10 ppt. After 1 week, some of the seedlings are moved up to a flood table which is irrigated with water having a salt content of 20 ppt. In the same way, plants are subsequently moved up to 30 ppt and finally to 40 ppt, with some being left behind at each level. After a suitable period of time, usually 3–4 weeks after the highest salinity level was reached, the percentage survival of the seedlings at each salinity level is recorded, and the weight of those plants is determined. All plants fall into one of two categories. Some have maximum growth at the lowest salinity level and decrease steadily with increasing salinity. The others have maximum growth at 10 ppt salinity and have poorer growth as salinity changes in either direction. Based on Chapman's (40) classification system, we have labeled these two groups as miohalophytes and euhalophytes, respectively. When all of those that fit in the euhalophyte category (15 species) and all of those in the miohalophyte category (12 species) from the first 27 species we have screened in this way are averaged, the general relationship is as shown in Figure 16.3. All 15 species in the euhalophyte category had 100% survival up to 40 ppt salinity, whereas the others were highly variable. Only two of the 12 miohalophytes had 100% survival at 40 ppt, and most had far less then 50% survival.

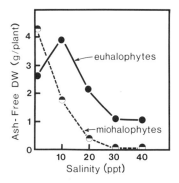

Figure 16.3. Dry weight of seedlings from salinity tolerance screening trials, harvested 8 weeks after seeding. Salinity of water used is given in parts per thousand (ppt) of total dissolved salts. See text for complete explanation of treatments. Average of 15 species for euhalophytes and 12 species for miohalophytes.

As a result of our experience so far, we have decided to streamline, and make even more rapid, the screening procedure as follows. Since the big difference between the two categories is the relative growth at zero versus 10 ppt salinity, we will henceforth only compare growth at those two salinity levels, and on that basis, they will be classified as euhalophytes or miohalophytes for our purposes.

3.3.   Screening for Beneficial Properties

Obviously, some preconceived uses must be in mind when one decides for which characteristics any halophyte should be analyzed, since it is unrealistic to expect that every one should be analyzed for every conceivable attribute. For example, historically, and even up to the present time, there has been considerable emphasis on use of halophytes as forage or fodder crops (see Section 2.1). As a result, routine screening for nutritional value has been done. We have found that over the salinity range from 0 to 40 ppt, the protein level of the foliage typically falls in the range of 15–20% of the dry weight. However, we have found in one of the euhalophytes that had 100% survival up to 40 ppt a protein content of 25–30%. This is a plant collected in Argentina that tentatively has been identified as *Atriplex triangularis*.

We also have measured vegetable oil content in seeds whenever we have had enough material for analysis, and some of the seeds have oil contents of 40–50%. As mentioned before (see Section 2.2), the possibility of a halophytic seed crop is particularly attractive due to the low ash content of seeds, even when the plants are grown at extremely high salinity.

As part of the screening process, we also have examined digestibility of the plant tissues. Standard *in vitro* digestibility techniques using buffered rumen fluid are used, and a standard alfalfa hay sample is included in each run. So far, we have looked at 45 species of halophytes, and the *in vitro* organic matter disappearance in 85% of them was equal to, or better than, the alfalfa standard. Furthermore, the salinity content of the tissue had no detrimental effect on the digestibility.

What is needed now are large-scale feeding trials with live animals to intensively examine feed conversion ratios, palatability, and acceptability of the material by the animals, and other aspects of practical use of the crops. Large-. scale production of halophytes for use in such trials has not yet been accomplished. Small-scale feeding trials conducted at our laboratory have been very encouraging (41).

3.4   Productivity Measurements

Even if the protein content of the leaves or oil content of the seeds is high, it is of little value if the total leaf or seed production per plant is low, or if the yield decreases significantly with increasing salinity. Thus, as soon as a halophyte is identified as having promise following the screenings outlined above, that plant should be grown outdoors during the normal crop season at different levels of salinity in order to determine the productivity or yield as a function of the salinity of the irrigation water, just as has been done for contemporary crop plants at the U.S. Salinity Laboratory (42) and elsewhere. We have constructed a facility for this purpose, in which crops can be grown in sand beds at several levels of salinity up to seawater concentration and beyond. Results from one of the first crops to be tested (*Atriplex barclayana*) are presented in Figure 16.2. As shown by this example, so far the field results validate the results of the quick-screening greenhouse technique. This example also emphasizes the point made earlier (Section 1.3) that productivity of certain halophytes should be very high when grown on brackish water.

3.5 Increasing Seed Supply

One of the most limiting aspects of any accelerated process of domestication is the availability of enough seeds for the necessary work. Typically, only small amounts of seed are available for most new accessions to the germ plasm collection. Thus, it is necessary to make plantings of these candidates for the purpose of harvesting a seed crop. This is a laborious procedure, and so far, very little of this type of work has been done. It takes a considerable amount of work just to produce enough seeds for the work outlined above, especially if numerous different species are involved; in our case, we still do not have large enough quantities of seeds available for general distribution. Since seed companies will not become interested in any of the plants until a potential market is established, it will remain for researchers in this area to develop their own seed increase programs.

3.6 Growth of Potential Crops Under Actual Agriculture Conditions

Once the feasibility of using some halophyte for a potential crop is demon-strated, it is necessary to actually grow it under agricultural conditions where methods of seeding, cultivating, and harvesting can be investigated. This is the step that must be accomplished before interest is generated among the agricultural industry hierarchy. Not until this step is accomplished will seed companies become interested in producing seed of that crop, for example. It must be demonstrated that this plant can be grown, handled, and marketed as a crop.

There is no halophyte yet ready for cultivation as an irrigated crop. However, with the current level of interest and effort, this point should be reached in 5–10 years. If there is a significant increase in effort, we could see large-scale planting of a potential halophyte crop in less than 5 years.

3.7 Genetic Improvement

As soon as any halophyte starts being seriously considered as a potential crop plant, there should be efforts toward improvement of the crop characteristics, just as there is with any crop. If the seed is the desired product, selection and breeding for increased seed size and / or number should be initiated, for example. The exact characteristics that should be the object of genetic improvement, and the specific genetic techniques to be used, will depend on the particular plant that is the object of these efforts.

3.8 Economic Analysis

No matter how attractive the products of any potential halophytic crops may be, if there is no economic benefit to be derived from cultivation of that crop, there probably will be no crop. Thus, once serious consideration is being given to a potential halophytic crop, there should be economic analyses of the particular crop and location combinations. Depending on the product, some crops may be judged economically feasible for some locations, but not for others. This sort of analysis will inevitably be initiated for any candidate by the time it reaches steps 4 or 5 in the process outlined previously.

4. MAJOR PROBLEMS TO BE OVERCOME

The major problems to be overcome before a wild halophyte becomes a domesticated, beneficial, economically feasible, irrigated crop plant are not directly plant related. That is, I think there is abundant evidence for halophytes having high productivity under saline irrigation conditions and for having reasonably high contents of valuable products, and so on, and relatively little work is needed in the area of plant improvement. Continued searching and screening, as outlined above, are necessary to find the best possible candidates for use as crop plants, but the major limitations are soil-related problems. Since our focus here is on plant improvement, I will not discuss those problems in detail, but I would like to at least mention what I see to be the major limitations to domestication of halophytes.

Based on well-known responses of soils to salinity (43, 44), it is clear that the higher the salinity of the irrigation water to be used, the sandier the soil should bc, and if seawater is to be used, then almost pure sand is required. In addition, the soil must be irrigated frequently enough, and with sufficient volume of water, to ensure adequate leaching, as well as to prevent the soil moisture from becoming too low. The important point is that, if highly saline water is used, there must be frequent irrigation of highly permeable, well-drained sandy soil to prevent salt accumulation. There are two obvious problems that arise in such a situation.

The first problem concerns the application of water. Since the sandy soil drains so well, if surface flooding is used, there will be a considerable amount of water that is lost to downward flow before enough surface flow occurs to adequately wet the soil at the point farthest from the point of water release. That is, an extremely large quantity of water will be required for flood irrigation. Even if the water were free, the pumping and delivery costs of the water could be prohibitive. The ideal situation would be to lay a thin layer of water on the surface and then release it into the soil, whenever irrigation is necessary. Sprinkler irrigation can be used to accomplish that, but so far this has proved to be unsuitable when brackish water is used, due to foliar burning (45). The tolerance of halophyte foliage to salt spray is not well known, so research in this area might prove profitable. Limiting the periods of sprinkling to nighttime should minimize the foliar burning problem, but it is not known whether large plants growing in pure sand in extremely arid areas would have enough water available to last the entire day. Drip irrigation with extremely saline water has not proved feasible either. Large, linear travel rigs with drip tubes might be used to irrigate with saline water, but would require a combination of plastic pipe to carry the water with metal framework for support. All these possibilities, and other new irrigation technologies, need investigation.

The other problem concerns fertilizer application. If well-drained sandy soil is irrigated continually to leach or flush soluble salts out of the root zone, then desirable salts such as nitrates and phosphates also will be leached away. That is, there will be no storage of nutrient salts in the soil. In that case, the nutrients would have to be supplied in the irrigation water. Since much of the irrigation water would pass quickly through the root zone and provide little opportunity for absorption of salts by the roots, there would be a considerable waste of costly fertilizer salts. Extremely dilute concentrations of some salts may be sufficient. Maybe there will have to be fertilizer irrigations (relatively low volume) alternated with leaching irrigations (comparatively higher volume). Since many coastal halophytes have very efficient mechanisms for harvesting nutrients from dilute solutions of those nutrients (46, 47, 48), the requirements may be met with such a schedule. On the other hand, it may be possible to use some form of slow-release fertilizer than can be immobilized at or near the soil surface and provide small amounts of soluble nutrient salts at every irrigation.

There is a need for much research before halophytes occupy a significant role in irrigated agriculture. Nevertheless, I feel it is inevitable that they will. It is important, though, that such a role neither be prematurely expected nor promised.

REFERENCES

1. S. Struever. *Prehistoric Agriculture*, 1st ed. Natural History Press, Garden City, New York, 1971.
2. R. Tannahill. *Food in History*, 1st ed. Stein and Day, New York, 1973.
3. G. D. Barney. *The Global 2000 Report to the President*, 1st ed. U. S. Government Printing Office, Washington, D.C., 1981.
4. L. T. Evans. *Crop Physiology. Some Case Histories*, 1st ed. Cambridge University Press, London, 1975.
5. M. M. Caldwell. Root structure: The considerable cost of belowground function. In O. Solbrig, S. Jain, G. Johnson, and P. Raven, eds., *Topics in Plant Population Biology*, 1st ed. Columbia University Press, New York, 1979, p. 408.
6. H. A. Mooney. *Bot. Rev.* **38**, 315 (1972).
7. P. R. Crosson. *The Cropland Crisis. Myth or Reality?*, 1st ed. Johns Hopkins University Press, Baltimore, 1982.
8. T. J. Flowers. Halophytes. In D. A. Baker and J. L. Hall, eds., *Ion Transport in Plant Cells and Tissues*, 1st ed. North-Holland, Amsterdam, 1975, p. 309.
9. T. J. Flowers, P. F. Troke, and A. R. Yeo, *Ann. Rev. Plant Physiol.* **28**, 89 (1977).
10. R. L. Jefferies, *BioScience* **31**, 42 (1981).
11. R. Jones, ed., *Biology of Atriplex*. CSIRO, Canberra, 1970.
12. C. B. Osmond, O. Bjorkman, and D. J. Anderson. *Physiological Processes in Plant Ecology. Toward a Synthesis with Atriplex*, 1st ed. Springer-Verlag, Berlin, 1980.
13. D. N. Sen and K. S. Rajpurohit. *Contributions to the Ecology of Halophytes*, 1st ed. Junk, the Hague, 1982.
14. I. A. Ungar, *Bot. Rev.* **44**, 233 (1978).
15. J. R. Goodin. In J. R. Goodin and D. K. Northington, eds., *Arid Land Plant Resources*. ICASALS, Lubbock, Texas, 1979, p. 418.
16. J. R. Goodin. Atriplex as a forage crop for arid lands. In G. A. Ritchie, eds., *New Agricultural Crops*. Westview Press Boulder, Colorado, 1979, p. 133.
17. C. M. McKell. *Science* **187**, 803 (1975).
18. P. Mudie. In R. J. Reimold and W. H. Queen, eds., *Ecology of Halophytes*. Academic Press, New York, 1974, p. 565.
19. G. F. Somers. In A. Hollaender, J. C. Aller, E. Epstein, A. San Pietro, and O. R. Zaborsky, eds., *The Biosaline Concept*. Plenum, New York, 1979, p. 101.
20. G. F. Somers. In A. San Pietro, ed., *Biosaline Research*. Plenum, New York, 1982, p. 127.
21. G. F. Somers, M. Fontes, and D. M. Grant. In J. R. Goodin and D. K. Northington, eds., *Arid Land Plant Resources*. ISACALS, Lubbock, Texas, 1979, p. 402.
22. M. Z. Zahran and A. A. Abdel Wahid. In D. N. Sen and K. S. Rajpurohit, *Contributions to the Ecology of Halophytes*, 1st ed. Junk, The Hague, 1982, p. 233.

23. J. H. Leigh and A. D. Wilson. In R. Jones, ed., *The Biology of Atriplex.* CSIRO, Canberra, 1970, p. 97.

24. C. V. Malcolm. *J. Agric. W. Austr.* **15,** 69 (1974).

25. C. J. Dawes. *Marine Botany,* 1st ed. Wiley, New York., 1981.

26. J. H. Day. Estaurine flora. In J. H. Day, ed., *Estuarine Ecology,* 1st ed. Balkema, Rotterdam, 1981, p. 77.

27. S. P. Long and H. W. Woolhouse. Primary Production in Spartina marshes. In R. L. Jefferies and A. J. Davy, eds., *Ecological Processes in Coastal Environments.* Blackwell, Oxford, 1979, p. 333.

28. W. A. Niering and R. S. Warren. *BioScience* **30,** 301 (1980).

29. E. P. Glenn, M. Fontes, and N. Yensen. In A. San Pietro, ed., *Biosaline Research.* Plenum, New York, 1982, p. 491.

30. D. Pasternak. In A. San Pietro, ed., *Biosaline Research.* Plenum, New York, 1982, p. 39.

31. M. G. Barbour, J. H. Burk, and W. D. Pitts. *Terrestrial Plant Ecology,* 1st ed. Benjamin/Cummings, Menlo Park, California, 1980.

32. J. W. O'Leary. In J. R. Goodin and D. K. Northington, eds., *Arid Land Plant Resources.* ICASALS, Lubbock, Texas, 1979, p. 574.

33. M. E. Ensminger and C. G. Olentine. *Feeds and Nutrition,* 1st ed. Ensminger Publishing, Clovis, California, 1978.

34. P. J. Schaible. *Poultry: Feeds and Nutrition,* 1st ed. AVI, Westport, Connecticut, 1970.

35. R. S. Felger. In A. San Pietro, ed., *Biosaline Research.* Plenum, New York, 1982, p. 473.

36. H. J. Teas. In A. San Pietro, ed., *Biosaline Research.* Plenum, New York, 1982, p. 369.

37. H. R. Bungay. *Energy, the Biomass Options,* 1st ed. Wiley, New York, 1981.

38. D. O. Hall, G. W. Barnard, and P. A. Moss. *Biomass for Energy in the Developing Countries,* 1st ed. Pergamon, Oxford, 1982.

39. Y. Waisel. *Biology of Halopytes,* 1st ed. Academic Press, New York, 1972.

40. V. J. Chapman, *Q. Rev. Biol.* **17,** 291 (1942).

41. E. P. Glenn, M. R. Fontes, S. Katzen, and L. B. Colvin. In A. San Pietro. ed., *Biosaline Research.* Plenum, New York, 1982, p. 485.

42. E. V. Maas and G. J. Hoffman. *J. Irrig. Drainage Div. ASCE* **103,** 115 (1977).

43. E. Bresler, B. L. McNeal, and D. L. Carter. *Saline and Sodic Soils,* 1st ed. Springer-Verlag, Berlin, 1982.

44. R. L. Hausenbuiller. *Soil Science,* 2nd ed. Brown, Dubuque, Iowa, 1978.

45. R. S. Ayers and D. W. Westcot. *Water Quality for Agriculture.* FAO, Rome, 1976.

46. R. L. Jefferies. *J. Ecol.* **65,** 847 (1977).

47. I. Valiela and J. M. Teal. *Nature* **280,** 652 (1979).

48. I. Valiela and J. M. Teal. Inputs, outputs, and interconversions of nitrogen in a salt marsh ecosystem. In R. L. Jefferies and A. J. Davy, eds., *Ecological Processes in Coastal Enviornments.* Blackwell, Oxford, 1979, p. 399.

17

PHYSIOLOGICAL GENETICS OF SALT RESISTANCE IN HIGHER PLANTS: STUDIES ON THE LEVEL OF THE WHOLE PLANT AND ISOLATED ORGANS, TISSUES AND CELLS

Moshe Tal

Department of Biology
Ben-Gurion University of the Negev
Beer Sheva, Israel

Evaluation of the prospects for increasing the quality and quantity of the world's food supply, while applying the energetically least costly methodologies, is of immediate priority because of the increase of the population pressure and the cost of energy (1). The increasing demands of the expanding population for food and energy necessitate the increase of arable land by exploiting marginal areas such as arid and semiarid lands, which comprise about 40% of the world's land surface (2). Such areas are characterized by high salinity in the soil and in the major water resources, which include underground brackish water and seawater (3–6). Increasing salinity of soil and water threatens agriculture in the arid and semiarid zones (7) and in other regions of the world. About one-third of the world's irrigated land is already affected by excess of salinity (8, 9).

The technological practices used by modern agriculture to manipulate the environment to make it suitable for the plant depends on a large input of energy and should, therefore, be supplemented by the biological approach, that is, production of altered plant varieties to suit the environment. While this approach has been used extensively to increase yield and the resistance to diseases or to improve the chemical composition, very little has been done to improve the adaptation of plants to stress conditions, such as extreme temperatures, drought, and high salinity. The feasibility of exploiting natural or induced genetic variability for developing plants better adapted to stresses is being increasingly appreciated in recent years (7, 10–13). The genetic approach for altering the plant to suit the conditions characteristic of saline environments may range from conventional methods of breeding (7, 14) to genetic engineering (9, 15, 16). One of the major prerequisites for this approach is the availability of genetic variability. In some species, genetic diversity of salt resistance occurs quite extensively among their cultivars (3, 10, 17, 18). In species in which such variation is limited or lacking, genes can be transferred from their wild salt-resistant relatives (19, 20). The existing germ plasm for salt resistance in crop plants may also be increased by selecting induced or spontaneous mutations for salt resistance in tissue culture. These subjects are discussed in Chapter 13 by Shannon (conventional breeding), Chapter 18 by Stavarek and Rains (cell selection in tissue culture), and Chapter 19 by Hanson (use of cell culture and genetic engineering for breeding).

Due to our ignorance of the regulation of gene action during the process of differentiation and in the differentiated multicellular organisms, we tend to describe the relation between genes and their product, that is, the phenotype, in such organisms by generalized statements, such as "the phenotype of a plant is the result of all the genes of the plant interacting with each other and with the environment" (6). The knowledge on the genetic basis of salt resistance, as on many other characteristics in higher plants, is very limited and it is usually described, therefore, as a complex or intricate attribute (6, 21). At present, most, if not all, of the occupation in the genetics of salt resistance is limited to aspects of breeding crop plants better adapted to high salinity. The efforts of the plant breeder are concentrated in achieving the highest possible yield, which is a product of the system as a whole, under saline conditions. The main, possibly the only, expectation from the plant physiologist of the present time is the identification of "salinity markers," as defined by Epstein et al. (18). Such markers include characters that are associated with salt resistance and are easily identified and, thus, can be used for screening salt-resistant plants in large breeding populations. The lack of such markers in most crop plants is one of the biggest problems of the conventional plant breeder at the present time (9, 22). This may result from (i) the tendency of the physiologist to disregard the whole plant, due to its complexity, and to regard the functions involved as independent parts (23); (ii) the lack of a comprehensive understanding of the principal mechanisms of salt resistance; or (iii) the lack of a sufficient integration between the physiological and genetic methodologies in the study of salt resistance.

The recognition of the advantages and the potential existing in studying isolated components, which can be manipulated, encourage the present tendency to disintegrate the plant system into its organizational components (isolated tissues and cells) and genome components (single genes).

The physiological and genetic basis of salt resistance as well as some of the accomplishments and potential contribution of the interdisciplinary approach and the disintegration of the whole plant system to the understanding and improvement of salt resistance in whole plants are discussed in the following sections.

1. PHYSIOLOGY OF SALT RESISTANCE ON DIFFERENT ORGANIZATIONAL LEVELS

Although appreciable progress has been made in the general understanding of the physiology of the plant–salt relationships, the lack of a comprehensive understanding of the principal physiological mechanisms of salt damage and resistance limits the application of the physiological information to breeding (22). Salinity can damage the plant through its osmotic effect, which is equivalent to a decrease in water activity, through specific toxic effects of ions and by disturbing the uptake of essential nutrients. These modes of action may operate on the cellular and on higher organizational levels (23, 24).

Salt resistance includes both avoidance and tolerance mechanisms (25). The former may operate through active extrusion of ions by specific pumps, passive exclusion of ions due to membrane impermeability, or dilution by rapid growth associated with an increase in water content. An unavoidable consequence of growth in a solution containing a high salt concentration is the development of osmotic stress, which is followed by a loss of turgor. Tolerance to osmotic stress may operate either through dehydration tolerance, which permits the cell to survive without growing when the turgor decreases, or by avoiding dehydration through osmoregulation, which includes an increase of solute concentration in the cell and consequently rehydration. The solutes may be either salt ions, which can be sequestered in the vacuole and osmotically balanced by organic solutes in the cytoplasm, or organic substances. The latter occurs when salt ions are prevented from entering the cells. The first strategy, typical to halophytes (26), was found to characterize the wild relatives of the cultivated tomato (13, 27). The second strategy seems to characterize the salt-resistant varieties of barley, grapes, and soybean (28, 29). According to Greenway and Munns (29), salt sensitivity in nonhalophytes may result from (i) inability of osmoregulation, which may result from either an insufficient uptake of salt ions or a lack of synthesis of organic solutes being used as osmotica; or (ii) injury caused by inorganic ions which are absorbed by the cell and are not compartmentalized. The latter situation characterizes the sensitive varieties of barley, grapes, and soybean. According to Yeo (30), the energetical cost of osmotic adjustment by the inorganic ions absorbed is much lower than that conferred by organic molecules synthesized in the cell. Greenway and Munns (29) recommended to look, when breeding for salt resistance, for exceptional genotypes having a halophytic way to cope with salinity, that is, a high rate of transport of inorganic ions to the shoot synchronized with ion compartmentation in the leaf cells.

The largest areas of saline soils in the world are saline because they occur in the arid and semiarid zones of the world (3). In such areas long periods of drought coincide with high temperatures. Wild species, which evolved in dry areas, for example, the wild relatives of the cultivated tomato and potato, may have, therefore, common mechanisms of resistance against several stresses. A support for the unified concept of stress resistance is provided by parallelism found between the variation in resistance to several stresses and by cases in which hardening of plants to one stress results in increased resistance in other stresses. Such a situation may result from an effect that is common to different stresses. Dehydration, for example is a strain common to water, salinity, heat, and freezing stress (25). Levitt (25) suggested the existence of "a common factor able to induce a partial tolerance to several kinds of different primary stresses." Possible candidates, which may play the role of such a factor, are the ability to prevent membrane changes caused by lipid peroxidation and/or protein aggregation and the accumulation of growth factors, such as abscisic acid (25).

The importance of extending the research on plant–stress interrelationships to the molecular level, including membrane structure and kinetics of enzymes, and to the identification of the regulatory mechanisms controlling adaptation is increasingly appreciated in recent years by scientists studying plant response to stresses (31, 32). The following are central targets, which may have a primary role in salt resistance and for which the recovery of the controlling major genes is recommended: (i) membrane characteristics, which include quality and quantity of lipids and proteins and functions connected with the membrane, for example, ATPase activity (23, 33; Chapter 5); (ii) balance of abscisic acid whose participation in a wide regulatory system of both salt and water balance in the plant is extensively supported (34); (iii) osmoregulation, whose significance in the tolerance against different stresses was emphasized in a recent symposium (35); (iv) the kinetics of enzymes that play a central role in the regulation of intermediary metabolism (36) in the presence of high ionic concentrations and factors stabilizing macromolecules (37).

Tissue culture technique has been recognized during the last decade as a powerful tool for breeding work (38, 39, 40) and physiological studies. It allows the selection of mutations and the potential to cross different species by fusion of protoplasts. These manipulations can only make use of genes which act on the cellular level. Some of these aspects are presented in this conference by Stavarek and Rains, and Hanson.

Physiological studies, in general, and studies of the mechanisms of stress resistance, in particular, can also benefit from tissue culture methodology. Some of the advantages in using tissue culture for physiological studies, which are limited only to mechanisms operating on the cellular level, are as follows (41–44): (i) experiments can be performed year-round since the growth of tissue culture is independent of seasonal fluctuations; (ii) tissue culture can be treated uniformly in a controlled way; (iii) relatively homogenous populations of cells can be developed in tissue culture, as compared with the heterogeneous (developmentally and metabolically) whole plant, thus serving as a potential tool in studying the effect of stress on various components of growth; (iv) tissue culture can be used as a model system for the whole plant but only in studying mechanisms which operate on both the cellular and the whole plant level; (v) the contribution of different parts to the response of the whole plant can be determined by studying their response in culture; (vi) cells in culture are more responsive to environmental stimuli than the whole plant; (vii) mechanisms operating on the cellular level only can be distinguished from those operating on other levels of organization; (viii) naked protoplasts are especially suited for studying the involvement of the surface membrane in stress injury. However, the use of tissue culture may be hampered by the appearance of spontaneous genetic variability (45). Such variability may be advantageous for breeding but disadvantageous for physiological research. Many of the basic mechanisms involved in salt tolerance operate at the cellular level (25, 40). Rains et al. (41) recommend, therefore, the application of cell culture for the identification and quantification of central physiological attributes to salinity, including compartmentation, osmoregulation and, most important, the cost of energy, that is, the energetical price to be paid for the operation of the resistance mechanisms. Lerner and Reuveni (46) suggested a method employing tissue culture, which may help to quantify compartmentation and thus to better understand the process of osmoregulation. The method is based on the estimation of the size of the solute pool in the cytosol by the induction of selective pore formation in the plasmalemma of cells growing in suspension culture and measurement of solutes leaking from the cytosol.

The application of tissue culture methodology to physiological studies of plant response to salinity stress is limited yet. First studies on the effect of salinity on isolated organs were reported by Strogonov (41). He compared the response to salinity of isolated roots and whole young plants of alfalfa and found that the isolated roots were more sensitive to salt than whole young plants. Strogonov (41) also showed that pieces isolated from phloem and xylem parenchyma and cambium tissue of carrot root responded differently to salt. The difference, however, progressively decreased upon subsequent passages of calli developed from these tissues. He suggested, therefore, that the salt tolerance of the plant cell depends on its functional and structural organization. At present efforts are carried out in our laboratory to find whether the difference in the resistance of whole plants of the cultivated tomato and its wild relatives to salinity is reflected differently in the responses of isolated fully differentiated and meristematic tissues to salt in sterilized medium.

Attempts to estimate the relative contributions of primary (toxic) and secondary (osmotic) effects of salt to growth inhibition or injury were made by growing isolated roots of alfalfa on media containing salt or mannitol (41). A pronounced toxic effect of Cl$^-$ was clearly demonstrated. Goldner et al. (47) compared the effects of seawater, solutions of different inorganic salts, and mannitol on the growth and coloration of callus obtained from carrot root. They concluded that growth inhibition resulted mainly from the increase of the osmotic pressure, whereas discoloration and necrosis resulted from the toxicity of salt.

The possible contribution of mechanisms of salt resistance operating on the cellular level to the resistance of the whole plant was studied in different species. The results of these studies are not consistent. No correlation was found between the response of calli originated from halophytic species [glasswort (41); *Atriplex undulata* and *Suaeda australis* (48)] and the whole plants. They concluded that the salt resistance in these cases depends not on cellular aspects but on the physiological and anatomical integrity of the whole plant. However, positive correlation between the response of the whole plant and callus derived from it was found in some glycophytes [cabbage, tobacco, sweetclover, and sorghum (41); *Lycopersicon esculentum*, *L. peruvianum* and *Solanum pennellii* (49); *Hordeum vulgare* (50); *Phaseolus vulgaris* and *Beta vulgaris* (48)] and in a halophyte [*Suaeda maritima* (51)], suggesting the operation of mechanisms of salt resistance on the cellular level in these species. Smith and McComb (52) suggested that when such a positive correlation is found, callus tissue can be used as an indicator of the level of salt tolerance in the whole plant. Rosen and Tal (53) found that, similarly to the whole plant and callus tissue, naked protoplasts isolated from leaves of *L. peruvianum* divided and grew better on saline medium than those of the cultivated species. Some of our results from the comparative study of the responses of *L. peruvianum* and the cultivated tomato to salinity on different organizational levels (13, 49, 53) are presented in Table 17.1. Additional uses of tissue culture for physiological studies of salt resistance are summarized by Stavarek and Rains in Chapter 18.

2.   GENETICS OF SALT RESISTANCE

According to the present state of knowledge, the inheritance of salt resistance per se does not seem to be different in principle from that of other characteristics of economic importance, which are usually specified in terms of metrics and referred to as quantitative or continuous characters. These include, for example, height, weight, time of maturity, and cold, heat, or drought resistance. One of the ideas that dominated the mind of geneticists for some time in the past, and that is still a widely accepted assumption on which the biometrical methods are based, was that the inheritance of quantitative characters depends on many genes with relatively small effects (54, 55). The present prevailing idea contends that the distinction between quantitative and qualitative characters is artificial

Table 17.1.   Sodium Accumulation and Growth Parameters in Different Organizational Levels in the Cultivated Tomato (CV) and its Wild Relative *Lycopersicon peruvianum* **(Lp) under NaCl Salinity**

Level of Organization		Concentration of Na under Salinity (meq/g dry wt)			Relative Growth under Salinity (% of control)		
Whole plant (0.58% NaCl)		Root	Stem	Leaf	*Growth Rate*[a]		
	CV	1.00	2.65	0.15	50.0		
	Lp	1.15	2.05	1.30	68.7		
Callus (0.50% NaCl)		Root	Stem	Leaf	*Dry Weight* Root	Stem	Leaf
	CV	—	—	0.7	13.4	59.1	59.1
	Lp	—	—	1.2	32.1	122.0	84.0
Protoplast (0.35% NaCl)					*Plating efficiency*[b]		
	CV				0		
	Lp				58		

[a]Additional growth of stem as percent of initial size/3 days.
[b]Dividing protoplasts as percent of total.

and results mainly from considerations of convenience, since: (i) "Regardless of the number of genes that may affect a polygenic character, there are nearly always a few genes that make the major contribution to the phenotype"(56). The combined control by a small number of major genes with continuous variation due to modifier effects was termed by Murfet (57) as an oligopolygenic system of control. (ii) No gene acts as a soloist (56), that is, all genes are members in an intricate network of interactions in which the product of a gene may affect many characters (plieotropy), and any character may be affected by many genes (polygeny). The complexity of the interactions increases gradually from Mendelian genes, which are relatively individualistic in their action as expressed phenotypically, to those controlling the inheritance of quantitative characters. (iii) The expression and contribution of any gene to the phenotype cannot be considered absolute since it may change depending on the genetic background and the physical or biotic environment. A character can, therefore, be considered under some conditions in Mendelian terms, that is, in clear-cut phenotypic categories, but it can also appear as continuous under different genetic or environmental circumstances (56–59). This situation can be demonstrated even in cases where the phenotypic expression measured is almost the first metabolic product of the gene. The inheritance of nitrate reductase activity in the common wheat can exemplify this situation. The enzyme, which plays a major role in the regulation of nitrogen metabolism in cereals, is very sensitive to genetic and environmental changes. It was found to be inherited as a continuous character in plants grown under field conditions (60) but as a Mendelian character in a different genetic background and while growing in a solution culture and under more controlled conditions (61). It seems, therefore, that by manipulating the genetic background and the environmental conditions, it may be possible to identify individual major genes in systems currently described as being continuous.

Several facts demonstrate the existence of an appreciable amount of intraspecific and interspecific genetic variability, which appears continuous phenotypically with respect to various aspects of mineral nutrition. These aspects include the rate of absorption and translocation, efficiency of utilization, and tolerance to high concentrations of specific elements in the medium (3, 28, 62). Although spectacular results were obtained in breeding for several continuous characters, such as various components of yield, very little has been done in improving the efficiency of mineral nutrition (10). The reason for that are as follows: (i) relative to other characters, phenotypic aspects of mineral nutrition are not easily distinguished; (ii) the technological solutions to problems of nutrient deficiency or excess of salts were preferred to the biological solution since the former were economically feasible in the past.

According to Ramage (6), salt tolerance is a complex character and its expression largely depends on the genetic background; hence, its successful breeding, that is, obtaining plants with the highest possible yield under saline conditions, will best be achieved by recurrent selection. The method enables the simultaneous selection of the character desired with the favorable genetic background.

Although statistical techniques have provided the breeder with adequate tools to manipulate continuous characters in his breeding work, they have not provided him with any clue for the mechanisms of action of the genes involved (63). As with other continuous characters, salt resistance may also be controlled mainly by a few major genes. Recovery of such major genes can be achieved through spontaneous or induced single gene mutations in whole plants or in cell culture. The latter technology seems much more promising for that matter (63). Major genes can possibly be recovered also by manipulating the environment or the genetic background when genotypes differing in their resistance to salt are crossed. Such genes can be used in genetic and physiological studies. A major promising development in plant breeding in the future will be the application of genetic engineering. One of the major prerequisites for such an application is the choice and isolation of adequate genes to be engineered. The plant physiologist can use the major genes involved in the control of salt resistance for the identification and understanding of the relevant mechanisms. This subject will be discussed in more detail in Chapter 18.

The successful application of biochemical mutants to solve many biological problems in microorganisms stimulated the imagination of plant scientists, who began selecting specific metabolic mutants to be employed for the elucidation of mechanisms of metabolic regulation. However, higher plants present many difficulties which might delay the progress in such an application (3, 64). Among the difficulties are the following: (i) in higher plants, which are diploid or polyploid multicellular organisms, recessive mutations easily escape detection; (ii) higher plants have a long generation time; (iii) mutation screening requires large populations of plants; (iv) induction and selection of auxotrophic mutant plants are difficult because of the possible existence of alternative biosynthetic pathways; (v) the nature of the metabolic lesion caused by the mutation is usually difficult to identify since the phenotypic change in the higher plant is usually removed from the gene by several levels of organization.

Although several ion transport mutants have been described in microorganisms (65), only few instances of such mutations are known in higher plants (3, 28, 62). The information presented in these three sources is summarized in Table 17.2.

Single gene mutations involved in the regulation of either uptake, transport, or utilization of the ions of the most prevailing salt (i.e., NaCl) in saline soils and water resources are very rare. A recessive single gene mutation that confers salt sensitivity was described in soybean (66). Plants homozygous for this mutation are sensitive to salt since they are not able to prevent the transport of Cl^- ions from the root to the shoot. Another recessive mutation related to ion balance, *scabrous diminutive* (*sd*), was reported in the pepper. Na^+ exclusion and K^+ uptake in the mutant root are less efficient than in the root of the normal genotype (67, 68). The identification of *sd* as a mutation in a gene that controls the balance of monovalent cations can exemplify the difficulties encountered in recovering such mutations in whole plants. This subject and the partial epigenetic sequence of action of *sd* gene will be discussed in the next section.

Table 17.2. Single Gene Mutations of Mineral Nutrition

Element Affected	Mutation and Expression[a]	Processes Affected[a]	Species
B	*btl*, recessive	Transport from root to shoot	*Lycopersicon esculentum*
B	?, recessive	?	*Apium graveolens*
Mg	*mg*, recessive	Uptake and transport to shoot	*Apium graveolens*
Fe	*fe*, recessive	Uptake by roots	*Glycine max*
Fe	*fer*, recessive	Uptake by roots	*Lycopersicon esculentum*
Fe	*ys*, recessive	Uptake by roots	*Zea mays*
P	*np*, intermediate	?	*Glycine max*
K	*Ke*, dominant	Efficiency of utilization	*Phaseolus vulgaris*

[a]The question marks signify lack of information.

3. PHYSIOLOGICAL GENETICS OF SALT RESISTANCE

Evolution is opportunistic in the sense that it favors any variation that provides a competitive advantage for survival. Consequently, multiple solutions can evolve for the same functional problem (69). The plant breeder, similarly to the evolutionary process, is also opportunistic but in the sense that he is usually interested in the final performance and not so much in the question of "how," an aspect belonging to the territory of the physiologist. However, unlike the evolutionary process, the breeder can, at least in part, rationally plan the final product by selecting the appropriate organism, by influencing the quantity and sometimes also the quality of mutations, and by being himself the selecting agent. This powerful ability can be elaborated with the development of science, including the method of genetic engineering and the information gathered by the plant physiologists on the mechanisms of salt resistance.

Tomato is one of the genera in which extensive physiological and genetic studies of salt tolerance have been performed in parallel. The wild relatives of the cultivated tomato which evolved in dry habitats in South America may be used as a source for genes for improving the salt tolerance of the cultivated tomato, in agreement with the general recommendation of Greenway and Munns (29). *Lycopersicon cheesmanii*, *L. peruvianum*, and *Solanum pennellii*, the latter bearing a much stronger affinity to *Lycopersicon* genus than to *Solanum* (70), seem to be good candidates for this purpose. The physiological studies reveal that the growth of the wild plants relatively to the control is less inhibited under salinity as compared with that of the cultivated species (13, 71–73). The wild plants, especially *S. pennellii*, are distinguished from the cultivated species by their high content of Na^+ and by a lower K^+ content under salinity. The increase of Na^+ content and the decrease of K^+ content in plants of the wild tomato species under NaCl salinity may be due to a low efficiency of Na^+ exclusion and K^+ uptake in these plants. A lower K^+ uptake was found in excised root tips of plants of the wild species (unpublished data). This difference between the tomato species is reminiscent of the difference between the *sd* mutant and the normal genotype of pepper to be discussed later. Among the three species studied, *S. pennellii* is the most succulent under both control and saline conditions. It was suggested that a better osmotic adjustment, mainly by Na^+ ions, is responsible, at least in part, for the superior performance of the wild species under salinity. Proline, which was suggested to play a role in the adaptation to various stresses in some plants (74), was found not be involved in this case (75). The inheritance of some of the differences between the wild and the cultivated species under salinity is being studied by comparing their expression, including growth rate, succulence, and accumulation of Na^+, K^+, and Cl^- ions in the main parts of the plant, in the parental, F_1, and F_2 generations (13).

An investigation such as that described above may lead to the discovery of "salinity markers" to be used for screening salt-resistant tomato plants in breeding populations. A screening test suggested in rice and discussed by Greenway and Munns (29) is based on the selection of individuals having the unusual favorable combination of high Na^+ in the leaf and high salt tolerance due, probably, to efficient compartmentation of this ion. The test consists of an analysis of Na^+ content in the third leaf during early tillering, when yield potential of the individual plant can already be evaluated.

A central problem that should be clarified by further research arises upon transferring genes from the genetic background of salt-resistant wild species to the background of their related cultivars. According to Nieman (76), the energy required for the operation of the salt-resistance mechanisms is obtained mainly at the expense of growth. While the genetic background in the wild species has been selected for maximal survival, which is not necessarily related to growth, in the cultivated ones it is being selected for maximal yield, which is frequently dependent on growth. Thus, while such an energy price may be tolerated in the former species, it may be critical in the latter. Such a situation seems to exist in the tomato, whose wild salt-tolerant species grow, in absolute terms, slower than the cultivated ones, in which yield is related to growth rate in the early stages of development, as suggested by Pasternak et al. (5).

Differences due to single gene mutations can provide a most powerful tool for investigating controlling mechanisms (3, 64). However, since most mutations in higher plants are morphological variants, in which the underlying lesion is removed from the phenotype by several levels of organization, they can be used as a tool for physiological research only if the lesion is identified and the epigenetic sequence between the gene and the phenotype is reconstructed (63). Single gene mutations can also be used for the discovery of unknown interrelations between different functions. Many mutations in higher plants are pleiotropic, that is, they cause several changes on the phenotypic level. Contrary to changes caused by a chemical, those caused by a single gene mutation always stem from a common source and are, therefore, necessarily interrelated.

Only few single gene mutations that may be involved in the control of mechanisms related to salt resistance are known. Mutations affecting aspects possibly related to osmoregulation include (i) a low cellular osmotic pressure mutation in *Arabidopsis thaliana* (77); and (ii) a proline-deficiency mutation, in *Zea mays* (78). It is worth mentioning in this context a salt-resistant mutation in *Salmonella typhimurium*, which induces the production of high amounts of proline (16). Mutations in major genes involved in the regulation of abscisic acid metabolism include *flacca*, *sitiens*, and *notabilis* in tomato (79, 80) and *droopy* in potato (81), all leading to wiltiness of the plant due to deficiency in the hormone. A relatively detailed study of the epigenetic sequence of gene action was performed in *flacca* (82). A scheme summarizing this study is presented in Figure 17.1. This study comprises the best evidence for the direct involvement of abscisic acid in the regulation of stomatal closure (83) and root resistance to water flow (84), as well as other interrelations which are depicted in the scheme.

The few instances in which mineral nutrition has been shown to be governed

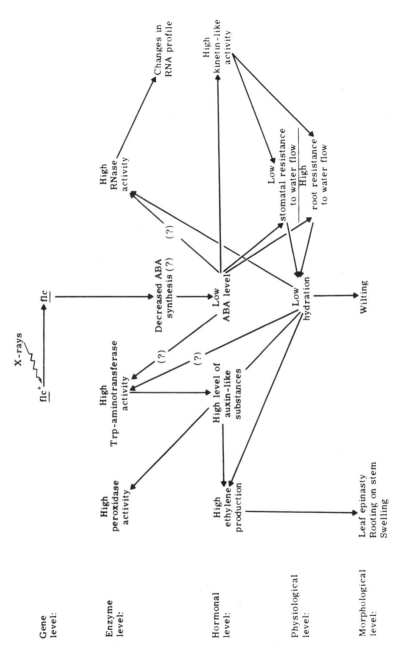

Figure 17.1. Possible epigenetic sequence of gene action in *flacca*, a wilty mutant of tomato. From ref. 82.

by a single gene mutation were discussed by Epstein and Jefferies (62), Epstein (3), and Lauchli (28) and summarized in Table 17.2. Fewer mutations in genes controlling the uptake or translocation of Na^+ and Cl^-, which are the ions of the prevalent salt in saline soils and water resources. They include *ncl* in soybean (66) and *scabrous diminutive* (*sd*) in pepper (85). A relatively extensive investigation is being performed in our laboratory on the epigenetic sequence of gene action in the *sd* mutant, which is mainly characterized phenotypically by excessive wilting due to excessive opening of stomata (85). Tal et al. (86) found that the difference in stomatal behavior between the mutant and the normal genotypes was correlated with differences in K^+ content in the guard cells. Na^+ substituted for K^+ in its function in stomatal movement in mutant plants grown in saline medium. Tal and Benzioni (67) and Benzioni and Tal (68) found a higher Na^+ content and a lower K^+ level in leaves of the mutant in saline medium as compared with those of wild type plants. Net Na^+ uptake into excised root tips was much higher whereas uptake of Rb^+ (representing K^+) was much lower in the mutant. They suggested that the defect in the mutant results from either an impaired Na^+/K^+ carrier system and/or an impaired membrane permeability. There are indications for a lower activity of a membrane-bound Ca-ATPase in the mutant root (87). Extensive experimental data suggest that plasm membrane $(Na^+ + K^+)$-activated ATPases play a role in cellular tolerance by mediating the energy required for cation uptake across the membrane of root cells (23, 33). Whether the difference in ATPase activity indicated between the mutant and the normal plants explains the difference in the uptake of the monovalent ions between them is still an open question. As suggested before, there is an apparent similarity between the situation in the mutant and normal pepper genotypes and the wild and cultivated tomato species with respect to Na^+ and K^+ balance. Whether a gene (or genes) similar to *sd* is responsible for the difference between the wild and the cultivated tomato with respect to Na^+/K^+ balance is still to be elucidated. It is worth mentioning in this context the preliminary finding of Sacher et al. (88). Based on the analysis of leaf sodium content in F_2 and backcross populations originated from the cross *L. esculentum* $\times$ *S. pennellii*, they suggested that a small number of genes, perhaps only one, is involved in the inheritance of sodium accumulation in these plants. A discovery of a major gene controlling ATPase activity can be of a great value for the study of energy transduction for cation uptake and for breeding for salt tolerance.

4. CONCLUDING REMARKS

The potential of genetic manipulation of the plant for extending agriculture into marginal areas is increasingly appreciated in recent years. Also acknowledged is the significance and potential of combining the methodologies of genetics and physiology in plant breeding, in general, and for the improvement of stress resistance, in particular. The unity of experimental approach may enable the scientist to focus his efforts on important questions rather than on disciplines (63).

At the present time, the most important problems confronting the conventional plant breeder in his efforts to improve salt resistance in plants, problems that are also relevant to other stresses, are (i) sources of germ plasm for salt resistance and the organization of the material and knowledge on readily available and potential sources; and (ii) markers for salt resistance, that is, characters associated with salt resistance which can be applied to screen salt-resistant plants in breeding populations. The identification of such markers is at present the main requirement of the plant breeder from the physiologist.

The feasibility of breeding for salt resistance can be increased by the determination and understanding of (i) the principal mechanisms of salt resistance; (ii) the factors involved in the energy cost of the operation of such mechanisms in terms of yield as well as the quantification of this cost; (iii) the organizational and genetic components relevant to salt resistance in the whole plant. The lack of, or the gap in, a comprehensive understanding of the principal mechanisms of salt resistance and the factors involved in the energy cost, which delays the progress in applying the existing extensive physiological information to the increase of the efficiency of breeding, can be overcome by extending the physiological research to molecular targets, such as membranes, growth regulators, osmoregulation, kinetics of key enzymes, and stabilizers of macromolecules, including the identification and understanding of the regulation of plant adaptation to salt stress. The use of the basic elements of the genome (single genes) and the whole plant body (isolated tissues and cells) in such a study will offer many advantages, which were proved to be very valuable in the study of microorganisms. It is recommended (Lüttge, in a discussion in this conference) to concentrate such combined physiological and genetic efforts on a small number of groups of plant species. It seems that the ideal criteria for selecting such groups of species should be (i) representation of the major strategies for salt resistance; (ii) economic significance; (iii) existence of background physiological knowledge; (iv) simple and well-mapped genome.

ACKNOWLEDGMENTS

I thank Ms. Dorot Imber for her valuable comments and Ms. Ruth Massil for her skillful typing. The research on tomato performed in our laboratory was partially supported by a grant from the United States–Israel (Binational) Agricultural Research and Development Fund (BARD).

REFERENCES

1. D. Pimentel, L. E. Hurd, A. C. Bellotti, M. J. Forster, I. N. Oka, O. D. Sholes, and R. J. Whitman. Food production and the energy crisis. *Science* **182,** 443 (1973).

2. R. A. Fisher and N. C. Turner. Plant productivity in the arid and semiarid zones. *Annu. Rev. Plant Physiol.* **29,** 277 (1978).

3. E. Epstein. *Mineral Nutrition of Plants: Principles and Perspectives.* Wiley, New York, 1972, p. 142.

4. L. Bernstein. Effects of salinity and sodicity on plant growth. *Annu. Rev. Plant Pathol.* **13,** 295 (1975).

5. D. Pasternak, M. Twersky, and Y. De Malach. Salt resistance in agricultural crops. In H. Mussell and R. C. Staples, eds., *Stress Physiology in Crop Plants.* Wiley, New York, 1979, pp. 127–142.

6. R. T. Ramage. Genetic methods to breed salt tolerance in plants. In D. W. Rains, R. C. Valentine, and A. Hollaender, eds., *Genetic Engineering of Osmoregulation.* Plenum, New York, 1980, pp. 311–318.

7. E. Epstein, J. D. Norlyn, D. W. Rush, R. W. Kingsbury, D. B. Kelley, G. A. Cunningham, and A. F. Wronn. Saline culture of crops: A genetic approach. *Science* **210,** 399 (1980).

8. E. V. Maas and G. J. Hoffman. Crop salt tolerance: Evaluation of existing data. In H. E. Dregne, eds., *Managing Saline Water for Irrigation.* Proceedings of the International Salinity Conference. Texas Technical University Press, Lubbock, Texas, 1977, pp. 187–198.

9. J. L. Marx. Plants: Can they live in salt water and like it? *Science* **206,** 1168 (1979).

10. E. Epstein. Genetic potentials for solving problems of soil mineral stress: Adaptation of crops to salinity. In M. J. Wright, eds., *Proceedings of a Workshop on Plant Adaptation to Mineral Stress in Problem Soils.* Cornell University Press, Ithaca, New York, 1976, pp. 73–82.

11. D. Atsmon. Breeding for drought resistance in small grains: In search for selection criteria and genetic sources. In J. H. J. Spiertz and T. Kramer, eds., *Crop Physiology and Cereal Breeding.* Proc. Eucarpia Workshop Wageningen, The Netherlands, 1979, pp. 126–131.

12. A. Bloom. Genetic improvement of drought resistance in crop plants: A case for sorghum. In H. Mussell and R. C. Staples, eds., *Stress Physiology in Crop Plants.* Wiley, New York, 1979, pp. 429–446.

13. M. Tal and M. C. Shannon. Salt tolerance in the wild relatives of the cultivated tomato: Response of *Lycopersicon esculentum, L. cheesmanii, L. peruvianum* and *Solanum pennellii,* and F_1 hybrids to high salinity 10, 109 (1983).

14. D. B. Kelly, J. D. Norlyn, and E. Epstein. Salt-tolerant crops and saline water: Resources for arid lands. In J. R. Godin and D. K. Northington, eds., *Arid Land Plant Resources.* Texas Technical University Press, Lubbock, 1979, pp. 326–334.

15. K. Mielenz, K. Anderson, R. Tait, and R. C. Valentine. Potential for genetic engineering for salt tolerance. In A. Hollaender, J. C. Aller, E. Epstein, A. San Pietro and O. R. Zaborsky, eds., *The Biosaline Concept? An Approach to the Utilization of Under-Exploited Resources.* Plenum, New York, 1979, pp. 361–370.

16. L. N. Csonka. The role of L-proline in response to osmotic stress in *Salmonella typhimurium:* Selection of mutant with increased osmoregulation as strains which over-produced L-proline. In D. W. Rains, R. C. Valentine, and A. Hollaender, eds., *Genetic Engineering of Osmoregulation.* Plenum, New York, 1980, pp. 35–52.

17. E. Epstein and J. D. Norlyn. Seawater-based crop production; a feasibility study. *Science* **197**, 249 (1977).

18. E. Epstein, R. W. Kingsburg, J. D. Norlyn, and D. W. Rush. Production of food crops and other biomass by seawater culture. In A. Hollaender, J. C. Aller, E. Epstein, A. San Pietro, and O. R. Zaborsky, eds., *The Biosaline Concept: An Approach to the Utilization of Under-Exploited Resources.* Plenum, New York, 1979, pp. 77–99.

19. J. D. Norlyn. Breeding salt-tolerant crop plants. In D. W. Rains, R. C. Valentine, and A. Hollaender, eds., *Genetic Engineering of Osmoregulation.* Plenum, New York, 1980, pp. 293–309.

20. D. W. Rush and E. Epstein. Breeding and selection for salt tolerance by the incorporation of wild germplasm into a domestic tomato. *J. Am. Soc. Hort. Sci.* **106**, 699 (1981).

21. Mary-Dell Chilton. Agrobacterium II plasmids as a tool for genetic engineering in plants. In D. W. Rains, R. C. Valentine, and A. Hollaender, eds., *Genetic Engineering of Osmoregulation.* Plenum, New York, 1980, pp. 23–31.

22. M. C. Shannon. Testing salt tolerance variability among tall wheat grass lines. *Agron. J.* **70**, 719 (1980).

23. A. Kylin and R. S. Quatrano. Metabolic and biochemical aspects of salt tolerance. In A. Poljakoff-Mayber and J. Gale, eds., *Plants in Saline Environments.* Springer-Verlag, Berlin, 1975, pp. 147–167.

24. A. Poljakoff-Mayber and J. Gale, eds. *Plants in Saline Environments.* Springer-Verlag, Berlin, 1975, 213 pp.

25. J. Levitt. *Responses of Plants to Environmental Stresses*, Vol. 2, 2nd ed. Academic Press, New York, 1980, p. 606.

26. T. J. Flowers, P. F. Troke, and A. R. Yeo. The mechanism of salt tolerance in halophytes. *Annu. Rev. Plant Physiol.* **28**, 89 (1977).

27. D. W. Rush and E. Epstein. Comparative studies on the sodium, potassium and chloride relations of a wild halophytic and a domestic salt sensitive tomato species. *Plant Physiol.* **68**, 1308 (1981).

28. A. Lauchli. Genotypic variation in transport. In U. Lüttge and M. G. Pitman, eds., *Encyclopedia of Plant Physiology*, New series, Vol. 2, Transport in plants. Springer-Verlag, Berlin, 1976, pp. 372–393.

29. H. Greenway and R. Munns. Mechanisms of salt tolerance in nonhalophytes. *Annu. Rev. Plant Physiol.* **31**, 149 (1980).

30. A. R. Yeo. Salt tolerance in the halophyte *Suaeda maritima L.* Dum.: Intracellular compartmentation of ions. *J. Exp. Bot.* **32**, 487 (1981).

31. P. J. Kramer. Drought stress and the origin of adaptation. In N. C. Turner and P. J. Kramer, eds., *Adaptation of Plants to Water and High Temperature Stress.* Wiley, New York, 1980, pp. 7–20.

32. J. S. Boyer. Physiological adaptations to water stress. In N. C. Turner and P. J. Kramer, eds., *Adaptation of Plants to Water and High Temperature Stress.* Wiley, New York, 1980, pp. 443–444.

33. T. K. Hodges. ATPase associated with membranes of plant cells. In U. Lüttge and M. G. Pitman, eds., *Encyclopedia of Plant Physiology*, New Series, Vol. 2, Transport in plants. Springer-Verlag, Berlin, 1976, pp. 260–283.

34. R. F. M. van Steveninck. Effect of hormones and related substances on ion transport. In U. Lüttge and M. G. Pitman, eds., *Encyclopedia of Plant Physiology*, New series, Vol. 2, Transport in plants. Springer-Verlag, Berlin, 1976, pp. 307–342.

35. D. W. Rains, R. C. Valentine, and A. Hollaender, eds. *Genetic Engineering of Osmoregulation*. Plenum, New York, 1980, p. 381.

36. P. W. Hochachka and G. N. Somero. *Strategies of Biochemical Adaptations*. Saunders, Philadelphia, 1973, p. 358.

37. A. Poljakoff-Mayber. Biochemical and physiological responses of higher plants to salinity stress. In A. San Pietro, ed., *Biosaline Research. A Look to the Future*. Plenum, New York, 1982, pp. 245–269.

38. R. S. Chaleff and P. S. Carlson. *In vitro* selection of mutants of higher plants. In L. Ledaux, ed., *Genetic Manipulations with Plant Material*. Plenum, New York, 1975, pp. 351–363.

39. H. E. Street. Plant cell cultures: Present and projected applications for studies in genetics. In L. Ledoux, ed., *Genetic Manipulations with Plant Material*. Plenum, New York, 1975, pp. 231–244.

40. D. W. Rains, T. P. Croughan, and S. J. Stavarek. Selection of salt-tolerant plants using tissue culture. In D. W. Rains, R. C. Valentine, and A. Hollaender, eds., *Genetic Engineering of Osmoregulation*. Plenum, New York, 1980, pp. 279–292.

41. B. P. Strogonov. Structure and function of plant cells in saline habitats: New trends in the study of salt tolerance. Israeli program for scientific translations, Jerusalem, 1973, 284 pp.

42. T. P. Croughan, S. J. Stavarek, and D. W. Rains. Selection of NaCl tolerant line of cultured alfalfa cells. *Crop Sci.* **18,** 959 (1978).

43. A. K. Handa, R. A. Bressan, S. Handa, and P. M. Hasegawa. Characteristics of cultured tomato cells after prolonged exposure to medium containing polyethylene glycol. *Plant Physiol.* **69,** 514 (1982).

44. M. Tal. Adaptation of plants to extreme environments. In D. A. Evans, W. R. Sharp, P. V. Ammirato, and Y. Yamada, eds., *Applications of Plant Tissue Culture Methods for Crop Improvement*, Vol. 2. MacMillan, New York, In press.

45. P. J. Larkin and W. R. Scowcroft. Somaclonal variation—A novel source of variability from cell cultures for plant improvement. *Theor. Appl. Genet.* **60,** 197 (1981).

46. H. R. Lerner and M. Reuveni. Induction of pore formation selectively in the plasmalemma of plant cells by poly-L-lysine treatment: A method for the direct measurement of cytosol solutes in plant cells. In D. Marme, E. Marre and R. Herte, eds., *Plasmalemma and Tonoplast: Their Functions in the Plant Cell*. Elsevier, Amsterdam, 1982, pp. 49–52.

47. R. Goldner, N. Umiel, and Y. Chen. The growth of carrot callus cultures at various concentrations and compositions of saline water. *Z. Pflanzenphysiol.* **85,** 307 (1977).

48. M. K. Smith and J. A. McComb. Effect of NaCl on the growth of whole plants and their corresponding callus cultures. *Aust. J. Plant Physiol.* **8,** 267 (1981).

49. M. Tal, H. Keikin, and K. Dehan. Salt tolerance in the wild relatives of the cultivated tomato: Responses of callus tissues of *Lycopersicon esculentum, L. peruvianum* and *Solanum pennellii* to high salinity. *Z. Pflanzenphysiol.* **86,** 231 (1978).

50. T. J. Orton. Comparison of salt tolerance between *Hordeum Vulgare* and *H. jubatum* in whole plants and callus cultures. *Z. Pflanzenphysiol.* **98,** 105 (1980).

51. H. Von Hendenstrom and S. W. Breckle. Obligate halophytes? A test with tissue culture methods. *Z. Pflanzenphysiol.* **74,** 183 (1974).

52. M. K. Smith and J. A. McComb. Use of callus cultures to detect NaCl tolerance in cultivars of three species of Pasture Legumes. *Aust. J. Plant Physiol.* **8,** 437 (1981).

53. A. Rosen and M. Tal. Salt tolerance in the wild relatives of the cultivated tomato: Responses of naked protoplasts isolated from leaves of *Lycopersicon esculentum* and *L. Peruvianum* to NaCl and proline. *A. Pflanzenphysiol.* **102,** 91 (1981).

54. R. W. Allard. *Principles of Plant Breeding.* Wiley, New York, 1960, p. 485.

55. D. S. Falconer. *Introduction to Quantitative Genetics.* Ronald Press, New York, 1961, p. 365.

56. E. Mayr. *Animal Species and Evolution.* Harvard University Press, Cambridge, Massachusetts, 1966, p. 797.

57. I. C. Murfet. Environmental interaction and the genetics of flowering. *Annu. Rev. Plant Physiol.* **28,** 253 (1977).

58. C. F. Wehrhahn and R. W. Allard. The detection and measurement of the effects of individual genes involved in the inheritance of a quantitative character in wheat. *Genetics* **51,** 109 (1965).

59. M. Tal. Genetic differentiation and stability of some characters that distinguish *Lycopersicon esculentum* Mill. from *Solanum pennellii* Cor. *Evolution* **21,** 316 (1967).

60. R. D. Duffield, L. I. Croy, and E. L. Smith. Inheritance of nitrate reductase activity, grain protein, and straw protein in a hard red winter wheat cross. *Agron. J.* **64,** 249 (1972).

61. L. W. Gallagher, K. M. Soliman, C. O. Qualset, R. C. Huffaker, and D. W. Rains. Major gene control of nitrate reductase activity in common wheat. *Crop Sci.* **20,** 717 (1980).

62. E. Epstein and R. L. Jefferies. The genetic basis of selective ion transport in plants. *Annu. Rev. Plant Physiol.* **15,** 169 (1964).

63. T. B. Rice and P. S. Carlson. Genetic analysis and plant improvement. *Annu. Rev. Plant Physiol.* **26,** 279 (1975).

64. G. Scholz and H. Bohme. Biochemical mutants in higher plants as tools for chemical and physiological investigations—A survey. *Kulturpflanze* **28,** 11 (1980).

65. C. W. Slayman. The genetic control of membrane transport. In F. Bronner and A. Kleinzeller, eds., *Current Topics in Membranes and Transport,* Vol. 4. Academic Press, New York, 1973, pp. 1–174.

66. G. H. Abel. Inheritance of the capacity for chloride inclusion and chloride exclusion by soybeans. *Crop Sci.* **9,** 697 (1969).

67. M. Tal and A. Benzioni. Ion imbalance in *Capsicum annuum, scabrous diminutive,* a wilty mutant of pepper. I. Sodium fluxes. *J. Exp. Bot.* **28,** 1337 (1977).

68. A. Benzioni and M. Tal. Ion imbalance in *Capsicum annuum scabrous diminutive,* a wilty mutant of pepper. II. rubidium flux. *J. Exp. Bot.* **29,** 879 (1978).

69. G. G. Simpson. *The Meaning of Evolution,* 4th ed. Yale Univ. Press, New Haven, Connecticut, 1969, 368 pp.

70. C. M. Rick. Biosystematic studies in *Lycopersicon* and closely related species of *Solanum*. In J. G. Hawkes, R. N. Lester, and A. D. Skelding, eds., *The Biology and Taxonomy of the Solanaceae*. Academic Press, London, 1979, pp. 667–678.

71. M. Tal. Salt tolerance in the wild relatives of the cultivated tomato: Responses of *Lycopersicon esculentum*, *L. peruvianum*, and *L. esculentum* minor to sodium chloride. *Aust. J. Agric. Res.* **22**, 631 (1971).

72. D. W. Rush and E. Epstein. Genotypic responses to salinity. Differences between salt-sensitive and salt-tolerant genotypes of the tomato. *Plant Physiol.* **57**, 162 (1976).

73. K. Dehan and M. Tal. Salt tolerance in the wild relatives of the cultivated tomato: Response of *Solanum pennellii* to high salinity. *Irrig. Sci.* **1**, 71 (1978).

74. L. G. Paleg, T. J. Douglas, A. van Daal, and D. B. Keech. Proline, betaine and other organic solutes protect enzymes against heat activation, *Aust. J. Plant Physiol.* **8**, 107 (1981).

75. M. Tal, A. Katz, H. Heikin, and K. Dehan. Salt tolerance in the wild relatives of the cultivated tomato: Proline accumulation in *Lycopersicon esculentum* Mill., L. peruvianum Mill, and *Solanum pennellii* Cor. treated with NaCl and polyethylene glycole. *New Phytol.* **82**, 349 (1979).

76. R. H. Nieman. Panel discussion: Osmoregulation in higher plants. In D. W. Rains, R. C. Valentine, and A. Hollaender, eds., *Genetic Engineering of Osmoregulation*. Plenum, New York, 1980, 264 pp.

77. J. Langridge. An osmotic mutant of *Arabidopsis thaliana. Aust. J. Biol. Sci.* **11**, 457 (1958).

78. G. Gavazzi, M. N. Racchi, and C. Tonelli. A mutant causing proline requirement in *Zea mays. Theor. Appl. Genet.* **46**, 339 (1975).

79. D. Imber and M. Tal, 1970. Phenotypic reversion of *flacca*, a wilty mutant of tomato, by abscisic acid. *Science* **169**, 591 (1970).

80. M. Tal and Y. Nevo. Abnormal stomatal behaviour and root resistance, and hormonal imbalance in three wilty mutants of tomato. *Biochem. Genet.* **8**, 291 (1973).

81. S. A. Quarries. A wilty mutant of potato deficient in abscisic acid. *Plant Cell* **5**, 23 (1982).

82. M. Tal, D. Imber, A. Erez, and E. Epstein. Abnormal stomatal behaviour and hormonal imbalance in *flacca*, a wilty mutant of tomato. V. The effect of abscisic acid on indoleacetic acid metabolism and ethylene evolution. *Plant Physiol.* **63**, 1044 (1979).

83. B. V. Milborrow. The chemistry and physiology of abscisic acid. *Annu. Rev. Plant Physiol.* **25**, 259 (1974).

84. M. Tal and D. Imber. Abnormal stomatal behaviour and hormonal imbalance in *flacca*, a wilty mutant of tomato. III. Hormonal effects on the water status in the plant. *Plant Physiol.* **47**, 849 (1971).

85. M. Tal, A. Witztum, and C. Shifriss. Abnormal stomatal behaviour and leaf anatomy in *Capsicum annuum*, *scabrous diminutive*, a wilty mutant of pepper. *Ann. Bot.* **38**, 983 (1974).

86. M. Tal, A. Eshel, and A. Witztum. Abnormal stomatal behaviour and ion imbalance in *Capsicum scabrous diminutive. J. Exp. Bot.* **27**, 111 (1976).

87. J. Morton, Comparative investigation of the activity of membrane-bound adenosine triphosphatase of roots of normal and mutant (*scabrous diminutive*) pepper. M.Sc. thesis, Ben Gurion University of the Negev, Beer-Sheva, Israel, 1979 (in Hebrew).

88. R. F. Sacher, R. C. Staples, and R. W. Robinson. Saline tolerance in hybrids of *Lycopersicon esculentum* & *Solanum pennellii* and selected breeding lines. In A. San Pietro, ed., *Biosaline Research. A Look to the Future.* Plenum, New York, 1982, pp. 325–336.

18

CELL CULTURE TECHNIQUES: SELECTION AND PHYSIOLOGICAL STUDIES OF SALT TOLERANCE

S. J. Stavarek and D. W. Rains

Plant Growth Laboratory
Department of Agronomy and Range Science
University of California at Davis
Davis, California

A significant portion of the terrestrial environment is affected by high levels of salt. Without including the major deserts, there are approximately 4 million km^2 of salt-affected land in the world (1). Loss of land to salt accumulation through irrigated agriculture has been estimated to be at least several hundred km^2 a year in India alone (1). In California, 20% of the irrigated agricultural land is affected to some extent by salt (2).

Management of the water and land can be successful in reclamation of salt-affected soils (3). The impact of salinity on plant productivity can also be managed through biologically manipulating the plant (4, 5). Identification of plant genotypes capable of increased tolerance to salt and incorporation of these desirable traits into economically useful crop plants may reduce the effects of salinity on productivity.

An understanding of the mechanisms by which plants adjust to a saline environment is important in programs attempting to establish salt-tolerant crop species. The identification of specific characteristics related to salt tolerance would also provide potential biological markers useful in selection systems.

Plants exposed to saline environments encounter three basic problems: (i) a reduction in water potential of the surrounding environment results in water becoming less available; (ii) toxic ions can interfere with the physiological and biochemical processes of the organism; and (iii) required nutrient ions must be obtained despite the predominance of other ions (6).

Plants have evolved several strategies to overcome the problems of saline environments. Plants may avoid the stress by growing during months of more favorable conditions (i.e., high rainfall months). Exclusion is another strategy used by plants. The exclusion may be at a whole plant level in which salts are not allowed to enter the shoots. This may be accomplished by active pumping of salts back into the roots. Exclusion mechanisms also may involve salt glands located on the surface of leaves which extrude salts that reach the shoots. Cellular level exclusions can be accomplished by either not permitting ions into the cells or pumping them out once they are in. A highly significant strategy is a physiological tolerance that may involve compartmentation and osmotic adjustment using inorganic and organic constituents. All these mechanisms used by plants have been discussed in detail in many reviews (1, 7–13)

Whichever mechanisms account for salt tolerance, there must be an energy cost. The energy required to make the organic compounds used for osmotic regulation and the carbon skeletons required may limit the growth of the organism. The energy required for regulation of inorganic ions may limit the potential for plant productivity. Respiratory energy for maintenance of processes competing with growth may increase when the organism is stressed, so that a greater portion of the energy is used for processes resulting in salt tolerance rather than for growth.

In development of new crop varieties with increased tolerance to salinity, the mechanism involved may greatly influence the potential success of that crop. Potential yield, not just the ability to survive, is important in determining the feasibility of using soils and/or waters with salinity problems. The ability of the plant to produce well in conditions with and without stress is critical, as the amount of salinity can vary across a field and during the growing season. The plant must be able to adapt to changing levels of stress and not require a specific concentration of salt for growth. Rosielle and Hamblin (14) discuss the theoretical aspects of selecting for yield in stress and nonstress environments. They suggest that selection should be made for increased mean productivity (the average yield in stress and nonstress environments) instead of tolerance to stress (yield difference between stress and nonstress environments). If selection is for increased mean productivity then mean yields in both environments will generally increase making the crop more adaptable to varying environments.

This chapter will discuss the use of cell culture techniques in developing salt-tolerant crop plants, as well as using cell culture systems for studying cellular level features of salt tolerance.

1. CELL CULTURE

Cell culture techniques offer several advantages for use in selection and basic physiological studies. With cell culture the media is rigidly defined and the environment controlled allowing a uniform and precise treatment. The relatively undifferentiated nature of the cultured cells reduces the complications of differences in morphology and stages of development. Cellular physiology can be studied directly. In cell culture systems literally millions of cells can be screened and evaluated for their performance in a relatively small area. Mutagenic agents can be used to increase variability.

There are several potential problems with cell culture systems that should also be considered. After selection, plants must be regenerated from the selected cells. In most cell culture systems the capacity for regeneration rapidly decreases with time. This, however, may be a technical problem which can be overcome by a better understanding of the processes involved in regeneration and the development of improved media sequences. In several species, plants have been regenerated from relatively old callus [*Lilium longiflorum* (15); *Daucus carota*, (16); *Medicago sativa* (17)]. The expression of the desirable traits in the regenerated plants as well as the heritability is very important. The plants regenerated should maintain the characteristics of the variety and incorporate the new genetic traits that were selected for in the cells without incorporating any inimical genes.

1.1. Selection for Salt Tolerance

The use of cell culture techniques has proved effective in several systems selecting for salt tolerance. Cells are exposed to normally lethal levels of salt (usually NaCl), and the resistant cells are visually isolated.

Zenk (18) reported the isolation of a resistant cell line from haploid cells of *Nicotiana sylvestris*. At 1% NaCl, the resistant cells grew at 50% the rate of that of the control without NaCl. No growth occurred at 1% NaCl for the nonselected cells. Dix and Street (19) were able to select cell lines of *Nicotiana sylvestris* and *Capsicum annuum* resistant to 1 and 2% (w/v) NaCl. The cells were selected in suspensions which contained NaCl. After several passages on salt they were able to isolate several cell lines whose growth equaled or exceeded the growth observed on 0% NaCl.

Nabors and co-workers (20) obtained *Nicotiana tabacum* cells which showed tolerance to 0.16% NaCl and then to 0.52%. Salt stress was applied stepwise to the suspension cultures. Later they produced lines tolerant to 0.88% NaCl (21).

Hasegawa et al. (22) reported the selection of *Nicotiana tabacum* cells tolerant to 1% NaCl. These cells however lost their tolerance to NaCl once they had been grown away from salt for five cell mass doublings. Kochba and co-workers (23) recently reported cell lines of *Citrus sinensis* and *C. aurantium* tolerant to NaCl. The cells were selected on 0.5 and 0.7% NaCl, respectively, and were able to grow on NaCl up to 1%.

In our laboratory, we have selected several cell lines of *Oryza sativa* and *Medicago sativa* which are tolerant to NaCl. Large amounts of cells were grown in suspension cultures. The cells are plated onto media which contain lethal levels of NaCl (1% for alfalfa, 2–3% for rice). Tolerant cell lines are isolated after several transfers on the salt. Alfalfa cell lines have been isolated which grow on

1% NaCl (24), whereas rice cells were isolated which grow on NaCl as high as 2% (25).

Regeneration of plants from the salt-tolerant cells has been limited. Nabors et al. (21) regenerated tobacco plants from cells tolerant to NaCl. F_1 and F_2 generation plants were tested for tolerance to NaCl. If watered with 2.62% NaCl, the plants regenerated from salt-tolerant cells had a higher survival rate than plants regenerated from nontolerant cells. Kochba et al. (23) obtained embryos from their salt-tolerant citrus cell lines. The embryos grew better in saline media but no production of plants was reported.

In our laboratory, we were able to regenerate hundreds of plants from the salt tolerant alfalfa callus once a new media sequence was developed (17). These plants regenerated from the salt-tolerant callus, which had been in culture for several years, proved to have many lethal genetic changes. The plants were stunted, slow growing and susceptible to many diseases. The amount of tolerance expressed in the plants was never fully determined (26).

Several studies have compared the response of callus to that of the whole plant of different species under increasing salt stress. Von Hendenström and Breckle (27) studied the growth of *Suaeda maritima* and *Salicornia europaea*. Both the plant and callus showed improved growth with the addition of NaCl. Tal et al. (28) looked at the response of *Lycopersicon esculentum, L. peruvianum*, and *Solanum pennellii* to 0-2% NaCl. They found that the callus responded in a similar fashion as the whole plant, the wild species (*L. peruvianum* and *Solanum pennellii*) had better growth under salinity than the cultivated species (*L. esculentum*).

Orton (29) compared the salt tolerance of *Hordeum vulgare* and *H. jubatum* in whole plants and callus cultures. The callus response paralleled the plant as NaCl was increased to 0.17 M.

Smith and McComb (30) compared plants and their corresponding callus cultures of a salt-sensitive glycophyte (*Phaseolus vulgaris*), a salt-tolerant glycophyte (*Beta vulgaris*), and two halophytes (*Atriplex undulata* and *Suaeda australis*) to increasing levels of NaCl (0-250 mM). For *P. vulgaris*, both the plant and callus decreased in growth with increasing salt, whereas *B. vulgaris* (both callus and plant) showed an increase in growth at intermediate salt levels and a decrease at higher levels. For both halophytes, the plants showed increased growth with increasing levels of salt, while the callus growth decreased. The halophytes have whole plant mechanisms (salt glands, succulence) for dealing with salt so tolerance is not expressed in their callus. *Beta vulgaris* is suggested to have a cellular level mechanism for salt tolerance.

Smith and McComb (31) have also examined the response to increasing NaCl of three pasture legumes and the corresponding callus cultures. They found very similar responses between the plant and callus and suggest that the use of callus cultures to screen for NaCl tolerance is a valid system.

The salt used in the majority of salinity studies is NaCl. Several researchers have compared the response of callus to different types of salt. Goldner et al. (32) using *Daucus carota* cultures and Chen et al. (33) using *Nicotiana tabacum*

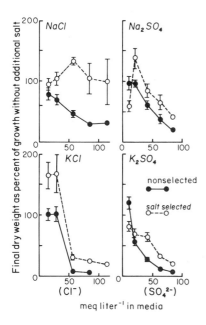

Figure 18.1 Final dry weight (as percent of growth without additional salts) as a function of increasing Cl^- or SO_4^{2+} in the media for nonselected alfalfa cells (●) and NaCl-selected alfalfa cells (○). Bars depict standard error.

found different responses when they compared growth on seawater, synthetic seawater, mannitol, NaCl, and other Cl^- and SO_4^{2-} salts. They suggest the use of multiple salts or synthetic seawater as a selection pressure to better parallel the salinity under field conditions.

Kochba and co-workers (23) tested their salt-tolerant citrus callus cultures, which were selected on NaCl, on other salts. The cells were sensitive to KCl, but able to grow well on Na_2SO_4, K_2SO_4, and $MgSO_4$.

We have compared the growth of the NaCl-selected alfalfa on different salts (NaCl, KCl, Na_2SO_4, and K_2SO_4,) (Fig. 18.1). The growth of the NaCl-selected alfalfa line was inhibited at high levels of all salts tested except NaCl (34). Other data (unpublished) show that the NaCl-selected cells can tolerate an osmotic stress, applied as mannitol to the medium. The nonselected alfalfa cells cannot tolerate an osmotic or salinity stress. It appears then that the NaCl-selected cells can tolerate osmotic stress and NaCl stress, but are sensitive to high levels of K^+ and SO_4^{2-}. Expression of tolerance at the whole plant level, however, will determine the feasibility of using just NaCl to select for salt tolerance.

When the same growth studies were performed on the NaCl-selected rice cells, different results were obtained (Fig. 18.2). The NaCl-selected rice cells can tolerate high levels of all salts tested. The nonselected cells did not grow on any salts tested. It appears then that the NaCl-selected rice cells have adapted a different mechanism for dealing with growth in a saline environment than the NaCl-selected alfalfa cells. Further studies are needed comparing the specificity of the cell lines to regulate their internal ionic compartments.

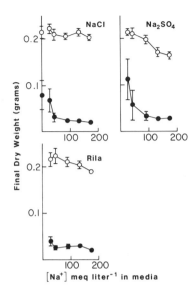

Figure 18.2 Final dry weight (in grams) as a function of increasing Na⁺ in the media for non-selected rice cells (●) and NaCl-selected rice cells (○). Bars depict standard error.

1.2. Physiological Studies

Cell culture systems are being used more frequently for studying cell physiology. Some researchers have looked at the effects of NaCl on different processes. Kulieva et al. (35) working with *Crepis capillaris* found that low concentrations of NaCl inhibit growth and depress mitotic activity. Also observed was increased polyploidization. Higher concentrations of NaCl cause immediate cell death.

K omizerko and Khretonova (36) examined the effect of NaCl on embryogenesis and regeneration in carrot cultures. Growth of callus was reduced on NaCl and embryogenesis was inhibited in callus tissues from phloem and xylem. Embryoids were formed in smaller numbers than in the control in callus originating from cambial tissue. Plants regenerated from these embryoids however were more resistant to salinization than plants from seeds.

A few studies have compared the differences between cell lines which differ only in the ability to tolerate NaCl. Heyser and Nabor (37) found few differences in the response of their adapted and nonadapted *Nicotiana tabacum* cell lines to NaCl. Dix and Pearce (38) looked at proline accumulation in their NaCl-resistant and-sensitive cell lines of *Nicotiana sylvestris*. Both cell lines rapidly accumulated free proline when placed on medium containing 1.5% NaCl; however, the amount was not sufficient to act as a osmoticum even if localized in the cytoplasm.

Tal and Katz (39) and Rosen and Tal (40) have studied the effects of proline on the salt tolerance of callus and protoplasts from *Lycopersicon esculentum* and *L. peruvianum*. The cultivated tomato *L. esculentum*, which is less salt tolerant, showed some stimulation of growth with proline. Under salt stress, proline had no effect on growth for either cell line.

Ionic Regulation

We have been studying the ionic differences between the NaCl-selected and nonselected cells of alfalfa. The differences may explain the mechanism used by the NaCl-selected cells for growing on normally lethal levels of NaCl.

Early studies showed that the NaCl-selected cells maintain a higher K^+ content at all levels of NaCl than the nonselected (24). This trait is known to be highly correlated with salt tolerance in both halophyllic bacteria and halophytic plants (6, 7, 41). Some researchers however have observed decreases in K^+ content of NaCl-tolerant cells as NaCl was increased in the media (26). These cells may use a different mechanism for growing on salt.

The NaCl-selected alfalfa cells accumulate approximately the same amount of Na^+ and Cl^- as the nonselected tissue when grown on different levels of NaCl. It appears then that the salt-selected cells do not exclude salts as the mechanism of tolerance. Further studies showing where Na^+ and Cl^- is located in the cells is needed to determine if the salt-selected cells are more efficient at compartmentation than the nonselected alfalfa cells.

Elevated levels of NO_3^- in the NaCl-selected cells at low salt concentrations is also consistent with observations in halophytic plants (24). Halophytic plants tend to accumulate NO_3^- when grown in the absence of Cl^-.

The Ca^{2+} concentration of the NaCl-selected and nonselected alfalfa cells decreases as NaCl is increased in the media (42). This decrease in Ca^{2+} is seen also in whole plants grown in salt (43). The salt-selected tissue however always maintained a higher Ca^{2+} concentration than the nonselected tissue. LaHaye and Epstein (44) found in *Phaseolus vulgaris* that when Ca^{2+} was present the plants grew as well with salt as the unsalinized control plants.

All these differences seen in the NaCl-selected alfalfa cells suggest that changes have occurred in the transport properties. Currently, uptake studies are being carried out to determine differences between the salt-selected and nonselected tissue. Many studies have used cell culture systems for studying uptake processes, but to our knowledge no studies have been published in which comparisons were made between selected and nonselected cells.

In our preliminary experiments looking at K^+ uptake, mannitol instead of NaCl was used to adjust the osmotic potential of the uptake solutions. This eliminated any affects Na^+ might have on the K^+ uptake system. It was found that with increasing osmotic stress the capacity of the nonselected cell line to take K^+ up decreased markedly (Fig. 18.3). The salt-selected cells, however, were able to maintain K^+ uptake at all osmotic stress levels. The salt-selected cells had higher rates of K^+ uptake as the osmotic stress increased. This ability to maintain K^+ uptake under an osmotic stress is an important mechanism which allows the salt-selected cells to survive in a stressful environment (42).

Further studies are being carried out in which the effects of competing

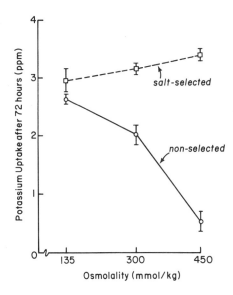

Figure 18.3 Potassium uptake after 72 hr as a function of increasing osmotic stress (applied as mannitol to media) for nonselected alfalfa cells (solid line) and salt-selected alfalfa cells (dashed line). Bars depict standard error.

cations, uptake of Na^+, and the effects of the anion have on the regulation of ion transport. These results should assist in elucidating the mechanisms used by the salt-selected alfalfa cells.

Metabolic Energy Costs

Another question being addressed in our laboratory is the metabolic cost imposed on a plant by the presence of salinity in its environment. Tissue culture techniques offer a unique opportunity to investigate the energy cost involved in dealing with salt and possibly other plant stresses. Plant cells can be cultured in media containing precise concentrations of sugar and salt, thereby exposing the cells to quantified amounts of energy and stress.

A similar approach has been used by Kato and Nagai (45) to determine the maintenance and growth yield values for sugar and oxygen in batch cultures of *Nicotiana tabacum* cells. Their growth yield value for sugar was 0.6 g cell dry weight per g glucose. Hunt and Loomis (46) obtained growth yields of 0.61 ± 0.04 g dry tissue per g sucrose supplied for *Nicotiana rustica* callus.

To examine the energy consumption associated with stress, liquid culture medium containing sugar and salt is inoculated with a weighed amount of tissue. Following a period of growth respiration is measured on an aliquot of cells, and the remaining suspension is filtered and growth is measured by the increase in weight. Cells are dried for 2 days at 70°C to determine dry weights. Filtered cells can also be pressed in a French cell press and expressed sap measured on a vapor pressure osmometer. Changes in osmotic potential of the cells can be followed. The medium from which the cells have been filtered is analyzed for the remaining sugar content, and the amount consumed during the growth of the culture is the difference between the starting and final concentrations of sugar.

Although there are a number of ways to express efficiency, for our purposes we have defined it as the weight of tissue produced per g of sugar consumed as follows:

$$\text{efficiency} = \frac{\Delta \text{ dry weight of tissue (g)}}{\text{weight of sugar consumed (g)}}$$

where Δ dry weight of tissue is the final dry weight of the cells filtered from the culture medium minus the inoculum dry weight and the weight of sugar consumed is the difference between the starting amount and the amount left in the medium after the cells have grown.

Growing the salt-tolerant and nontolerant alfalfa cells as suspension cultures using glucose as the energy source and various levels of salinity, we determined efficiency values as shown in Figure 18.4. The Y axis is efficiency as we defined it earlier. The solid line is the nonselected cells, which show a decrease in efficiency as the salt level rises. The dotted line is the salt-selected cell line, which shows less efficiency in the absence of salt, but a substantial improvement in efficiency as the salt level is raised. In the absence of salt, the nonselected cells are better adapted having twice the efficiency as the salt-selected cells. Under salt stress though, the salt-selected cells are better adapted to this environment and have approximately three times the efficiency of the nonselected cells. The maximum efficiency for both cell lines are similar, but for the nonselected cells that efficiency is reached in the absence of salt, whereas the salt-selected cells, with an apparent requirement for salt, reach maximum efficiency in the presence of salt. These values for the alfalfa cells are somewhat lower than that found by Kato and Nagai (45) and Hunt and Loomis (46) for tobacco. These lower values

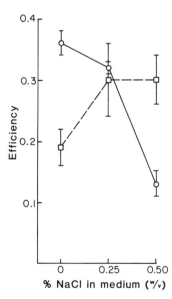

Figure 18.4 Efficiency of alfalfa suspension culture cells as a function of NaCl concentration in the nutrient medium, comparing a salt-selected cell line (□) to the nonselected cell line (○). Bars depict standard error.

are due more to the measurement procedures and not necessarily the result of physiological differences in these two species.

By comparing the respiration rates and the energy consumed by cells grown in the absence of salt with that of cells exposed to various levels of salinity, any additional energy consumption associated with the salinity stress can be determined. By including both nontolerant and salt-selected cell lines, any difference in the stress-related energy requirements of the two cell types can be examined.

Respiration was followed over time for the two cell lines growing on 0, 0.5, and 1.0% NaCl. After 8 days the rate of respiration for the nonselected cells was less than the salt-selected cells at all NaCl levels (Table 18.1). The rate of respiration dropped 60–64% when salt was present for the nonselected cells. The salt-selected cells showed little effect of salt on respiration. There was no difference in respiration rate between cells grown at 0% or 0.5% NaCl, whereas the rate for the cells on 1.0% NaCl dropped only 26%.

Evaluation of respiration and energy use efficiency provides some insight into the effect of salinity on the coupling of respiration with the metabolic process leading to biomass formation. The salt-selected cells, when exposed to increasing salinity, showed increasing energy use efficiency with little change in respiratory activity. This could be interpreted as a biological system that efficiently converts energy sources to growth and development in stress environments. In contrast, the unselected cells showed a considerable effect of salinity on respiration and efficiency. Respiration and efficient use of glucose decreased as salinity increased. It is possible that reduction in respiration of these cells also reduces metabolic conversion of glucose to metabolites and to growth and development. This could result in reduced efficiency and in the inability of these cells to respond to increasing stress.

Although we can only conjecture from these preliminary experiments, perhaps the efficiency and respiration differences between the salt-selected and nonselected cells could be a significant component of survival, and more importantly, productivity for plants exposed to salt stress. A major limitation to yield in stress environments such as salinity may be the efficient use of metabolic

Table 18.1   Respiration of Alfalfa Cells after 8 Days Growth in Suspension Containing NaCl

% NaCl	Respiration (nmoles O_2 min^{-1} g fr. wt^{-1})	
	Nonselected	Salt-Selected
0	286.19 ± 27.03	353.71 ± 27.35
0.5	113.37 ± 15.05	334.69 ± 15.35
1.0	103.31 ± 22.79	261.92 ± 18.47

energy for growth. A better understanding of this efficiency of carbohydrate utilization may led to improved crop production in saline environments.

2. CONCLUSIONS

A prime objective of cell selection procedures is the isolation of cell lines characterized by properties favoring growth in chemical or physical environments that are commonly inimical to growth of unselected lines. The selected lines are presumed to be genetic variants that upon regeneration to whole plants will express the selected characteristic and provide genetic material for plant improvement programs. There has been very limited success for this approach; however, these materials may provide the basis for physiological studies.

Selection of cell lines for tolerance to salinity provides an opportunity for comparative studies on the cellular mechanisms associated with salt tolerance.

Salt-selected cells evaluated for physiological differences demonstrated a number of processes that could be identified with salt-tolerant mechanisms. Salt-selected cells were characterized by ion regulation processes which maintained higher intracellular levels of potassium and calcium than nonselected cells when the two cell lines were exposed to salinity. The selected cell line was found to have a greater energy use efficiency and to maintain higher levels of respiration than the nonselected line when these lines were exposed to salinity. This could be related to the apparent requirement for diversion of metabolic products and energy for osmoregulation processes essential in plant cells exposed to salt stress.

REFERENCES

1. T. J. Flowers, P. F. Troke, and A. R. Yeo. The mechanism of salt tolerance in halophytes. *Annu. Rev. Plant Physiol.* **28,** 89–121 (1977).
2. T. P. Croughan and D. W. Rains. Terrestrial halophytes: habitats, productivity, and uses. In A. Mitsui and C. C. Black, eds., *Handbook of Biosolar Resources*, Vol. 1, Part 2, Basic Principles. CRC Press, Boca Raton, Florida, 1982, pp. 245–255.
3. L. Bernstein. Effects of salinity and sodicity on plant growth. *Annu. Rev. Phytopathol.* **13,** 295–312 (1975).
4. E. Epstein. Responses of plants to saline environments. In D. W. Rains, R. C. Valentine, and A. Hollaender, eds. *Genetic Engineering of Osmoregulation: Impact on Plant Productivity for Food, Chemicals, and Energy.* Plenum, New York, 1980, pp. 7–21.
5. D. W. Rains. Salt tolerance—New Developments. In J. T. Manassah and E. J. Briskey, eds., *Advances in Food Producing Systems for Arid and Semiarid Lands.* Academic Press, New York, 1981, pp. 431–456.

6. D. W. Rains. Salt tolerance of plants: strategies of biological systems. In A. Hollaender, ed., *The Biosaline Concept: An Approach to the Utilization of Saline Environments*. Plenum, New York, 1979, pp. 47–67.

7. H. Greenway and R. Munns. Mechanisms of salt tolerance in non-halophytes. *Annu. Rev. Plant Physiol.* **31**, 149–190 (1980).

8. J. A. Hellebust. Osmoregulation. *Annu. Rev. Plant Physiol.* **27**, 485–505 (1976).

9. R. L. Jefferies. The role of organic solutes in osmoregulation in halophytic higher plants. In D. W. Rains, R. C. Valentine, and A. Hollaender, eds., Genetic Engineering of Osmoregulation; Impact on Plant Productivity for Food, Chemicals, and Energy. Plenum, New York, 1980, pp. 135–154.

10. J. Levitt, *Responses of Plants to Environmental Stresses*, Vol. 2. Academic Press, New York, 1980.

11. D. W. Rains. Salt transport by plants in relation to salinity. *Annu. Rev. Plant Physiol.* **23**, 367–388 (1972).

12. G. R. Stewart and F. Larher. Accumulation of amino acids and related compounds in relation to environmental stress. In P. K. Stumpf and E. E. Conn, eds., *The Biochemistry of Plants*, Vol. 5. Academic Press, New York, 1980, pp. 609–635.

13. R. G. Wyn Jones. An assessment of quaternary ammonium and related compounds as osmotic effectors in crop plants. In D. W. Rains, R. C. Valentine, and A. Hollaender, eds., *Genetic Engineering of Osmoregulation: Impact on Plant Productivity for Food, Chemicals, and Energy*. Plenum, New York, 1980, pp. 155–170.

14. A. A. Rosiella and J. Hamblin. Theoretical aspects of selection for yield in stress and non-stress environments. *Crop Sci.* **21**, 943–946 (1981).

15. W. F. Sheridan. Plant regeneration and chromosome stability in tissue culture. In L. Ledoux, ed., *Genetic Manipulations with Plant Material*, NATO Advanced Study Institutes Series A, Vol. 3. Plenum, New York, 1975, pp. 263–295.

16. Z. S. Wochok and D. F. Wetherell. Restoration of declining morphogenetic capacity in long term tissue cultures of *Daucus carota* by kinetin. *Experientia* **28**, 104–105 (1972).

17. S. J. Stavarek, T. P. Croughan, and D. W. Rains. Regeneration of plants from long-term cultures of alfalfa cells. *Plant Sci. Lett.* **19**, 253–261 (1980).

18. M. H. Zenk. Haploids in physiological and biochemical research. In K. J. Kasha, ed., *Haploids in Higher Plants*. University of Guelph Press, Guelph, Canada, 1974, pp. 339–353.

19. P. J. Dix and H. E. Street. Sodium chloride-resistant cultured cell lines from *Nicotiana sylvestris* and *Capsicum annuum*. *Plant Sci. Lett.* **5**, 231–237 (1975).

20. M. W. Nabors, A. Daniels, L. Nadolny, and C. Brown. Sodium chloride tolerant lines of tobacco cells. *Plant Sci. Lett.* **4**, 155–159 (1975).

21. M. W. Nabors, S. E. Gibbs, C. S. Bernstein, and M. E. Meis. NaCl-tolerant tobacco plants from cultured cells. *Z. Pflanzenphysiol. Bd.* **97**, 13–17 (1980).

22. P. M. Hasegawa, R. A. Bressan, and A. K. Handa. Growth characteristics of NaCl-selected and non-selected cells of *Nicotiana tabacum* L. *Plant Cell Physiol.* **21**, 1347–1355 (1980).

23. J. Kochba, G. Ben-Hayyim. P. Spiegel-Roy, S. Saad, and H. Neumann. Selection

of stable salt-tolerant callus cell lines and embroyos in *Citrus sinensis* and *C. aurantium*. *Z. Pflanzenphysiol. Bd.* **106**, 111–118 (1982).

24. T. P. Croughan, S. J. Stavarek, and D. W. Rains. Selection of a NaCl tolerant line of cultured alfalfa cells. *Crop Sci.* **18**, 959–963 (1978).

25. T. P. Croughan, S. J. Stavarek, and D. W. Rains. *In vitro* development of salt resistant plants. *Environ. Exp. Bot.* **21**, 317–324 (1981).

26. T. P. Croughan. The application of cell culture techniques to the selection and study of salt tolerance. Ph.D. dissertation, University of California, Davis, 1981.

27. H. von Hedenström and S. W. Breckle. Obligate halophytes? A test with tissue culture methods. *Z. Pflanzenphysiol Bd.* **74**, 183–185 (1974).

28. M. Tal, H. Heikin, and K. Dehan. Salt tolerance in the wild relatives of the cultivated tomato: Responses of callus tissue of *Lycopersicon esculentum, L. peruvianum* and *Solanum pennellii* to high salinity. *Z. Pflanzenphysiol. Bd.* **86**, 231–240 (1978).

29. T. J. Orton. Comparison of salt tolerance between *Hordeum vulgare* and *H. jubatum* in whole plants and callus cultures. *Z. Pflanzenphysiol. Bd.* **98**, 105–118 (1980).

30. M. K. Smith and J. A. McComb. Effect of NaCl on the growth of whole plants and their corresponding callus cultures. *Aust. J. Plant Physiol.* **8**, 267–275 (1981).

31. M. K. Smith and J. A. McComb. Use of callus cultures to detect NaCl tolerance in cultivars of three species of pasture legumes. *Aust. J. Plant Physiol.* **8**, 437–442 (1981).

32. R. Goldner, N. Umiel, and Y. Chen. The growth of carrot callus cultures at various concentrations and composition of saline water. *Z. Pflanzenphysiol. Bd.* **85**, 307–317 (1977).

33. Y. Chen, E. Zahavi, P. Barak, and N. Umiel. Effects of salinity stresses on tobacco. I. The growth of *Nicotiana tabacum* callus cultures under seawater, NaCl, and mannitol stresses. *Z. Pflanzenphysiol. Bd.* **98**, 141–153 (1980).

34. S. J. Stavarek and D. W. Rains. Growth and ionic characteristics of alfalfa cells exposed to different salts. Proceedings, 58th Annual Meeting of the American Society of Plant Physiologists, 1982.

35. F. B. Kulieva, Z. B. Shamina, and B. P. Strogonov. Effect on high concentrations of sodium chloride on multiplication of cells of *Crepis capillaris in vitro*. *Fiziologiya Rastenii* **22**, 131–135 (1975).

36. E. I. Komizerko and T. I. Khretonova. Effect of NaCl on the process of somatic embryogenesis and plant regeneration in carrot tissue cultures. *Fiziologiya Rastenii* **20**, 268–276 (1973).

37. J. W. Heyser and M. W. Nabors. Osmotic adjustment of cultured tobacco cells (*Nicotiana tabacum* var. Samsum) grown on sodium chloride. *Plant Physiol.* **67**, 720–727 (1981).

38. P. J. Dix and R. S. Pearce. Proline accumulation in NaCl-resistant and sensitive cell lines of *Nicotiana sylvestris*. *Z. Pflanzenphysiol. Bd.* **102**, 243–348 (1981).

39. M. Tal and A. Katz. Salt tolerance in the wild relatives of the cultivated tomato: The effect of proline on the growth of callus tissue of *Lycopersicon esculentum* and *L. peruvanum* under salt and water stresses. *Z. Pflanzenphysiol. Bd.* **98**, 283–288 (1980).

40.   A. Rosen and M. Tal. Salt tolerance in the wild relatives of the cultivated tomato: Responses of naked protoplasts isolated from leaves of *Lycopersicon esculentum* and *L. peruvianum* to NaCl and proline. *Z. Pflanzenphysiol. Bd.* **102,** 91–94 (1981).

41.   E. Epstein. Mineral metabolism of halophytes. In I. H. Rorison, ed., *Ecological Aspects of the Mineral Metabolism of Plants.* Blackwell, Oxford, 1969, pp. 345–355.

42.   S. J. Stavarek and D. W. Rains. Ion uptake of salt-selected plant cells. Proceedings, 57th Annual Meeting of the American Society of Plant Physiologists, 1981.

43.   Y. Waisel. *Biology of halophytes.* Academic Press, New York, 1972.

44.   P. A. LaHaye, and E. Epstein. Salt toleration by plants: Enhancement with calcium. *Science* **166,** 395–396 (1969).

45.   A. Kato and S. Nagai. Energetics of tobacco cells, *Nicotiana tabacum* L., growing on sucrose medium. *Eur. J. Appl. Microbiol. Biotechnol.* **7,** 219–225 (1979).

46.   W. F. Hunt and R. S. Loomis. Carbohydrate-limited growth kinetics of tobacco (*Nicotiana rustica* L.) callus. *Plant Physiol.* **57,** 802–805 (1976).

19

CELL CULTURE AND
RECOMBINANT DNA METHODS
FOR UNDERSTANDING
AND IMPROVING
SALT TOLERANCE OF PLANTS

Maureen R. Hanson

Department of Biology
University of Virginia
Charlottesville, Virginia

This chapter will review current and potential applications of plant cell culture techniques for improvement of crop salt tolerance. Plant cell culture is just one among the battery of methods we have for studying how plants tolerate stress and for producing or selecting genetically superior plants. Potentially much progress in this area may be made by an integrated approach, using classical plant breeding, physiological, biochemical, cell culture, and recombinant DNA methods. The focus of this chapter will be to point out the advantages as well as current limitation of cell culture technology for salt-tolerant crop synthesis.

Plant tissue culture is often praised as a means to perform selections in several petri dishes which would take hundreds of acres of performed at the whole plant level. It is indeed true that millions of potential plants can exist in the form of protoplasts, callus, or suspension cells in one culture dish. However, there are important differences between the plant breeder's evaluation of relative stress tolerance in a large field population of plants and the cell culturist's selection for stress tolerance of the cell *in vitro*. For one, certain stresses are more practically applied at the whole plant level (e.g., wind, insect pests) while others (e.g., rare toxins, chemicals) are often easier to incorporate into cell culture selection. Salt stress is one that is reasonable to apply both at the whole plant and cell level, although salt inclusion in cell culture media is certainly simpler than in whole plant substrates.

A more important consideration with regard to salt tolerance is whether the degree of genetic diversity is greater in whole plant or in cell populations on which selection is applied. For some crop plants, little genetic diversity for salt tolerance may be available among conventionally propagated material. In such a case, the plant breeder has no more initial resources than the cell culturist, who normally starts cultures from one plant. However, for some crop plants, the breeder can construct highly heterozygous material with unexploited genetic variation to compare, with many genotypes and phenotypes represented in the population. If somatic cells of one plant are not significantly genetically variable *in vivo*, the cell culturist must rely on *in vitro* generation of diversity to produce a diverse cell population from which variants can be selected. Thus, the question becomes, how can we manipulate a cell culture so as to maximize the probability of producing and detecting a stress tolerant variant from a stress-sensitive initial cell population?

I will consider the prospects for using cell cultures, with emphasis on protoplast culture, to produce salt-tolerant plants from initially more salt-sensitive material. In the case of protoplasts, genetic diversity has been demonstrably generated by (i) passage through cell culture, (ii) mutagenesis, (iii) fusion with other protoplasts, and (iv) tumorigenic transformation by *Agrobacterium tumefaciens*. Active research is currently being carried out on two other conceptually promising methods to generate diversity, namely, chromosome and DNA-mediated transformation of cell cultures. I will therefore also discuss the prospects for genetic engineering of salt-tolerant crops via recombinant DNA technology. Finally, I will point out how cell culture and nucleic acid techniques can be used to further our understanding of the mechanism of salt tolerance and plant salt stress reactions. This fundamental knowledge is required to apply certain strategies for the deliberate genetic engineering of salt tolerance.

1. SOMACLONAL VARIATION

The term "somaclone," a plant derived from any form of cell culture, was defined by Larkin and Scowcroft in 1981 in a review destined to become a classic. For several years, the scientific community had begun to reconsider the concept of tissue culture as a neutral pathway from which to produce copies of an initial genotype. Larkin and Scowcroft (1) presented an integrated view of dispersed observations of variability from cell culture, one that has generated many testable hypotheses.

From the definition of somaclone, somaclonal variation is the variation among plants derived from cell culture. Many observations of such variation have been made. But these were misfits to the prevailing general concept that cell cultures reproduced the germ line of the starting genotype faithfully, unless the chromosomal number changed and/or mutagenesis occurred. Also puzzling have been the unexpectedly high recovery of new phenotypes in cell culture without deliberate mutagenesis. The observations of variability have more coherence if viewed as results of uncharacterized phenomena which alter the genome of somatic cells such that novel genomes and thereby new original phenotypes appear in regenerated plants. With future study, a reversal of the view of culture-derived plants may emerge. We may eventually ask, not why so much variation appears, but how plant genomes remain so stable.

Larkin and Scowcroft (1) review reports of somaclonal variation in the crop plants sugar cane, potato, tobacco, rice, oats, maize, barley, and *Brassica* species. The defining of the phenomenon has spurred further comparisons and inquiries into the molecular mechanisms of variability generation should follow. Larkin and Scowcroft suggest a number of possible genetic mechanisms, many or all of which may prove to contribute to somaclonal variation (Table 19.1).

Thus, somaclonal variation may provide an unexpected source of genetic variability for such useful characteristics as stress tolerance. Thomas et al. (2) have demonstrated somaclonal variation among 10 different plants regenerated from the same initial plant cell, which indicates that the variability arose during culture. This does not eliminate the possibility that variability also preexists in somatic cells outside of the germ line, which is now detected because of our recent ability to produce quantities of plants from somatic cells. If substantial numbers of somatic cells have genomes differing from those destined to form gametes, then the cell culturist does not start with just one genotype in a petri dish. Instead the petri dish is more comparable to a plant breeder's field of heterozygous plants than previously suspected. Even if all of the variability arises from cell culture itself, then we must modify the previous concept of cell cultures as stable populations from which to select simple single-step alterations, analogous to bacterial mutagenesis. Instead, phenomena may be occurring which produce multiple changes within the same cell at high frequencies. Secor and Shepard (3) found single somaclones with as many as 17 different phenotypic alterations. Such multiple changes may be critical for obtaining plants via cell culture which are improved for polygenic traits such as salt tolerance.

Table 19.1.   Possible Causes of Somaclonal Variation

Karyotype changes
Cryptic changes associated with chromosome rearrangement
Transposable elements
Somatic gene rearrangements
Gene amplification and depletion
Cryptic virus elimination
Rearrangements in organelles genomes

Of the possible types of variation mechanisms (Table 19.1), which are likely to be most relevant for salt tolerant genotype breeding? All such changes have the potential to alter the expression of genes affecting salt tolerance, or to alter the structure of the gene products. Chromosomal loss, amplification, or gross chromosomal rearrangements are likely to be the least useful, since such changes are often detrimental and may not be transmitted stably through either vegetative or sexual propagation. Other gross chromosome changes, such as somatic crossing-over, may result in more stable changes. If cell culture does increase the frequency of sister chromatid exchange, then for optimal generation of diversity, highly heterozygous material should be cultured. Culturing interspecific hybrids that have sterility problems may also be a way to obtain hybrids with increased fertility. If tissue culture enhances chromosome breakage and exchange, then alien chromosomal material may become attached to crop plant-derived chromosomes. Orton (4, 5) already has some interesting evidence of such occurrences in hybrids between barley and the salt-tolerant species *Hordeum jubatum*. Some regenerates were obtained in which all *H. jubatum* chromosomes had been eliminated, but a few *H. jubatum* iszyme bands were still present (6).

Culture-induced alterations in smaller units of genetic material (such as could ensue from mobilization of transposable elements, somatic gene rearrangement, or gene amplication or depletion) also could affect a cell line's response to salt stress. Such changes must be carried to the whole plant level and be stable following propagation for such variation to be useful. A critical ongoing area for research is the degree of stability of somaclonal variation. From the fact that some somaclonal variants have chromosome number alterations but others do not, we already know that more than one mechanism generates the variability. We do not yet know how stability of somaclonal variation will differ according to the generating mechanism. Current results for practical exploitation are promising (1, 7), but much further evaluation of somaclones and their progeny is needed.

Even if no increases in salt tolerance are found among regenerated plants themselves, such plants may still be useful in a breeding program. Variation may be present in somaclones, but require combination by sexual crosses to obtain the expression of a salt-tolerant phenotype. Furthermore, we may discover that only certain crop species will exhibit useful somaclonal variation.

1.1. Somaclonal Variation and Selection of Salt Tolerance *In Vitro*

The mechanisms underlying somaclonal variation are likely also operating during growth of cell lines *in vitro* and may be responsible for producing phenotypic changes of cells in culture. Selection of cell lines *in vitro* with increased salt tolerance has already been achieved by several laboratories and is reviewed by Stavarek and Rains in Chapter 18. I will therefore restrict myself to discussing how knowledge of mechanisms underlying salt stress response can affect strategies for production of salt-tolerant plants from salt-tolerant cultures.

Diversity is the most striking aspect emerging from a survey of our current knowledge of the plant kingdom's responses and adaptations to salt stress. Many apparent contradictions are likely eventually to be resolved as real genotypic differences in strategy. Only one general principle has actually surfaced: a negative one. Plants do *not* evolve salt tolerance by alteration of cellular enzymes to accommodate high cytoplasmic salt concentrations (8–10). How this "cytoplasmic avoidance" is achieved differs among species, among related genotypes, among tissues, and among development stages of the same plant. In designing cell culture or recombinant DNA engineering strategies for improving salt tolerance, we should be aware that a plant, depending on its genetic constitution, might be best improved by affecting (i) how much and which ions are transported to the shoot (11, 12), (ii) the reexport of ions to roots, (iii) the cellular compartmentalization of ions, (iv) the concentration and type of osmotic cytoplasmic solutes (13, 14), (v) the efficiency of ion pumps, (vi) lipid composition of membranes, and (vii) other physiological characteristics (15–17).

Can we predict from type of tolerance mechanism used by a plant whether it will be tolerant *in vitro* as well as *in vivo*? Can we predict mechanism and *in vivo* tolerance from the observation of *in vitro* tolerance? Although relevant studies are increasing in number, so far there are insufficient data on this important point. Results from my laboratory are consistent with the hypothesis than any relative salt tolerance of the xerophyte *Solanum pennellii* (18), reported by Dehan and Tal (19) to be more salt tolerant than *Lycopersicon esculentum*, requires whole plant structure. We have not found any evidence for differences in NaCl sensitivity of organ cultures of roots and shoots of these two species (Fig. 19.1; Table 19.2). We also did not observe convincing reproducible differences in callus growth (Table 19.3), contrary to other reports (20, 21). Since Dehan and Tal (19) reported higher Na ion concentrations in *S. pennellii* compared to *L. esculentum*, we might have expected to find cellular salt resistance *in vitro*. Conversely, we might not expect cellular salt resistance in a genus such as *Hordeum* in which salt tolerance is thought to derive from inhibition of salt diffusion or transport into the xylem sap (22). However, Orton (23) concluded that *H. jubatum* callus was more NaCl tolerant than *H. vulgare in vitro*. Our ability to make predictions is also not expanded by the comparisons of *in vitro* salt tolerance made by Smith and McComb (24). Two halophytes were not more resistant than a salt-sensitive glycophyte in culture, whereas a salt-tolerant glycophyte was as tolerant *in vitro* as *in vivo*.

One problem that has plagued comparisons of salt sensitivity and tolerance of different genotypes in culture is a frequently observed disparity in growth at control, low concentrations of NaCl. Genotypic-specific growth responses on

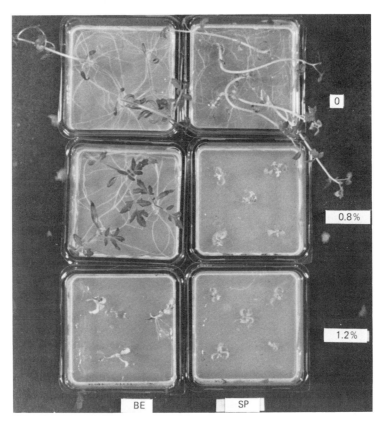

Figure 19.1. Growth of shoot culture of *L. esculentum* cv. Big Early (BE) and *S. pennellii* (SP) on media containing 0 added NaCl, or 0.8% or 1.2% NaCl. Seedling shoot tips were excised after germination *in vitro* and placed on shoot culture media for 2 weeks of growth.

Table 19.2. Effect of NaCl on Dry Weight (mg) of Root Cultures of Three Related Genotypes[a]

Added NaCl (mM)	*S. pennellii*	BE × SP	Big Early
0	29.9	27.8	26.9
100	4.0	9.7	9.3
130	3.8	5.3	3.3
160	2.1	1.7	2.3
200	1	1	1

[a]Cultures were grown in liquid media for 3 weeks. Big Early is a *Lycopersicon esculentum* cultivar. BE × SP is the F₁ hybrid between the cultivar and *S. pennellii*.

Table 19.3.   Growth of Tomato Callus on Media Containing NaCl

	Red Cherry		Red Cherry × S. pennellii		S. pennellii	
Added NaCl (%)	FWI[a]	%C[b]	FWI	%C	FWI	%C
0	108.2	100	314.4	100	650.7	100
0.8%	30.1	27.8	75.8	24.1	148.0	22.7
1.2%	17.7	16.4	97.2	30.9	104.6	16.1

[a]Percent increase in fresh weight from inoculum.
[b]Percent increase compared to control FWI.

standard media are well known (25). When one genotype grows much better than the other on the control medium, then a higher apparent salt tolerance of the rapidly growing genotype might actually result from a nonspecific better toleration of any stress compared to a poorly growing tissue that is already stressed by a nonoptimal medium. In our own comparisons of *S. pennellii* and *L. esculentum*, we are most confident of our results with shoots and roots, because we were able to formulate control media on which both genotypes grew well (Fig. 19.1; Table 19.2).

1.2. Strategies and Predictions

Although further work is needed on *in vivo* versus *in vitro* expression of salt tolerance, the following point is already evident. Even though salt selection *in vitro* has produced regenerated plants with increased tolerance (26), selection *in vitro* for salt tolerance will not detect all potentially salt-tolerant plants. This reaffirms the importance of screening somaclonal variants at the whole plant level. The observations also suggest that salt-tolerance mechanisms of relatives of crops should be considered before deciding whether to (i) select cell lines or (ii) screen somaclones for salt tolerance. Crops in which wild relatives have halophytic adaptions which are not expressed in culture may be less suitable targets for *in vitro* selection than for improvement via breeding (conventionally or with somaclonal variants). Finally, we should exploit the ability to isolate tissues and organs in culture to study mechanisms of salt tolerance as well as to use culture techniques for direct improvement (27).

2. NEW HYBRIDS VIA *IN VITRO* TECHNIQUES

Breaking fertilization barriers is one achievement of cell and tissue culture technology which has potential to be utilized for salt-tolerant plant breeding. Both somatic cell fusion and sexual gamete fusion or embryo culture in vitro have been successfully applied for producing hybrids unobtainable through traditional means.

The less complex techniques of *in vivo* fertilization and embryo culture are reasonable first approaches to producing hybrids where sexual incompatibilities have prevented cross-breeding. Such techniques have been reviewed elsewhere (28–30). Here, I will consider the current state of somatic hybrid plant production.

2.1. Nuclear Hybrids

Some cell hybrids are produced by isolating and fusing protoplasts from two different sources, followed by culture and regeneration to whole plants. In order to produce a somatic cell hybrid, the following techniques must be possible: production of protoplasts, induction of cell fusion to form heterokaryons, selection or identification of somatic hybrid cells or plants, and culture of protoplasts to plant via embryogenesis or apical meristem organization from callus. The specifics of plant protoplast fusion have been extensively reviewed (31–34). With moderate effort, production of protoplasts and their fusion are now possible for a wide range of different plant species. However, inability to culture protoplasts to plants is the primary barrier to somatic cell hybrid production in a number of genera. The active research being carried out in this area has already resulted in some progress with crop plants since 1975 (35).

Cell culture techniques may be useful not only for conventional crops, but for improvement of halophytes not previously exploited commercially (36). This may especially be true if a less tolerant but more economically useful halophyte has close relatives with high tolerance. How amenable halophytes will be to cell culture techniques cannot be judged from the small amount of work yet performed (24, 37).

The limiting factor for somatic hybrid production in those genera where protoplasts will regenerate into plants has been means to select hybrids from unfused parental cells. In most cases thus far, genetic selection methods, whereby hybrid callus or plants have phenotypes distinguishing them from their parents, have been used for somatic hybrid selection (Table 19.4) (38–80). Genetic selection requires construction of appropriate parental genotypes or screening of genotypes for existing differences which can be exploited in a selection scheme. More versatile methods, permitting synthesis of hybrids between any two genotypes from which protoplasts can be regenerated, may become routine within the next 5 years. One promising strategy that has been successfully applied is chemical inhibition of two parental cell types by two different poisons, followed by complementation of function in fused cells but death of unfused parental cells (81). This method can also be used to select against one parent which brings a selectable marker to the heterokaryon. For example, streptomycin-resistant tobacco cells were chemically poisoned before

Table 19.4.   Heterokaryon-Derived
Plants[a]: Production and Characterization

Genus	References
Intrageneric	
Petunia	38–48
Nicotiana	49–67
Datura	68–73
Solanum	43
Daucus	74
Intergeneric	
Daucus-Aegopodium	75
Atropa-Datura (leaves only)	76
Solanum-Lycopersicon	77–80
Arabidopsis-Brassica	57

[a]See ref. 14.

fusing with streptomycin-sensitive cells, and somatic hybrid callus could be selected on streptomycin medium (82).

A second versatile heterokaryon selection method is based on differential fluorescence labeling of parental cells (83). Fused cells have fluorescences characteristic of both types of parental cells and can be isolated for culture either manually with micropipets (84) or by processsing through fluorescence-activated cell sorters (85). This latter approach is particularly promising, because automated cell sorting will be able to provide large quantities of heterokaryons for regeneration into somatic hybrid plants. In most somatic hybridization experiments reported (references in Table 19.4) only small numbers of plants were regenerated. Rare is a report such as that of Izhar (86) in which over 3000 somatic hybrids were produced following a genetic selection. For breaking fertilization barriers, often only a small number of somatic hybrid plants per two genotypes fused will suffice for further breeding. However, for the asymmetric hybridization experiments discussed below, large numbers of hybrids would increase the possibility of transfer of the desired genes.

Currently somatic hybrid plants have been produced in a limited number of genera, primarily among the Solanaceae (Table 19.4). Regenerated hybrid plants are generally confined to the same genus, or closely related genera. Although a number of wider fusions of rather unrelated genotypes have been carried out and cell divisions obtained, most near-term applications of protoplast fusion are likely to be intrageneric because of the developmental aberrations resulting from combinations of evolutionarily diverged genotypes.

In order to assess the prospects of utilizing somatic hybrids in salt tolerance breeding, we must review what types of plants and variability has been produced following protoplast fusion. A number of characteristics have been found to vary among hybrid plants derived from the same fusion (Table 19.5). Much of

Table 19.5.   Types of Varability among Sibling Somatic Hybrids[a]

Characteristic	References
Chromosome number	50, 53, 58, 59, 65, 67, 69, 76, 77, 87–90
Leaf morphology	53, 56, 58, 59, 66, 67, 90
Plant height	53, 63
Fertility	50, 58, 59, 63, 81, 86, 90
Flower morphology	10, 50, 53, 58, 67, 87, 88, 91
Isozyme banding pattern	53, 91
Mitochondrial DNA restriction fragment pattern	92–95
RuBPCase subunits	6, 50, 52, 59, 60
Chloroplast DNA restriction fragment pattern	50, 65, 89, 92

[a]Somatic hybrid plants derived from the same fusion experiment.

this variability is likely due to differences in chromosome complement between sibling hybrids as well as to chromosome instability in the regenerated plants. Presently we do not know how to control variability in chromosome number and other characteristics of somatic hybrids. It is unknown why, using the same *N. glauca* and *N. langsdorfii* parents, one laboratory obtained somatic hybrids with a range of chromosome numbers (67) whereas others (51, 56) obtained all summation-number hybrids. In one remarkable experiment, 1114 somatic hybrid plants resulting from fusion of two sexually compatible *Petunia* species were complete tetraploids (86). Probably both details of culture technique as well as degree of nuclear compatibility will be found to determine chromosomal stability in hybrids.

In most of the experiments listed in Table 19.5, severely unbalanced chromosome complements were not obtained in the regenerated plants. Hexaploid or near hexaploid hybrids, which can be explained as triple cell fusion products, have been recovered in *Nicotiana* and *Datura* (67, 73). However, even in somatic hybrids containing more chromosomes of one parent than the other, often nearly a complete complement of each parental chromosome set is present. A fusion between *Arabidopsis* and *Brassica* which gave hybrids containing only a few chromosomes of one parent (88) is the exception rather than the rule thus far. Indeed, a large asymmetry in ratio of parental chromosome content in a heterokaryon-derived dividing tissue may well be inimical to the regeneration developmental pathway, particularly if the two parental lines are not closely related. Even so, some rather remarkable ranges of chromosome constitution have been obtained (refs. in Table 19.5). Chromosome segregation and loss has been observed to occur both in dividing cells (96, 97) and after hybrid plant regeneration (98).

Many somatic hybrids have chromosomal aberrations or incompatibilities sufficient to make the hybrid plants economically worthless or too sterile to be of use for transfer of introduced genes any further. For many purposes the cell culturist would like to produce hybrids with limited genetic information from one parent, so that developmental aberrations are minimized, and the ensuing plant either remains economically useful or sufficiently fertile for breeding via sexual genetics. For proper development, better than the presence of only one or two chromosomes from one parent may be the transfer of smaller pieces of genetic information from one parent's chromosomes to the other, with elimination of the "donor" chromosome set. Chromosomal and DNA-mediated gene transfer following fusion has been well studied in animal cells (99, 100).

2.2. Asymmetric Hybridization

There are three intriguing reports concerning putative chromosomal gene transfer in hybrids following protoplast fusion (75, 90, 101). When parental protoplasts of one *Nicotiana* species were irradiated with X-rays before fusion, most hybrids detected had chromosomes of the unirradiated parents and cytoplasmic genomes of the irradiated parent. Some plants were obtained, however, which carried some genes from the irradiated nucleus in addition to chromosomes from the unirradiated parent (90). When albino carrot plants were fused with wild-type *Aegopodium* protoplasts, green plants were obtained (101). These carried only carrot chromosomes and carrot chloroplast genomes, suggesting that the green plants were recipients of genetic information from the wild-type nuclear-incompatible parent. Finally, green plants with $2n=19$ chromosomes were produced when X-irradiated parsley protoplasts were fused with a nuclear albino carrot ($2n=18$) (75).

In two of these cases, "donor" nuclear chromosomes were fragmented by irradiation, whereas Dudits et al. (101) suggest that the mitotically inactive *Aegopodium* cells' chromosomes were fragmented during cell division as believed to occur in inactive nuclei in animal cell fusions (99).

The conceptualization that these experiments suggest for further experimentation is that of the protoplast as a possible carrier for chromosomal gene introduction into a second protoplast type. Because of the possibility of transfers of multiple genes by such a process, this approach may be particularly relevant for transfer of a polygenic complex controlling salt tolerance to protoplasts of a sensitive plant.

To clarify this point further, imagine an artificial vehicle to be used to inject foreign genes into a plant cell nucleus. One might theoretically design a sac contained within another vesicle. First, the outer membrane would fuse with the target plant cell. Then, within the target cell, the inner sac containing the foreign genes would move toward the target cell nucleus and fuse with it. This vehicle need not be artificially constructed, since current evidence suggests that this describes what follows fusion of two plant protoplasts.

Unfortunately, the reports thus far of putative chromosomal gene transfer suffer from the unavailability of sufficient linkage maps and cytogenetic knowledge to define the genetic events. My laboratory has been engaged (102) in development of protoplast techniques for tomato because of possibility to characterize any asymmetric protoplast-derived hybrids in this genetically well-known species. Although asymmetric hybridization is a conceptually promising approach, donation of nuclear genes from irradiated nuclei to heterokaryon-derived plants is certainly not the rule. Menczel et al. (62) X-rayed *N. tabacum* at four different doses before fusing with *N. plumbaginofolia.* Although plants carrying cytoplasmic genomes of *N. tabacum* and nuclear chromosomes of *N. plumbaginofolia* were identified, none of these hybrids had detectable nuclear genetic information from the irradiated parent.

2.3.   New Organelle–Nuclear Combinations via Protoplast Fusion

The discussion above has focused on making new combinations of nuclear genes, as these have clearly been shown to specify relative salt tolerance. However, protoplast fusion can also result in hybrids containing new combina-

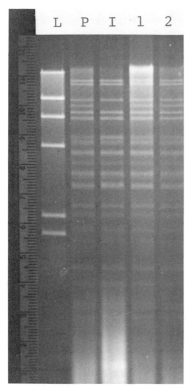

Figure 19.2.   Organelle genomes present in parental genotypes and somatic hybrids of *Petunia inflata* + *P. parodii.* L, λ HindIII molecular weight markers. I, *P. inflata* chloroplast DNA (cpDNA). P, *P. parodii* cp DNA. 1,2-cpDNA of progeny of two somatic hybrid plants. All cpDNAs were digested with enzyme BAM HI. Unpublished data of E. Clark and M. Hanson.

tions of nuclear and organelle genes, as well as novel organelle genomes. Fusion of two somatic cells circumvents the maternal inheritance that can prevent certain conbinations of nuclear and organelle genomes following sexual hybridization.

During the cell divisions following protoplast fusion, parental chloroplast genomes evidently segregate from one another, so that often regenerated somatic hybrid plants contain only one or the other parental genome (43, 65; Coney, Clark, Hanson, unpublished; Fig. 19.2). Mitochondrial genomes in somatic hybrid plants of *Nicotiana* and *Petunia* however, differ from either parent and from each other (93, 95, 103; Fig. 19.3). These results are consistent with the hypothesis that rearrangements, perhaps mediated by recombination of the two parental genomes, occurred in somatic hybrids.

Whether the ability to create new nuclear–organelle combinations and novel mitochondrial genomes is useful for salt-tolerant genotype production is presently not known. If differences in salt sensitivity were observed between genotypes derived from reciprocal sexual crosses of lines carrying different cytoplasmic genomes, then cytoplasmic genomes would be implicated in specification of salt tolerance. However, no systematic studies of salt tolerance have been reported on either isonuclear sexual or somatic hybrids carrying different organelle genomes. Investigation of this point may be worthwhile,

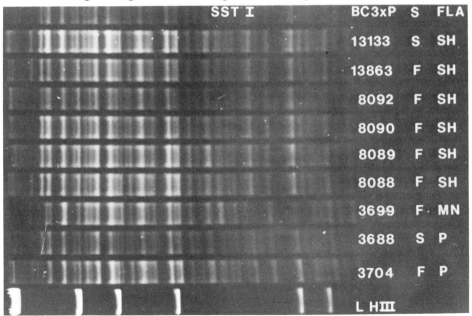

Figure 19.3. Diversity of mitochondrial genomes present in somatic hybrid *Petunia* plants. Mitochondrial DNA was isolated and digested with restriction enzyme SST I prior to gel electrophoresis. Lanes marked "P" contain mitochondrial DNA from the two parental lines, and lanes marked "SH" containing mitochondrial DNA of somatic hybrids. Unpublished data of M. Boeshore, I. Lifshitz, M. Hanson, and S. Izhar.

because organelle genomes can specify resistance to such stresses as a herbicide (104), toxin (105), and cold (106).

2.4. Somaclonal Variation among Somatic Hybrids

Somatic hybrid plants, as products of somatic cells, are by definition soma-clones. We have little data yet indicating whether the type of variability observed among unfused protoplast-derived, explant-derived, and callus-derived soma-clones is also found in somatic hybrids. Currently the variation caused by unequal chromosome numbers (66, 107) has obscured the question of what variation might have arisen otherwise. Comparison of sexual hybrids at the tetraploid level with comparable summation-number tetraploid somatic hybrids is an approach to this question. So far such a comparison has been reported only by Evans et al. (55) who indeed provide evidence that somatic hybrids were more variable than the corresponding *Nicotiana* sexual hybrids.

2.5. Strategies and Predictions

The above discussion has a number of implications for salt-tolerant crop production. First, somatic hybridization cannot currently be applied to a large number of genera, although this number will be increasing with protoplast regeneration technology as the rate-limiting step. Protoplast fusion is currently a reasonable approach for salt-tolerant genotype production only if (i) genetic variability for salt tolerance within a protoplast-culturable genus must be identified and (ii) this variability must be unaccessible to the cultivated plants by sexual hybridization *in vivo* or *in vitro*.

What of the prospects for salt-tolerant crop breeding among the most readily culturable family, the Solanaceae? Protoplast regeneration has been achieved for several solanaceous economic species, namely, tobacco, petunia, potato, eggplant, and tomato (102, 108, Table 2). Although some of these are grown in salt-affected areas such as irrigated land in California, the reservoir of genes for salt tolerance in sexually compatible species has barely been tapped. Tomato breeding programs are now in progress with the sexually compatible *L. cheesmanii* for increasing tomato salt tolerance (68, 109). Whether the less accessible genetic variation for salt tolerance in *L. peruvanium* is significantly different from that available in more sexually compatible wild species is not known. Certainly the use of bridging species and embryo culture of *L. peruvanium* × *L. esculentum* hybrids (110) would be simpler first resorts than protoplast fusion.

3. RECOMBINANT DNA TECHNOLOGY FOR SALT TOLERANCE

3.1. Predictions

Several predictions can be made regarding the near-term developments of recombinant DNA technology for plant genetic engineering. Within the next 5 years, one or more suitable vectors and techniques will likely be developed for introduction and expression of a foreign or altered DNA sequence into a plant cell. Whole plants constituted of cells containing copies of the introduced sequence will be obtainable. This will most likely be achieved by cell culture regeneration of a DNA-transformed cell. Another route may be microinjection into the germ line or apical meristem, and production of vegetative or sexual "progeny." Introducing several different genes into the same cell line and thereby plant will be possible provided that high-efficiency transforming vectors are available or repeated transformation cycles are possible. For a polygenic trait such as salt tolerance, the primary barrier to genetically engineering a more tolerant plant cell will thus not be merely the introduction of multiple new sequences.

In 5 years some progress will have been made in understanding tissue-specific and developmental stage-specific modulation of gene expression. But within this time, our knowledge will likely not be sufficient for us to direct an introduced gene to be expressed to a specified degree in a specified tissue. In fact, the ability to introduce DNA sequences in different environments and monitor their expression will open new methods for finding how a cell regulates gene expression. Only with this information will become possible the genetic engineering of traits which require developmentally regulated gene expression. But proper expression of introduced genes is not the only major barrier to the near-term improvement of crops via recombinant DNA technology. Identification and isolation of genes that specify salt tolerance are requisite for directed genetic engineering of crop plants.

Thus, two gaps in knowledge, namely, how to control introduced gene expression and what genes to introduce, make speculation on the actual genetic engineering of salt tolerance premature. Beginning to identify genes specifying salt tolerance, however, is not a premature endeavor. In fact, this effort is imperative not only for future engineering attempts, but more importantly to help us discover how plants respond to stress. Fortunately, we have some clues to sources of stress tolerance-specifying genes from comparative studies of different plant species and genotypes (111, 112).

Recombinant DNA technology adds an arsenal of new methods for uncovering basic mechanisms of stress tolerance (for methods, see ref. 113). Evident in current molecular biological studies of differentiation is a basic experimental strategy, which I will describe below in terms of its use for studying salt tolerance. For clarity, hypothetic experimental results will be presented, drawn from those obtained in studies of regulation of seed storage proteins and other developmentally regulated proteins (114).

3.2. Methods and Application to Salt Stress

The essential starting point is two different types of tissue for comparison. The range of tissues suitable for salt tolerance studies will be discussed later; we will begin our discussion with the example of root tissue of control versus root tissue of salt-stressed plants. Messenger RNA can be isolated from the initial salt-unstressed tissue and from tissue in the days following NaCl application, when growth inhibition is measurable. A DNA version of these mRNAs can be made *in vitro*. These so-called "cDNAs" have the same sequence as the corresponding mRNA except the appropriate DNA nucleotides are substituted. The cDNAs are cloned into plasmid vehicles, so that large amounts of individual cDNA sequences ("cDNA clones") can be obtained. These individual cDNAs, each representing a mRNA present in the salt unstressed or the stressed tissue, are then available for further characterization.

By comparative hybridization techniques, cDNA clones that represent genes present in salt-stressed, but not initial tissue, can be identified. These clones can be compared to one another by various techniques so that individual clones with DNA sequences in common are distinguished from ones that have completely different sequences. These distinct cDNAs can then be used to study either chromosomal DNA or the proteins they specify.

To study chromosomal DNA, the cDNA clones can be used to identify "genomic clones," which represent stretches of chromosomal DNA from which mRNA transcription was induced during salt stress. To accomplish this, chromosomal DNA is attached to plasmid or phage vehicles, creating a "library" of genomic DNA sequences. A DNA sequence corresponding to a salt stress-specific cDNA can be selected from the library by hybridization techniques. A collection of genomic clones, each representing a chromosomal DNA sequence transcribed during salt stress, can be made. By electron microscopic and DNA sequencing methods, the structural features of these genomic clones can be analyzed, with the goal of discerning any features common to salt stress-induced genes or their nearby nucleic acid environment.

DNA sequencing can also give information concerning proteins specified by salt stress-induced mRNA. The size and amino acid compositions of proteins can be derived from the DNA sequence. Furthermore, small amounts of cDNA-coded proteins can be made by using the cDNA clones to isolate mRNA which is translated *in vitro*. These proteins can be displayed on a gel for comparison to proteins isolated from the tissue. In some cases, such gel comparisons or reactions to specific antibodies will identify what protein is coded for by the induced mRNA.

The above is the current general strategy for analyzing inducible genes during a development process via recombinant DNA technology, and as such has potential for generating basic information on regulation of gene expression, gene structure, signals involved in initiation, and so on. What can such a study tell us more specifically about salt tolerance mechanisms? The techniques are standard; the amount of new information gained is determined largely by what tissues are compared and how the isolated genes are analyzed with reference to these tissues. The analyses will be most effective only with the plant physiologist's and biochemist's knowledge and hypotheses regarding salt stress and salt tolerance mechanisms.

For example what tissues are appropriate for comparisons of gene expression? In addition to the comparisons of control and stressed tissues, one might compare the same salt-tolerant genotype under low salt and high salt, or the same salt-sensitive genotype under low salt and high salt, or a salt-tolerant genotype and a salt-sensitive related genotype under the same condition of either low or high salt. Because the interesting combinations are vast, judgment must be exerted to decide which will give the most information. A plant physiologist interested, for example, in the effect of Ca^{2+} on NaCl tolerance, might compare a genotype under NaCl stress at different concentrations of Ca^{2+}. Others might test the hypothesis of Wyn Jones and Storey (115) who have suggested that proline accumulation is a major response to transient stress, whereas glycinebetaine accumulation is a response to more gradual long-term stress, by comparing short-term versus long-term changes in genes expressed. Analysis of cDNA clones could reveal whether enzymes responsible for proline synthesis are induced soon after salt stress. Indeed, since new proteins have been observed to be synthesized under salt and osmotic stress (116, 117), the methods described above could lead to isolation of the genes specifying these proteins.

Gene expression in salt-stressed versus heat-stressed, water-stressed, toxin-stressed, and so on, tissue could also be compared to separate salt-stress specific responses from general stress responses. This is one area where there already is some effort. J. Key's laboratory at U. Georgia has determined that a group of homologous RNAs are induced in soybean following several different types of stresses (Key, personal communication). These RNAs may represent genes mediating general developmental stress responses.

One final example will be given which is relevant to the cell culture studies discussed in this chapter. Suppose that the same tissue can be manipulated *in vitro,* for example, by exogenously supplied hormones, to vary in salt sensitivity at a given salt concentration. Is expression of salt-tolerance genes being turned on and off by the exogenous factor? Recombinant DNA methods have the power to settle this question, which is relevant to our previous consideration of genotypes which are tolerant at the whole plant level but not *in vitro.* Without molecular studies, we cannot choose between these three interpretations: (i) lack of *in vitro* tolerance represents repression of salt tolerance genes; (ii) the plant's salt tolerance mechanism requires an integrated whole plant structure, (iii) an existing fully expressed tolerance mechanism is unable to function because of a physiological deficit (such as insufficient K in the medium, for example). Possibility (ii) can be ruled out if a tissue or organ varies in salt tolerance *in vitro* depending on medium composition, but molecular studies are needed to distinguish (i) and (iii).

4. CONCLUSIONS: DIRECTED VERSUS UNDIRECTED GENERATION OF DIVERSITY

Plant breeding is a powerful technology that permits crop improvement without any basic understanding of the biological mechanisms producing a useful phenotype. Two types of cell culture applications to salt-tolerant crop improvement, namely, regeneration of protoplasts for somaclonal variation and production of somatic hybrids, are essentially plant breeding techniques, with variability obtained via culture methods rather than by conventional sexual crosses. Theoretically, little understanding of the mechanisms of salt tolerance is necessary to apply these techniques for crop improvement.

Other tissue culture technology, as well as recombinant DNA technology, is most effectively applied with some knowledge of mechanisms, which must direct how experiments are performed. *In vitro* selection of salt tolerant types and incorporation of foreign genes cannot be done in a vacuum of knowledge of the physiological and biochemical nature of adaptation to saline environments. Although directed genetic engineering of crop plants via introduced genes is premature, the present is the time for use of recombinant DNA methods and cell culture methods to help reveal the molecular mechanisms of the plant kingdom's solutions to salt stress.

ACKNOWLEDGMENTS

Experiments on organelle DNA described here were supported by the Binational Agriculture Research Development Fund and the National Science Foundation. Tomato tissue culture studies were supported by the USDA Competitive Grants Program, and were performed with the excellent technical assistance of Christine Snyder.

REFERENCES

1. J. Larkin and W. R. Scowcroft. Somaclonal variation—A novel source of variability from cell cultures for plant improvement. *Theor. Appl. Genet.* **60,** 197–214 (1981).

2. E. Thomas, S. W. J. Bright, J. Franklin, V. A. Lancaster, B. J. Miflin, and

R. Gibson. Variation amongst protoplast-derived potato plants (*Solanum tubero-sum* cv. 'Maris Bard') *Theor. Appl. Genet.* **62,** 65–68 (1982).

3. G. A. Secor and J. F. Shepard. Variability of protoplast-derived potato clones. *Crop Sci.* **21,** 102–105 (1981).

4. T. J. Orton. Chromosomal variability in tissue cultures and regenerated plants of *Hordeum. Theor. Appl. Genet.* **56,** 101–112 (1980).

5. T. J. Orton. Haploid barley regenerated from callus cultures of *Hordeum vulgare* $\times$ *H. Jubatum. J. Hered.* **71,** 780–782 (1980).

6. T. J. Orton and R. P. Steidl. Cytogenetic analysis of plants regenerated from colchicine-treated callus cultures of an interspecific *Hordeum* hybrid. *Theor. Appl. Genet.* **57,** 89–95 (1980).

7. J. F. Shepard, D. Bidney, and E. Shahin. Potato protoplasts in crop improvement. *Science* **207,** 17–24 (1980).

8. J. A. Hellebust. Osmoregulation. *Annu. Rev. Plant. Physiol.* **27,** 485–505 (1976).

9. R. L. Jefferies. The role of organic solutes in osmoregulation in halophytic higher plants. In D. W. Rains, R. C. Valentine, and A. Hollander, eds., *Genetic Engineering of Osmoregulation: Impact on Plant Productivity for Food, Chemicals and Energy.* Plenum, New York 1980, pp. 135–159.

10. P. H. Yancey, M. E. Clark, S. C. Hand, R. D. Bowlus, and G. N. Somero. Living with water stress: evolution of osmolyte systems. *Science* **217,** 1214–1222 (1982).

11. D. W. Rains. Salt transport by plants in relation to salinity. *Annu. Rev. Plant. Physiol.* **23,** 367–388 (1972).

12. A. Lauchli. "Genotypic variation in transport," in A. Pirson and M. H. Zimmerman, *Encyclopedia of Plant Physiology,* Vol. 2, Part B, 1976, pp. 372–393.

13. R. G. Wyn Jones. An assessment of quartenary ammonium and related com-pounds as osmotic effecters in crop plants. In D. W. Rains, R. C. Valentine, and A. Hollander, eds., *Genetic Engineering of Osmoregulation: Impact on Plant Productivity for Food, Chemicals and Energy.* Plenum, New York, 1980, pp. 155–170.

14. D. Aspinall and L. G. Paleg. Proline accumulation: physiological aspects. In L. G. Paleg and D. Aspinall, eds., *The Physiology and Biochemistry of Drought Resistance in Plants.* Academic Press, Sydney, 1981, 205–241.

15. T. J. Flowers, P. F. Troke, and A. R. Yeo. The mechanisms of salt tolerance in halophytes. *Annu. Rev. Plant. Physiol.* **28,** 89–121 (1977).

16. J. Levitt. *Responses of Plants to Environmental Stresses,* Academic Press, New York, 1972, pp. 489–530.

17. H. Greenway and R. Munns. Mechanisms of salt tolerance in nonhalophytes. *Annu. Rev. Plant. Physiol.* **31,** 149–170 (1980).

18. C. M. Rick. Potential genetic resources in tomato species: clues from observations in native habitats. In A. M. Srb, ed., *Genes, Enzymes, and Populations.* Plenum, New York, 1972.

19. K. Dehan and M. Tal. Salt tolerance in the wild relatives of the cultivated tomato: Responses of *Solanum pennellii* to high salinity. *Irrig. Sci.* **1,** 71–76 (1978).

20. A. Rosen and M. Tal. Salt tolerance in the wild relatives of the cultivated tomato: Responses of naked protoplasts isolated from leaves of *Lycopersicon esculentum* and *L. peruvianum* to NaCl and proline. *Z. Pflanzenphysiol. Bd.* **102,** 91–94 (1981)

21. M. Tal, H. Heikin, and K. Dehan. Salt tolerance in the wild relatives of the cultivated tomato: Responses of callus tissue of *Lycopersicon esuculentum, L. Peruvianum* and *Solanum pennellii* to high salinity. *Z. Pflanzenphysiol. Bd.* **86,** 231–240 (1978).

22. H. Nassery and R. L. Jones. Salt-induced pinocytosis in barley and bean. *J. Exp. Bot.* **27,** 358–367 (1976).

23. T. J. Orton. Comparision of salt tolerance between *Hordeum vulgare* and *H. jubatum* in whole plants and callus cultures. *Z. Pflanzenphysiol. Bd.* **98,** 105–118 (1980).

24. M. K. Smith and J. A. McComb. Effect of NaCl on the growth of whole plants and their corresponding callus cultures. *Aust. J. Plant Physiol.* **8,** 267–275 (1981).

25. R. C. Skvirsky, M. R. Hanson, and F. M. Ausubel. A genetic approach for studying plant regeneration: Analysis of cytokinin-controlled shoot morphogenesis from tissue explants of *Petunia hybrida.* In E. D. Earle and Y. Demarly, eds., *Variability in Plants Regenerated from Tissue Culture,* Praeger, New York, 1982, pp. 101–120.

26. M. W. Nabors, S. E. Gibbs, C. S. Bernstein, and M. E. Meis. NaCl-tolerant tobacco plants from cultured cells. *Z. Pflanzenphysiol. Bd.* **97,** 13–17 (1980)

27. D. W. Rains, T. P. Croughan, and S. J. Stavick. Selection of salt-tolerant plants using tissue culture. In D. W. Rains, R. C. Valentine, and A. Hollander, eds., *Genetic Engineering of Osmoregulation: Impact on Plant Productivity for Food, Chemicals and Energy.* Plenum, New York, 1980, pp. 279–292

28. J. M. Stewart. In vitro fertilization and embryo rescue. *Environ. Exp. Bot.* **21,** 301–316 (1981)

29. M. Zenkteler. Intraovarian and *in vitro* pollination. In I. K. Vasil, ed., International Review of Cytology, Suppl. 11B, *Perspectives in Plant Cell and Tissue Culture,* Academic Press, New York, 1981, pp. 137–155.

30. V. Raghavan. "Embryo culture," in *Int. Rev. Cytol.,* Suppl. 11B: *Perspectives in Plant Cell and Tissue Culture* I. K. Vasil. eds., 1981, pp. 209–236

31. T. Eriksson, K. Glimelius, and A. Wallin. Protoplast isolation, cultivation and development. In T. A. Thorpe, ed., *Frontiers of Plant Tissue Culture.* International Association for Plant Tissue Culture, Calgary, Canada, 1978, pp. 131–140.

32. O. Schieder and I. K. Vasil. Protoplast fusion and somatic hybridization. In I. K. Vasil, ed., International Review of Cytology, Suppl. 11B, *Perspectives in Plant Cell and Tissue Culture.* Academic Press, New York, 1980, pp. 21–46.

33. F. Constable. Development of protoplast fusion products, heterokaryocytes, and hybrid cells. In T. A. Thorpe, ed., *Frontiers of Plant Tissue Culture.* International Association for Plant Tissue Culture, Calgary, Canada, 1978, pp. 141–149.

34. E. C. Cocking. Selection and somatic hybridization. In T. A. Thorpe, eds., *Frontiers of Plant Tissue Culture.* International Association for Plant Tissue Culture, Calgary, Canada, 1978, pp. 151–158.

35. P. Carlson, O. Gamborg, and T. Murashige. Crop improvement through techniques of plant cell and tissue culture. Working paper, Rockefeller Foundation, 1975.

36. P. J. Mudie. The potential economic uses of halophytes. In R. J. Reinardt and W. H. Queen, *Ecology of Halophytes.* Academic Press, New York, 1974, pp. 565–598.

37. H. von Hedenstrom and S-W Breckle. Obligate halophytes? A test with tissue culture methods. *Z. Pflanzenphysiol. Bd.* **74,** 183–185 (1974).

38. C. Bergounioux-Bunisset and C. Perennes. Transfert de facteurs cytoplasmiques de la fertilite male entre 2 lignees de *Petunia hybrida* par fusion de protoplastes. *Plants Sci. Lett.* **19,** 143–149 (1980).

39. E. C. Cocking, D. George, M. J. Price-Jones, and J. B. Power. Selection procedures for the production of interspecies somatic hybrids of *Petunia hybrida* and *Petunia parodii.* II. Albino complementation selection. *Plant. Sci. Lett.* **10,** 7–12 (1977).

40. S. Izhar and J. B. Power. Somatic hybridization in *Petunia:* A male sterile cytoplasmic hybrid. *Plant Sci. Lett.* **14,** 49–55 (1979).

41. S. Izhar and Y. Tabib. Somatic hybridization in *Petunia* Part 2: Heteroplasmic state in somatic hybrids followed by cytoplasmic segregation into male sterile and male fertile lines. *Theor. Appl. Genet.* **57,** 241–245 (1980).

42. A. Kumar, D. Wilson, and E. C. Cocking. Polypeptide composition of Fraction 1 protein of the somatic hybrid between *Petunia parodii* and *P. parviflora. Biochem. Genet.* **19,** 255–261 (1981).

43. A. Kumar, E. C. Cocking, W. A. Bovenberg, and A. J. Kool. Restriction endonuclease analysis of chloroplast DNA in interspecies somatic hybrids of *Petunia. Theor. Appl. Genet.* **62,** 377–383 (1982).

44. J. B. Power, K. C. Sink, S. F. Berry, S. F. Burns, and E. C. Cocking. Somatic and sexual hybrids of *Petunia hybrida* and *petunia parodii.* A comparison of flower color segregation. *J. Hered.* **69,** 373–376 (1978).

45. J. B. Power, S. F. Berry, J. V. Chapman, and E. C. Cocking. Somatic hybridization of sexually incompatible *Petunias: Petunia parodii, Petunia parviflora. Theort. Appl. Genet.* **57,** 1–4 (1980).

46. J. B. Power, S. F. Berry, J. V. Chapman, E. C. Cocking, and K. C. Sink. Somatic hybrids between unilateral cross-incompatible *Petunia* species. *Theor. Appl. Genet.* **55,** 97–99 (1979).

47. J. B. Power et al. Somatic hybridization of *Petunia hybrida* and *P. Parodii. Nature* **263,** 500–502 (1976).

48. J. B. Power et al. Selection procedures for the production of interspecies somatic hybrids of *Petunia hybrida* and *Petunia parodii.* I. Nutrient media and drug sensitivity complementation selection. *Plant. Sci. Lett.* **10,** 1–6 (1977).

49. D. Aviv and E. Galun. Restoration of fertility in cytoplasmic male sterile (CMS) *Nicotiana sylvestris* by fusion with x-irradiated *N. tabacum* protoplasts. *Theor. Appl. Genet.* **58,** 121–127 (1980).

50. D. Aviv, R. Fluhr, M. Edelman, and E. Galun, Progeny analysis of the interspecific somatic hybrids *Nicotiana tabacum* (CMS) + *Nicotiana sylvestris* with respect to nuclear and chloroplast markers. *Theor. Appl. Genet.* **56,** 145–150 (1980).

51. P. S. Carlson, H. H. Smith, and R. D. Dearing. Parasexual interspecific plant hybridization. *Proc. Natl. Acad. Sci. USA* **69,** 2292–2294 (1972).

52. K. Chen, S. G. Wildman, and H. H. Smith. Chloroplast DNA distribution in parasexual hybrids as shown by polypeptide composition of fraction 1 protein. *Proc. Nat. Acad. Sci. USA* **74,** 5109–5112 (1977).

53. G. C. Douglas, L. R. Wetter, C. Nakamura, W. A. Keller, and G. Setterfield. Somatic hybridization between *Nicotiana rustica* and *tabacum* III. Biochemical, morphological, and cytological analysis of somatic hybrids. *Can. J. Bot.* **59**, 228–237 (1981).

54. D. A. Evans, C. E. Flick, and R. A. Jensen. Disease resistance: Incorporation into sexually incompatible somatic hybrids of the genus *Nicotiana*. *Science* **213**, 907–909 (1981).

55. D. A. Evans, C. E. Flick, and S. A. Kut. Comparison of *Nicotiana tabacum* and *Nicotiana nesophila* hybrids produced by ovule culture and protoplast fusion. *Theor. Appl. Genet.* **62**, 193–198 (1982).

56. D. A. Evans, L. R. Wetter, and O. L. Gamborg. Somatic hybrid plants of *Nicotiana glauca* and *Nicotiana tabacum* obtained by protoplast fusion. *Physiol. Plant.* **48**, 225–230 (1980).

57. Y. Y. Gleba, R. G. Butenko, and K. M. Sytnik. Fusion of protoplasts and parasexual hybridization in *Nicotiana tabacum*. *Dokl. Akad, Nauk, SSSR. Sen. Biol.* **221**, 1196–1198 (1975).

58. K. Glimelius and H. T. Bonnett. Somatic hybridization in Nicotiana: Restoration of photoautotrophy to an albino mutant with defective plastids. *Planta* **153**, 497–503 (1981).

59. K. Glimelius, T. Eriksson, R. Grafe, and A. J. Müller. Somatic hybridization of nitrate-deficient mutants of *Nicotiana tabacum* by protoplast fusion. *Physiol. Plant.* **44**, 273–277 (1978).

60. S. Iwai, T. Nagao, K. Nakata, N. Kawashima, and S. Matsuyama. Expression of nuclear and chloroplastic genes coding for fraction I protein in somatic hybrids of *Nicotiana tabacum* + *rustica*. *Planta* **147**, 414–417 (1980).

61. G. Melchers and G. Labib. Somatic hybridization of plants by fusion of protoplasts. I. Selection of light resistant hybrids of "haploid" light sensitive varieties of tobacco. *Mol. Gen. Genet.* **135**, (1974).

62. L. Menczel, G. Galiba, F. Nagy, and P. Maliga. Dose used to irradiate protoplasts influences the efficiency of chloroplast transfer by protoplast fusion in *Nicotiana*. *Genetics* **100**, 487–495 (1982).

63. T. Nagao. Somatic hybridization by fusion of protoplasts. I. The combinations of *Nicotiana tabacum* and *Nicotiana rustica*. *Jpn. J. Crop Sci.* **57**, 491–498 (1978).

64. T. Nagao. Somatic hybridization by fusion of protoplasts. II. The combinations of *Nicotiana tabacum* and *N. Glutinosa* and of *N. tabacum* and *N. alata*. *Jpn. J. Crop Sci.* **48**, 385–392 (1979).

65. W. R. Scowcroft and P. J. Larkin. Chloroplast DNA assorts randomly in intraspecific somatic hybrids of *Nicotiana debneyi*. *Theor. Appl. Genet.* **60**, 179–184 (1981).

66. H. H. Smith and I. A. Mastrangelo-Hough. Genetic variability available through cell fusion. In W. R. Sharp, P. O. Larsen, E. F. Paddock, and V. Raghavan, eds., *Plant Cell and Tissue Culture Principles and Applications*. Ohio State University Press, Columbus, 1979, pp. 266–285.

67. H. H. Smith, K. M. Kao, and N. C. Combatti. Interspecific hybridization by protoplast fusion in *Nicotiana*. *J. Hered.* **67**, 123–128 (1976).

68. D. W. Rush and E. Epstein. Breeding and selection for salt tolerance by the

incorporation of wild germplasm into a domestic tomato. *J. Am. Soc. Hort. Sci.* **106**(6), 699–704 (1981).

69. O. Schieder. Hybridization experiments with protoplasts from chlorophyll deficient mutants of some Solanaceous species. *Planta* **137**, 253–258 (1977).

70. O. Schieder. Genetic evidence for the hybrid nature of somatic hybrids from *Datura innoxia* Mill. *Planta* **141**, 333–334 (1978).

71. O. Schieder. Somatic hybrids of *Datura innoxia* Mill + *Datura discolor* Bernh. and of *Datura innoxia* Mill. + *Datura stramonium* L. var. tatula. I. Selection and characterization. *Mol. Gen. Genet.* **162**, 113–120 (1978).

72. O. Schieder. Somatic hybrids between a herbaceous and two tree *Datura* species. *Z. Pflazenphysiol.* **98**: 119–127 (1980).

73. O. Schieder. Somatic hybrids of *Datura innoxia* Mill. + *Datura discolor* Bernh. and of *Datura innoxia* Mill. + *Datura stramonium* L. var. tatula L. II. Analysis of progenies of three sexual generations. *Mol. Gen. Genet.* **179**, 387–390 (1980).

74. D. Dudits, G. Hadlaczky, E. Levi, O. Feger, Z. Haydu, and G. Lazar. Somatic hybridization of *Daucus carota* and *D. capillifolius* by protoplast fusion. *Theor. Appl. Genet.* **51**, 127–132 (1977).

75. D. Dudits, O. Fejer, G. Hadlaczky, C. Koncz, G. B. Lazar, and G. Horvath. Intergeneric gene transfer mediated by plant protoplast fusion. *Molec. Gen. Genet.* **179**, 283–288 (1980).

76. G. Krumbiegel and O. Schieder. Selection of somatic hybrids after fusion of protoplasts from *Datura innoxia* Mill and *Atropa belladonna* L. *Planta* **145**, 371–375 (1979).

77. G. Melchers and M. D. Sacristan. Somatic hybridization of plants by fusion of protoplasts. II. The chromosome numbers of somatic hybrid plants of four different fusion experiments. In R. J. Gautheret, ed., *La Culture Des Tissus et Des Cellules Des Vegetaux.* Masson, Paris, 1977, pp. 169–177.

78. G. Melchers. The somatic hybrids between tomatoes and potatoes (topatoes and pomatoes). In P. Sala et al., eds., *Plant Cell Cultures: Results and Perspectives.* Elsevier/North-Holland, Amsterdam, 1980, pp. 57–58.

79. G. Melchers, M. D. Sacristan, and A. A. Holder. Somatic hybrid plants of potato and tomato regenerated from fused protoplasts. *Carlsburg Res. Commun.* **43**, 203–218 (1978).

80. C. Poulsen, D. Porath, M. D. Sacristan, and G. Melchers. Peptide mapping of the ribulose bisphosphate carboxylase small subunit from the somatic hybrid of tomato and potato. *Carlsberg Res. Commun.* **45**: 259–267 (1980).

81. R. Nehls. The use of metabolic inhibitors for the selection of fusion products of higher plant protoplasts. *Mol. Gen. Genet.* **166**, 117–118 (1978).

82. P. Medgyesy, L. Menczel, and P. Maliga. The use of cytoplasmic streptomycin resistance: Chloroplast transfer from *Nicotiana tabacum* into *Nicotiana sylvestris*, and isolation of their somatic hybrids. *Mol. Gen. Genet.* **179**, 693–698 (1980).

83. D. W. Galbraith and J. E. C. Galbraith. A method for identification of fusion of plant protoplasts derived from tissue cultures. *Z. Pflanzenphys.* **93**, 149–1581 (1979).

84. G. Patnaik, E. C. Cocking, J. Hamill, and D. Pental. A simple procedure for the

manual isolation and identification of plant heterokaryons. *Plant. Sci. Lett.* **24,** 105–110 (1982).

85. D. Galbraith. Fluorescence activated cell sorting. Proceedings Fifth International Congress, International Association for Plant Tissue Culture, Japan, July 1982.

86. S. Izhar. Male sterility in petunia. In K. C. Sink and J. B. Power, eds., *Monograph: The Genus Petunia.* Springer-Verlag, Berlin, 1983.

87. Y. Chupeau, C. Missoner, M-C. Hommel, and J. Goujaud. Somatic hybrids of plants by fusion of protoplasts. Observations on the model system *Nicotiana glauca-Nicotiana langsdorffii. Molec. Gen. Genet.* **165,** 239–245 (1978).

88. Y. Y. Gleba and F. Hoffmann. "Arabidobrassica": A novel plant obtained by protoplast fusion. *Planta* **149,** 112–117 (1980).

89. L. Menczel. G. Lazar, and P. Maliga. Isolation of somatic hybrids by cloning *Nicotiana* heterokaryons in nurse cultures. *Planta* **143,** 29–32 (1978).

90. A. Zelcer, D. Aviv, and E. Galun. Interspecific transfer of cytoplasmic male sterility by fusion between protoplasts of normal *Nicotiana sylvestris* and X-ray irradiated protoplasts of male-sterile *N. tabacum. Z. Pflanzenphysiol.* **90,** 397–407 (1978).

91. P. Maliga, A. R. Kiss, A. H. Nagy, and G. Lazer. Genetic instability in somatic hybrids of *Nicotiana tabacum* and *Nicotiana Knightiana. Mol. Gen. Genet.* **163,** 145–151 (1978).

92. G. Belliard, G. Pelletier, F. Vedel, and F. Quetier. Morphological characteristics and chloroplast DNA distribution in different cytoplasmic para-sexual hybrids of *Nicotiana tabacum. Mol. Gen. Genet.* **165,** 231–237 (1978).

93. M. L. Boeshore, I. Lifshitz, M. R. Hanson, and S. Izhar. Novel composition of mitochondrial genomes in *Petunia* somatic hybrids derived from cytoplasmic male sterile and fertile plants. *Mol. Gen. Genet. 190:*459–467.

94. E. Galun, P. Arzee-Gonen, R. Fluhr, M. Edelman, and D. Aviv. Cytoplasmic hybridization in *Nicotiana:* Mitochondrial DNA analysis in progenies resulting from fusion between protoplasts having different organelle constitutions. *Mol. Gen. Genet.* **186,** 50–56 (1982).

95. F. Nagy, I. Török, and P. Maliga. Extensive rearrangements in the mitochondrial DNA in somatic hybrids of *Nicotiana tabacum* and *Nicotiana knightiana. Mol. Gen. Genet.* **183,** 437–439 (1981).

96. H. Binding and R. Nehls. Somatic cell hybridizations of *Vicia faba* and *Petunia hybrida. Molec. Gen. Genet.* **164,** 137–143 (1978).

97. K. N. Kao. Chromosomal behavior in somatic hybrids of soybean—*Nicotiana glauca. Molec. Gen. Genet.* **150,** 225–230 (1977).

98. P. Maliga, G. Lazar, F. Joo, A. H. Nagy, and L. Menczel. Restoration of morphogenetic potential in *Nicotiana* by somatic hybridization. *Mol. Gen. Genet.* **157,** 291–296 (1977).

99. S. Kit, W-C. Leung, G. Torgensen, D. Trkula, and D. R. Dubbs. Acquisition of chick cytosol thymidine kinase activity by thymidine kinase-deficient mouse fibroblast cells after fusion with chick erythrocytes. *J. Cell Biol.* **63,** 505–514 (1974).

100. C. M. Corsarao and M. L. Pearson. Competence for DNA transfer of ouabain resistance and thymidine kinase: Clonal variation in mouse L-cell recipeints. *Som. Cell Gen.* **7,** 617–630 (1981).

101. D. Dudits, G. Hadlaczky, G. Y. Bajszar, C. S. Koncz, G. Lazar, and G. Horvath. Plant regeneration from intergeneric cell hybrids. *Plant. Sci. Lett.* **15**, 101–112 (1979).

102. M. R. Hanson. Cell and tissue culture of *Lycopersicon.* In Proceedings of 5th Congress of the International Association for Plant Tissue Culture, Japan, July 1982.

103. G. Belliard, G. Vedel, and G. Pelletier. Mitochondrial recombination in cytoplasmic hybrids of *Nicotiana tabacum* by protoplast fusion. *Nature* **281**, 401–403 (1979).

104. C. J. Arntzen, C. L. Ditto, and P. E. Brewer. Chloroplast membrane alterations in triazine-resistant *Amaranthus retroflexus* biotypes. *Proc. Natl. Acad. Sci. USA* **76**, 278–282 (1979).

105. A. A. Fleming. Effect of cytoplasm on Southern leaf blight of corn. *Plant Dis. Reptr.* **55**, 473–475 (1971).

106. N. Gilboa, D. Lapushner, and R. Frankel. *Curr. Genet.* Submitted.

107. D. A. Evans. Genetic variability of somatic hybrid plants. *Newslett. Int. Assoc. Plant Tissue Culture,* No. 33, 6–11 (1981).

108. D. P. Bhatt and G. Fassuliotis. Plant regeneration from mesophyll protoplasts of eggplant. *Z. Pflanzenphysiol.* **104,** 81–89 (1981).

109. J. Norlyn. Breeding salt-tolerant crop plants. In D. W. Rains, R. C. Valentine, and A. Hollander, eds., *Genetic Engineering of Osmoregulation Impact on Plant Productivity for Food, Chemicals and Energy.* Plenum, New York, 1980, pp. 293–310.

110. B. R. Thomas and D. Pratt. Efficient hybridization between *Lycopersicon esavlintum* and *L. perivianum* via embryo callus. *Theor. Appl. Genet.* **9,** 215–219 (1981).

111. E. Epstein, J. D. Norlyn, D. W. Rush, R. W. Kingsbury, D. B. Kelley, G. A. Cunningham, and A. F. Wrona. Saline culture of crops: A genetic approach. Science **210**, 399–404 (1980).

112. E. Epstein. *Mineral Nutrition of Plants: Principles and Perspectives.* Wiley, New York, 1972, pp. 325–370.

113. T. Maniatis, E. F. Fritsch, and J. Sanbrook. *Molecular Cloning: A Laboratory Manual.* Cold Spring Harbor Laboratory, New York, 1982.

114. L. Dure and O. Ciferri. Proceedings, NATO Course on Plant Molecular Biology, Lake Garda, Italy, 1982.

115. R. G. Wyn Jones and R. Storey. Betaines. In *The Physiology and biochemistry of Drought Resistance in Plants,* L. G. Paleg and D. Aspinall, eds., Academic Press, Sydney, 1981, pp. 172–204.

116. A. K. Handa, R. A. Bressan, S. Handa, and P. M. Hasegawa. Tolerance to water and salt stress in cultured cells. Proceedings of the Vth International Congress of Plant Tissue and Cell Culture, IAPTC Tokyo, Japan, July 11–16, 1982.

117. J. Fleck, A. Durr, C. Fritsch, T. Vernet, and L. Hirth. Osmotic-shock "stress proteins" in protoplasts of *Nicotiana sylvestris. Plant. Sci. Lett.* **26**, 159–165 (1982).

PART THREE

CONTROLLED ENVIRONMENTS AND ECONOMIC ANALYSES

20

CULTIVATION OF PLANTS IN BRACKISH WATER IN CONTROLLED ENVIRONMENT AGRICULTURE

J. Gale

Department of Botany
The Hebrew University of Jerusalem
Jerusalem, Israel

M. Zeroni

Blaustein Institute for Desert Research
Ben-Gurion University of the Negev
Sde-Boquer, Israel

Within the general framework of the book, this chapter may be considered as somewhat singular. It will not discuss the "improvement" of plants for salt tolerance but rather an alternative strategy whereby other environmental factors may be altered so as to increase plant tolerance of saline or brackish waters. This strategy may also greatly reduce the quantity of water required per unit family farm, thus facilitating the economical use of desalinated water.

The general principles of controlled environment agriculture (CEA) have been reviewed before (1–3). Here, after a brief summary of the purpose of CEA in arid regions, emphasis is placed on aspects of environmental physiology in relation to salt tolerance under CEA conditions; on the germane possibility of combining CEA and desalination technologies and on economic considerations.

1. PURPOSE OF CEA IN ARID REGIONS

Most regions of the world where salinity is a problem are arid, frequently hot deserts, situated around 30° latitude. These deserts are characterized by low and highly erratic rainfall, high evapotransporation potentials, very high midday solar radiation flux densities, extremes of temperature, poor or nonexisting soil, high wind force often associated with sand and dust abrasion, and a paucity of good quality water for irrigation.

The purpose of CEA in arid regions is to minimize the disadvantages of this environment while making maximal use of the main desert advantages. The latter include the availability of abundant sunshine, especially in winter months. For example, much of the Middle-East enjoys about 8000 MJ m^{-2} yr^{-1} solar radiation versus some 4000 MJ in central western Europe (4). A second factor that may be used to advantage is the frequent availability of brackish water (5). As will be shown below, CEA is perhaps the only form of crop agriculture for which desalinated brackish water could be used economically. This is because the quantities required are small and the value of the crops high.

CEA for arid regions is an offshoot of the greenhouse industry. The particular aim is to devise a greenhouse that can be kept closed for as many hours of the day as possible, minimizing evergy and irrigation water expenditures. There are many possible engineering variants (1, 6). In some such systems the surplus heat energy accumulating during the daylight hours can be removed from the plant area, stored, and sent back or dissipated at night. In this way ventilation costs are reduced during daytime and solar energy, instead of fossil fuel, is used for nighttime heating of high cash out-of-season crops. Keeping the greenhouse as closed as possible (within the limitations of ventilation requirements for either cooling or removal of toxic gases) has a number of advantages in arid regions. For example, in a system in which exhaust ventilation (whether direct or via moist pads) is not used for cooling, carbon dioxide fertilization can be employed. This brings about an increased utilization of the high desert solar radiation in photosynthesis, resulting in higher yields (7–9).

The CEA advantage that constitutes the main argument of this chapter is that water use per farming family is very low, because transpiration is low in the high humidity and moderated temperature and radiation environment of CEA. Furthermore, plants in CEA may tolerate higher levels of salinity than under conditions of open field agriculture. The reason for this may be found in the otherwise less stressful environment. Factors of the CEA environment which favor increased salt tolerance are nutrient solution hydroponics, low evapotranspiration demand, moderated temperatures, reduced intensity of nonphotosynthetic radiation, and high levels of carbon dioxide.

Engineering aspects of CEA have been discussed by De Bivort et al. (1), Jensen (10), and Luft and Froechtenight (6). Descriptions of actual desert CEA systems have been given by Bettaque (11) and Hodges (12). By way of

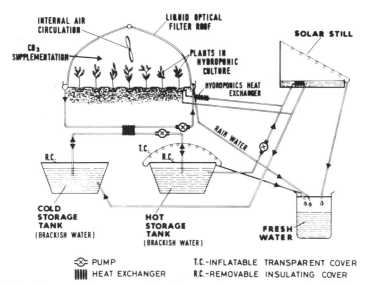

Figure 20.1. Diagrammatic representation of a combined controlled environment agriculture (CEA) and solar still installation for arid regions. From ref. 13. Courtesy of Plenum Press, New York.

illustration a diagrammatic representation of one possible CEA configuration is shown in Figure 20.1 (13).

2. A CEA SYSTEM COMBINING A LIQUID OPTICAL ROOF AND SOLAR STILL

The system shown in Figure 20.1 is based on the liquid optical filter (LOF) concept in combination with a solar still. The LOF design is based on an early idea of Canham (14) which has been developed by Chiapale et al, (15, 16) and Van Bavel et al. (17). The design shown in Figure 2 is a further advance incorporating an improved filter, a heat exchanger between the filter and the heat storage tanks, and solar desalination. A CEA greenhouse of this configuration is presently under construction at the Blaustein Desert Research Institute.

In the above system the greenhouse is covered by a hollow (6 mm thick) polycarbonate roof. Through the roof circulates a pigmented solution designed to absorb the near infrared and UV solar radiation bands, while transmitting nearly all of the 400- to 700-nm, photosynthetically active, waveband.

As the liquid filter circulates through the roof during the daylight hours, it heats up as it absorbs the 52% of the incident solar radiation energy within the nonphotosynthetic wavebands. It is also warmed by heat transferred to it from within the greenhouse by convection, conduction, and condensation. This heat is transferred via a heat exchanger to a brackish water heat storage tank. At night the process is reversed, and the stored heat transferred to the roof in the hot liquid filter then serves to prevent excessive cooling of the greenhouse. At the same time the liquid filter is cooled by radiation and sensible heat exchange and subsequently cools the water in the heat storage tank, via the heat exchanger.

An exact thermal balance during any 24-hr period of heat input and loss would be fortuitous. There may be either a surplus or a shortage of energy. Consequently various alternative dispositions of the available energy are planned: transfer of heat into the soil at root level, into a heat exchanger within the greenhouse, into the entire roof, or into the adjacent solar still. The operation control of the heat system and of other CEA parameters such as carbon dioxide level (8) and hydroponic system set points (18) will be governed by a microcomputer, preprogrammed with a dynamic simulation model of the system and inputted with the farmer's directives. The microcomputer will also receive information from the greenhouse itself (flow rates, temperatures, CO_2 levels, etc.) and from weather sensors (radiometers, anemometers, air temperature and humidity gauges, etc.) (17, 19). We believe that one such microcomputer and sensor installation could control some 10 CEA greenhouses each of about 0.2 hectares.

3. THE CEA CLIMATE

The CEA climate cannot be entirely or exactly controlled by the farmer. This is not because of any lack of engineering possibilities but because of the necessity for establishing a compromise between optimal environmental specifications and the cost of the system which can be borne by the final marketable product. Thus, somewhat paradoxically, presently conceived arid region CEA systems will produce tropical conditions. These will be characterized by yearlong high solar radiation, high but not extreme daytime temperatures, warm nighttime temperatures, and very high air humidities. Add to this a high CO_2 atmosphere and there exists a combination of environmental conditions which, as discussed below, has a marked influence on plant tolerance of salinity.

4. THE INTERACTION OF SALT WITH OTHER ENVIRONMENTAL
FACTORS AFFECTING PLANT GROWTH

Salinity is often differentiated into high levels of salt which result in chlorosis, necrosis, and other obvious signs of toxicity and low level salt, which brings about reduced growth but with fewer or no visible symptoms of toxicity. The present discussion relates mainly to the latter, low level salinity. The reason is practical. In an expensive CEA system it would never be worthwhile to use water of such low quality as would cause severe plant damage. We plan to use brackish water of as high a concentration as possible but which under the special CEA conditions will not adversely affect crop yield or quality. However, even under CEA conditions that alleviate salinity stress of many crop species, it would not be wise to cultivate those that are particularly salt sensitive.

Plants exposed to salt in the root environment tend to adjust their internal osmotic concentrations so as to maintain a constant water potential gradient from the rhizosphere to the roots and the plant shoots (20). The extent to which they are successful in achieving a complete osmotic adjustment, the level of salt to which they can adjust, and the nature of the internal osmoticum are extremely variable among plant species. A plant that is unable to achieve a complete water potential gradient adjustment between itself and the root medium becomes very sensitive to the level of evapotranspiration potential (ETP). Even moderate levels of ETP will tend to cause a water imbalance, resulting in loss of turgor and stomatal closure.

The hydraulic conductivity of the root system tends to decrease under saline conditions (21–24). This causes an even greater sensitivity to ETP. Furthermore, even with full osmotic adjustment gas diffusion resistance of the leaves to water vapor loss increases under conditions of salt (25–26). This affects water balance in the opposite direction to the decreased root permeability. Transpiration is reduced and this tends to maintain plant turgor. However, stomatal closure increases the resistance to the diffusion of CO_2 into the leaf, thus reducing photosynthesis and growth (20). It follows that there are many environmental factors that directly or indirectly modify the transpiration and growth responses to salinity.

4.1. The Root Environment under CEA Conditions in Relation to Salinity

Under field conditions, even the most carefully irrigated crop is exposed to temporal and spatial fluctuations in soil water content and salinity (27). Consequently, in order to survive plants must tolerate not the average but the most extreme salt concentrations encountered.

Lack of soil aeration is known to be one of the major indirect factors of salt damage to plants under field conditions. This is particularly acute where exchangeable sodium is present in alkali soils and is causing deflocculation (28). However, the additive deleterious effects of lack of aeration and salinity can be demonstrated even in solution culture (29).

The intensive nature of greenhouse and CEA agriculture makes it worthwhile to invest in one or other forms of hydroponics. In such hydroponic systems salt concentration can be maintained at a low and uniform level. Furthermore, other potential root environment stress factors, such as low oxygen (30) or low temperatures (31, 32), can be controlled.

4.2. Transpiration and Temperature under CEA Conditions in Relation to Salinity

At the single leaf level, the potential rate of transpiration can best be derived from the leaf energy budget equation (33):

$$S_t + L_d = L_e \pm C + LE \pm M \tag{1}$$

where S_t and L_d are the absorbed short (below 700 nm) and long (above 10 μm) wave radiation, respectively. L_e, C, E, and M are dissipation terms. L_e, the long wave radiation of the leaf, is mainly a function of leaf temperature; C is the sensible heat exchange; E is the evaporation (transpiration) from the leaf and L is the latent heat of evaporation; M is the metabolic energy exchange (photosynthesis and respiration) and although it is the "raison d'être" for the leaf, it is quantitatively small enough to be neglected within the energy balance equation.

Inspection of the various terms of the heat balance Equation (1) shows that most are affected by CEA conditions. S_t is usually reduced by about 30% in greenhouses. In a liquid filter greenhouse incoming solar radiation may be attenuated by as much as 65%, because of the approximately 30% reduction of radiation caused by shading from cladding and other structural materials and the almost complete absorption of the near infrared. L_d is a function of the temperature of the CEA roof or liquid filter. During the day it will not usually be so significantly different from the sky temperature. With the liquid optical roof system, it is the high roof temperature at night (versus the low desert night sky temperature) which prevents the cooling of the greenhouse and condensation on the leaves.

As shown in Equation (1), in the open field the C term can be negative. In other words, in an arid situation with dry winds hotter than the transpiring leaf, energy is actually added to the leaf from the air. This leads to higher leaf temperatures and transpiration rates. This does not occur under greenhouse conditions unless hot dry air is brought in from outside. In closed system CEA internal ventilation will tend to reduce leaf temperatures to the average temperature of the air within the greenhouse.

Leaf transpiration can be calculated from

$$L = \frac{e_1 - e_a}{r_a + r_s} \tag{2}$$

where e_1 and e_a are the vapor pressures of the leaf and air and r_a and r_s are the diffusional resistances to water vapor diffusion of the leaf boundary layer and of the stomata, respectively.

Although r_s will tend to be higher in CEA than in field crops due to the high carbon dioxide level and attenuated light (34) in the final analysis it will probably not be very much greater as water stress will be lower due to the good supply of soil (or hydroponic) water and to the high level of e_a. Boundary layer resistances, r_a, in an internally ventilated CEA system are not expected to differ significantly from open field conditions. If the CEA system is successful engineering-wise then air and leaf temperatures in the greenhouse and hence e_l will be as low or lower than in the field during hot summer days. Inspection of Equation (2) shows that this combination of constant or lower e_l, higher e_a, and moderately increased r_s will tend to greatly reduce transpiration, E.

From the above discussion it will be appreciated that water stress is much lower and leaf temperatures less extreme in CEA than under open field conditions.

For many years it has been observed that high temperatures are associated with lower salt tolerance (35). However it is often difficult to differentiate between the direct effects of temperature, per se, and the indirect effects of increased ETP (36). Lunt et al. (37) reported that under conditions of constant leaf to air humidity gradient either heat or salinity reduced growth of chrysanthemum. However, the combined effect of the two factors was no greater than their arithmetic product, suggesting a lack of any special interaction.

High humidity (presumably reducing ETP) with temperature held constant has been shown to alleviate salt stress [beans (38, 39) beets (40), barley and corn (42)]. This effect of high humidity on plants exposed to salinity is not always found, as reported for cotton (41) and wheat (42). Some of their results on the effect of salinity on plant growth at various humidity levels are presented in Table 20.1. Pitman (Chapter 6) has also reported the different effects of variations in the transpiration stream on the uptake and accumulation of minerals in a number of vegetable crops.

As shown in Table 20.1, for some plant species high humidity alone can go a long way to alleviate the salinity effect on growth.

4.3. The Carbon Balance of the Plant, under CEA Conditions, in Relation to Salinity

As previously noted, only low level salinity is discussed here. This is because of practical considerations. Brackish water would only be used for the intensively grown high cash crops of CEA if it can be demonstrated that there will be no adverse effects.

Under conditions of low level salinity many plants have reduced growth rates, but otherwise do not show any visual symptoms of toxicity. For many years it was suggested that this reduced growth could be largely ascribed to the increased use of assimilates for "maintenance respiration" (R_m). This R_m would supply the extra energy required for coping with salt: exclusion, secretion, and the more rapid turnover of damaged enzymes and organelles (43, 44).

Recently we have made some measurements of the rate of change of R_m in response to low levels of salt (45). The result for two plant species, the

Table 20.1.   Growth of Plants under Saline Conditions at Two Air Humidities[a]

Relative Humidity (%)	Root Medium Osmotic Potential (bars)	Dry Weight Yield (g/plant)
	Beet	
45	−0.4	6.9
	−5.0	4.4
	−10.0	2.4
90	−0.4	10.2
	−5.0	7.1
	−10.0	4.3
	Onion	
45	−0.4	47
	−2.0	33
	−4.0	19
90	−0.4	53
	−2.0	46
	−4.0	36
	Radish	
45	−0.4	0.56
	−2.0	0.46
	−3.5	0.22
90	−0.4	0.69
	−2.0	0.65
	−3.5	0.53

[a]Adapted from Hoffman and Rawlins (40).

salt-sensitive bean (*Phaseolus vulgaris*) and the more tolerant cocklebur (*Xanthium strumarium*), exposed to NaCl salinity in nutrient solution are shown in Figure 20.2.

As shown in Figure 20.2, the salt induced R_m increased with each rise in the concentration of NaCl in the external medium. However, above a certain level of salt (different for the two species) R_m declined. Our interpretation of these data is that the increase of R_m with salt concentration is evidence for the plants' ability to cope with this stress factor. However, above a certain salt level toxicity effects begin to dominate and R_m declines. The inflection points of these curves coincide with the level of salt which we have found to cause an increase in mesophyll resistance to carbon dioxide fixation in photosynthesis (r_m) (46). This "r_m" is defined as the reciprocal of the initial slope of the photosynthesis: $[CO_2]_i$ curve, where $[CO_2]_i$ is the carbon dioxide concentration within the leaf. An increase of r_m is an indication of damage to the basic photosynthetic apparatus.

It is interesting to note that Stavarek and Rains (Chapter 18) report that salt-selected lines of alfalfa cells show a much higher rate of respiration than nonselected cells, especially in the presence of salt.

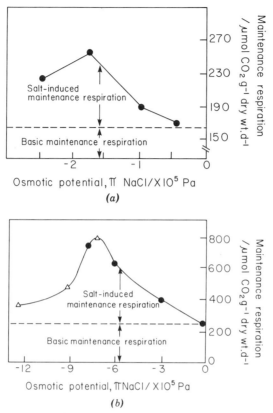

Figure 20.2. Maintenance respiration as affected by salinity. (a) *Phaseolus vulgaris.* Peak value of R_m is significantly different from that of the controls ($P < 0.01$). (b) *Xanthium strumarium.* Δ and $\bullet$ points are from different experiments. -8 and -6×10^5 Pa values of R_m are significantly different from that of control ($P < 0.1$). ref. 45.

Even with complete osmotic adjustment and high levels of turgor, leaf stomata are often partly closed when plants are exposed to salt(26, 27). This reduces photosynthesis by lowering the stomatal diffusion conductance for CO_2 gas.

Schwarz and Gale (45) calculated that when *Xanthium strumarium* plants were exposed to 5×10^5 Pa NaCl in the nutrient solution, approximately 75% of the reduction of growth (as compared to controls growing in normal culture solution) could be ascribed to reduced photosynthesis. The other 25% resulted from the increased expenditure of assimilates for R_m.

The reported values for R_m (45) were calculated from measurements of overall dark respiration as shown by the rate of carbon dioxide efflux. A complicating factor may be the engagement of the alternative, cyanide-resistant respiration pathway. Lambers (47) has proposed that under conditions of surplus carbohydrate, when the plant is unable to utilize available assimilates (e.g., because of salinity damage), this pathway may be operative as an overflow

Table 20.2 The Effect of Carbon Dioxide Supplementation on the Growth of Control and Salinized *Xanthium strumarium* Plants[a]

[CO_2] (μl/liter)	Increase in Dry Weight During 12 Days (mg)		Depression of growth
	Controls	Salinized Plants	
320	515 ± 7	242 ± 12	53%
2500	790 ± 28	573 ± 18	28%
Increase of growth	53%	136%	

[a]Figures for dry weight increase are each the average for six plants ±se. Initial dry weight of parallel plant sample: 558 ± 5 mg. Plants grown in half-strength Hoagland solution; salinized plants with the addition of 130 mM NaCl. Other environmental conditions: 12 hr light at 415 μE m^{-2} s^{-1} (400–700 nm). Air temperatures: 26°C day; 18°C night. Day/ night air humidity: $65 \pm 5\%$ RH. From Schwarz and Gale (in preparation).

that is not producing energy for metabolism. Evidence suggesting that this is not the case here is given below.

As suggested above, R_m is putting an increased burden on the carbon economy of the plant at the very time when photosynthate production is reduced. If the plant has not a surplus but a shortage of assimilate, then any factor that increases photosynthesis such as CO_2 supplementation should also increase plant tolerance of salt. Such indeed has been found in recent experiments of Schwarz and Gale (in preparation) for *Phaseolus, Xanthium*, and *Atriplex halimus* (a halophyte) and for roses by Zeroni and Gale (in preparation). Some representative data are presented in Tables 20.2 and 20.3.

As can be seen from the data of Tables 20.2 and 20.3, CO_2 supplementation

Table 20.3.   The Effect of Carbon Dioxide Supplementation on the Growth of Canes of Control and Salinized Rose Bushes[a]

[CO_2] (μl/liter)	Dry Weight of Canes per Branch at Harvest[b](g)		Percent Difference in Growth
	Controls[c]	Salinized Plants[d]	
320	5.9 ± 0.06	4.1 ± 0.1	−30%
600	7.4 ± 0.1	9.8 ± 0.2	+32%
Increase of growth	25%	140%	

[a]Plants grown in cinder hydroponic culture in growth chambers, with conditions simulating those projected for CEA. From Zeroni and Gale (in preparation).
[b]Representing 45 days growth. Each figure mean + se. of 17–29 branches.
[c]Standard culture solution.
[d]Standard culture solution with the addition of 1718 ppm TSS of a salt mixture similar to that found in Negev, fossil brackish water.

may much decrease or eliminate the effect of salt in reducing plant growth. These results indicate that an inability to utilize assimilate was probably not involved in the reduced growth of the salt-treated plants. Low water potential (48) and high temperature (5) are two other forms of stress which may be alleviated by an increase of carbon dioxide.

Adding together all the effects of CEA environmental conditions, that is, constant salt concentration and high water potential around the plant roots, low evapotranspiration potential, moderated temperatures, and carbon dioxide supplementation, we believe that many crop plants could be grown under CEA conditions at a relatively high level of salinity. By "relatively high" is meant about 1000–2000 ppm total soluble salts (TSS) as compated to about 250 ppm TSS, often considered as the upper limit for conventional open field agriculture. It is possible (but as yet unproved) that many greenhouse plants could be grown at this level of salt ($1-2 \times 10^5$ Pa osmotic potential) in CEA, without any adverse effects.

5.   WATER USE IN A CEA FARM UNIT AS COMPARED TO ONE BASED ON OPEN FIELD IRRIGATED AGRICULTURE

If the purpose of innovative desert agriculture with brackish water is to maximize plant productivity in a country or region, then CEA is obviously inappropriate, because of the high capital cost per unit land area. However, if the intention is to use the small amounts of available brackish water as sparingly as possible, and to minimize water use per unit farm, then CEA shows a two orders of magnitude advantage over conventional, open field, irrigated agriculture. Furthermore, owing to the small quantities of water required, solar desalinization of brackish water (50) may be both feasible and economically worthwhile.

Solar desalination appears to be particularly appropriate for supplying CEA fresh water requirements from a limited source of brackish water in arid regions. This is because of the relatively low cost of land and the intrinsic linkage between fresh water production and plant water requirements, both of which are driven by the same varying energy source, namely solar radiation.

Assumptions used for estimating water requirements for CEA, based on a combined CEA–solar still system, are given in Table 20.4. Estimates of these requirements are given in Table 20.5.

As can be seen from the assumptions and estimates presented in Tables 20.4 and 20.5 a very large difference is projected between the water use of conventional irrigated versus CEA agriculture. Using fresh water (less than 250 ppm TSS) a farm unit would require some 80,000 m^3 yr^{-1}. This figure would be even larger if brackish water were to be used, as yields would be lower and the required farm area proportionally greater. With the CEA and solar still system the only water input would be about 900 m^3 of brackish water per farm.

Table 20.4   Assumptions Used for Estimating Brackish Water and Solar Still Requirements for CEA in Arid Regions, and Water Requirements for Open field Agriculture

Required CEA area, per single family	0.2 ha (Israeli standard for greenhouse farm industry)
Available brackish water	2000–4000 ppm TSS
Solar still production	2.5 liters day^{-1} m^{-2} (yearly average)
Salt level tolerated by plants under CEA conditions	1000–2000 ppm TSS (estimate)
Rainfall	100 mm yr^{-1} (assumed)
Fresh water requirement in CEA	0.45 m^3 m^{-2} yr^{-1} (average for well-closed greenhouse)
per farm unit (0.2 ha)	900 m^3 yr^{-1}
Minimum area for single family unit of conventional open field, irrigated agriculture	4 ha (Israeli standard)
Fresh water requirement per conventional farm	20,000 m^3 ha^{-1} yr^{-1}
farm unit	80,000 m^3 yr^{-1}

6. A LOOK TO THE FUTURE

6.1. Economic Aspects of CEA

It is often assumed that underdeveloped countries should aim toward the local production of those basic commodities that are in shortest supply. These may be high calorie protein foods or clothing. However, a more sophisticated approach is to produce those crops or products that will yield the most profit under local or export market conditions. The income is then used to import local

Table 20.5.   Estimate of Sources of Water and Solar Still Area Required for a Single Family Farm Unit Based on CEA

350 m^2 solar still yield from brackish water	320 m^3 distilled water yr^{-1}
Extra nighttime distillation in solar still as a result of transfer of surplus heat from CEA	60 m^3 distilled water yr^{-1}
Rainfall harvest from 2000 m^2 roof of CEA and 350 m^2 solar still, at 50% efficiency	117 m^3 fresh water yr^{-1}
Total salt-free fresh water obtained	497 m^3 yr^{-1}
Total water yield after 1:1 dilution with 500 m^3 of 2000–4000 ppm TSS brackish water	$\sim$ 1000 m^3 yr^{-1}

requirements. Although making economic sense this approach has a number of drawbacks. In underdeveloped areas of the world the necessary technical skills, capital and market organization are often lacking.

Controlled environment agriculture combined with a solar or other desalinization system is a foremost example of very high agricultural technology. Presently conceived, CEA is basically suited to the production of high cash crops for local or foreign markets that can afford such luxury products. These may be out of season ornamental plants, fresh vegetables, fruits, and so on. It can only be considered where know-how, technical competence, and capital investment are either locally available or can be acquired. At 1982 prices, for example, the investment in a 0.2 h, single family farm in Israel is estimated to be of the order of $100,000. We have calculated that, at today's prices for fuel, this system will be profitable, given yields of roses 20% higher than in conventional greenhouses. This is considered reasonable in view of the response of roses to CO_2 supplementation (e.g., Table 20.3). Furthermore, relative profitability (as compared to conventional, fossil fuel heated greenhouses) should increase with each rise in the price of heating fuel.

In most arid regions of the world CEA cannot be considered as an immediate agriculture option, but rather as something to be introduced at a later stage of development. It does, however, hold out the promise of an agricultural activity capable of providing a satisfying and profitable livelihood for a modern, educated farmer living in an arid desert. This is in contrast to most projections for the use of marginal waters on marginal lands.

At the present time CEA technology is of greatest interest to more advanced countries, having substantial desert regions or to those rapidly advancing countries, blessed with the availability of sufficient capital (6).

6.2. Required CEA Research

Everything said here concerning CEA in arid regions relates not to an established technology, but to a potential agriculture option that could utilize brackish water. There are still many questions to be solved. Some are unique to CEA but most are common to the entire greenhouse industry. It cannot be over emphasized that there are few or no technological or scientific problems of growing plants in CEA, in any climatic regions, which cannot be solved by present-day technology. The problem as stated above, is to design a system that, by maximizing local advantages, will operate at a cost which can be borne by the marketable product. Research in many multidisciplinary and interrelated fields is required before CEA can become profitable. This includes the following:

1. Design of the simplest and most cost-effective engineering system, for cooling the CEA greenhouse during the day and for preventing heat loss from the plants at night.

2. Choice of the most appropriate system for desalination of available brackish or seawater, preferably also exploiting surplus energy from the CEA. This may be solar still, reverse osmosis, or any other desalination technology found to be cost effective for the quantities required and the saline water available.

3. Design of automated control systems for optimization of CEA performance, based on dynamic mathematical models of the system. Apart from real-time operational and meteorological data these models will be inputted by the farmers' program and by the relevant market data.

4. Selection and breeding of plants for maximum profitability taking into consideration the growth conditions that can be produced in CEA at reasonable cost, and the prospective market.

5. Development of response curves to different factors of the environment, for each of the target crop species. This information is essential for the optimization control program. It includes response of plants, during the day and night, to the following factors and to different combinations of these factors at different stages of growth: temperatures (root and shoot), light intensity and spectral quality, nutrient supply (including oxygen to roots), water quality, carbon dioxide concentration, and humidity.

6. Development of a system for the supply of inexpensive, pollutant-free carbon dioxide. Possible sources are from industrial effluents such as are produced by ammonia or cement factories, or from burning organic agricultural wastes.

7. Development of inexpensive, seawater-cooled enclosures, with simple controls, for the intensive production of more basic crops. This would be aimed at impoverished peoples living in coastal deserts (51) for the year-round production of protein crops such as alfalfa (*Medicago*) using brackish or brackish-diluted water. Potential production of protein crops in such a system is very high (52). Minimum cost estimates do not indicate that such a system would be profitable at present world market prices for protein. However with increasing demands from the American vegetable protein market (concomitant with the increasing world population) this may also change.

6.3 The Coming Years*

First experimental CEA systems have already been set up or are under construction in the United States, Australia, Israel, and the Persian Gulf Emirates. CEA systems were also tried, and later abandoned in Chile (51) and Mexico (12). It is to be expected that within the next 10 years there will be further

*"Since the destruction of the temple, prophecy has passed into the domain of fools and the very young." So said the Talmud some 1500 years ago, and yet we try!

developments in arid countries where capital is available, based on research in the areas noted above.

An optimistic view of the next 15–20 years suggests that, in the more backward arid lands, development of mining, industry, tourism, and biosaline technology (53) will so raise their technological standards as to make them ready for CEA. By then CEA will have reached an advanced stage of development. It will move into the arid regions because of their one natural advantage, namely, abundant sunshine for plant growth and for heating (during the winter season). Even should the advent of hydrogen fusion reduce the cost of electrical energy to close to zero, it is doubtful if CEA in more temperate or cold regions could compete with arid zone CEA. This is because of the cost of the lighting, heating, and cooling installations that would be required.

ACKNOWLEDGMENT

This work has been supported by a grant from the BMFT/KFT Germany and the National Council for Research and Development, Israel.

REFERENCES

1. L. H. De Bivort, T. B. Taylor, and M. Fontes. An assessment of controlled environment agriculture technology. National Technical Information Service. U. S. Dept. Commerce. PB-279-211, 462 pp. (1978).

2. J. Gale. High yields and low water requirements in closed system agriculture in arid regions: Potentials and problems. In L. Berkofsky, D. Faiman, and J. Gale, eds., *Settling the Desert.* Gordon and Breach, London, 1981, pp. 81–96.

3. J. Gale. Controlled environment agriculture for hot desert regions. In J. Grace, E. D. Ford, and P. G. Jarvis, eds., *Plants and Their Atmospheric Environment,* Blackwell, Oxford, 1981, pp. 391–402.

4. W. D. Sellers. *Physical Climatology.* University of Chicago Press, Chicago, 1965, p. 272.

5. A. Issar, ed. *Brackish Water as a Factor in Development.* Ben-Gurion University of the Negev, Isreal, 1975, 305 pp.

6. W. Luft and J. Froechtenigt. Solar energy controlled-environment agriculture in the United States and in Saudi Arabia. SERI Publication TP-270-1465. U.S. Dept. Commerce, Springfield, Virginia, 1981, 11 pp.

7. E. Enoch, I. Rylski, and Y. Samish. CO_2 enrichment to cucumber, lettuce and sweet pepper plants, grown in low plastic tunnels in a subtropical climate. *Isr. J. Agric. Res.* **20,** 63–69 (1970).

8. D. Rudd-Jones, A. Calvert, and G. Slack. CO_2 enrichment and light dependent temperature control in glasshouse tomato production. Abstr. Int. Symp. Potential Productivity in Protected Cultivation, Kyoto, Japan, pp. 35–37 (1978).

9. B. A. Kimball and S. T. Mitchell. CO_2 enrichment of tomatoes in unventilated

greenhouses in an arid climate. Abstr. Int. Symp. Potential Productivity in Protected Cultivation, Kyoto, Japan, pp. 30–32 (1978).

10. M. H. Jensen, ed. *Proceedings of the Solar Energy–Fuel and Food Workshop.* University of Arizona, 1976, 262 pp.

11. R. Bettaque. Agua para tierras sedientas. *El Campo* **61**, 630–638 (1977).

12. C. N. Hodges. Desert food factories. *Tech.* Rev., 33–39 (Jan. 1975).

13. J. Gale. Use of brackish and solar desalinated water in closed system agriculture. In A. San Pietro, ed., *Biosaline Research, a look to the future.* Plenum Press, New York, 1982, pp. 315–324.

14. A. E. Canham. Shading glasshouses with liquid films. British Electrical and Allied Industries Research Association Report W/T40 (1962).

15. J. P. Chiapale, J. Damagnez, P. Denis, and P. Jourdan. La serre solaire. 12th *Colloq. Natl. Plastiq. Agric.* 87–90 (1976).

16. J. P. Chiapale. La serre solaire INRA-CEA: Resultats physiques *Acta Horticult.* **115**, 387–393 (1981).

17. C. H. M. van Bavel, E. J. Sadler, and G. C. Heathman. Infra-red filters as greenhouse covers: Preliminary test of a model for evaluating their potential. Proc. ASAE National Energy Symposium. ASAE Publ. 4-81, 552–557 (1981).

18. D. Rudd-Jones and G. W. Winsor. Environmental control in the root zone: Nutrient film culture. Abstr. Int. Symp. Potential Productivity in Protected Cultivation, Kyoto, Japan, pp. 53–54 (1978).

19. J. van Zeeland, C. J. van Asselt, and W. C. Nuijen. A portable low cost micro-computer system for glasshouse climate control. *Acta Hort.* **115**, 347–350 (1981).

20. J. Gale. Water balance and gas exchange of plants under saline conditions. In A. Poljakoff-Mayber and J. Gale, eds., *Plants in Saline Environments.* Springer-Verlag, Heidelberg, 1975, pp. 168–185.

21. A. Kaplan and J. Gale. Effect of sodium chloride salinity on the water balance of *Atriplex halimus. Aust. J. Biol. Sci.* **25**, 895–903 (1972).

22. M. B. Kirkham, W. R. Gardner, and G. C. Gerloff. Leaf water potential of differentially salinised plants. *Plant Physiol.* **44**, 1378–1382 (1969).

23. J. J. Oertli and W. F. Richardson. Effects of external salt concentrations on water relation in plants IV The compensation of osmotic and hydrostatic water potential differences between root xylem and external medium. *Soil Sci.* **105**, 177–183 (1968).

24. J. W. O'Leary. The effect of salinity on permeability of roots to water. *Isr. J. Bot.* **18**, 1–9 (1969).

25. W. J. S. Downton and B. R. Loveys. Abscisic acid content and osmotic relations of salt-stressed grapevine leaves. *Aust. J. Plant Physiol.* **8**, 443–482 (1981).

26. J. Gale, H. C. Kohl, and R. M. Hagan. Changes in the water blance and photosynthesis of onion, bean and cotton plants under saline conditions. *Physiol. Plant.* **20**, 408–420 (1967).

27. A. Meiri and A. Poljakoff-Mayber. Effect of various salinity regimes on growth, leaf expansion and transpiration rate of bean plants. *Soil Sci.* **109**, 26–34 (1970).

28. I. Shainberg. Salinity of soils. Effects of salinity on the physics and chemistry of soils. In A. Poljakoff-Mayber and J. Gale, eds., *Plants in Saline Environments.* Springer-Verlag, Heidelberg, 1975, pp. 39–55.

29. Y. Mizrahi, A. Blumenfeld, and A. E. Richmond. The role of abscissic acid and salination in the adaptive response of plants to reduced aeration. *Plant Cell Physiol.* **13**, 15–21 (1972).

30. M. Zeroni, J. Gale and Y. Ben-Asher. Root aeration in a deep hydroponic system; growth and yield response to tomato. *Sci. Hortic.* **19**, (1983) (in press).

31. A. T. Moustafa and J. V. Morgan. Root zone warming of spray chrysanthemums in hydroponics. *Acta Hortic.* **115**, 217–226 (1981).

32. M. Zeroni and J. Gale. The effect of root temperature on the development, growth and yield of Sonia roses. *Sci. Hortic.* **18**, (1982) (in press).

33. K. Raschke. Uber die physikalischen Beziehungen zwischen Warmeubergangszahl, Strahlung, Austausch, Temperatur und Transpiration eines Blattes. *Planta* **48**, 200–237 (1956).

34. H. Meidner and T. A. Mansfield. *Physiology of Stomata.* McGraw-Hill, New York, 1968, p. 179.

35. S. M. Ahi and W. L. Powers. Salt tolerance of plants at various temperatures. *Plant Physiol.* **12**, 767–789 (1938).

36. J. Gale. The combined effect of environmental factors, and salinity on plant growth. In A. Polgatroff-Mayber and J. Gale, eds., *Plants in Saline Environments.* Springer-Verlag, Heidelberg, 1975, pp. 186–192.

37. O. R. Lunt, J. J. Oertli, and H. C. Kohl. Influence of certain environmental conditions on the salinity tolerance of *Chrysanthemum morifolium. Proc. Am. Soc. Hort. Sci.* **75**, 676–687 (1960).

38. G. J. Hoffman and S. L. Rawlins. Design and performance of sunlit climate chambers. *Trans. Am. Soc. Angric. Eng.* **13**, 656–660 (1970).

39. J. T. Prisco and J. W. O'Leary. Effects of humidity and cytokinin on growth and water relations of salt-stressed bean plants. *Plant Soil* **39**, 263–276 (1973).

40. G. J. Hoffman and S. L. Rawlins. Growth and water potential of root crops as influenced by salinity and relative humidity. *Agric. J.* **63**, 877–880 (1971).

41. G. J. Hoffman, S. L. Rawlins, M. J. Garber, and E. M. Cullen. Water relations and growth of cotton as influenced by salinity and relative humidity. *Agric. J.* **63**, 822–826 (1971).

42. G. J. Hoffman and J. A. Jobes. Growth and water relations of cereal crops as affected by salinity and relative humidity. *Agric. J.* **70**, 765–769 (1978).

43. H. Greenway. Salinity, plant growth and metabolism. *J. Aust. Inst. Agric. Sci.* **1**, 24–34 (1973).

44. A. Poljakoff-Mayber and J. Gale, eds. *Plants in Saline Environments.* Springer-Verlag, Berlin, 1975.

45. M. Schwarz and J. Gale. Maintenance respiration and carbon balance of plants at low levels of sodium chloride salinity. *J. Exp. Bot.* **130**, 933–941 (1981).

46. J. Gale and A. Poljakoff-Mayber. Interrelations between growth and photosynthesis of saltbush (*Atriplex halimus*) grown in saline media. *Aust. J. Biol. Sci.* **23**, 937–945 (1970).

47. H. Lambers. Cyanide-resistant respiration: A non-phosphorylating electron transport pathway acting as an energy overflow. *Physiol. Plant* **55**, 478–485 (1982).

48. N. Sionit, B. R. Strain, H. Hellmers, and P. J. Kramer. Effects of atmospheric

carbon-dioxide concentration and water stress on water relations of wheat. *Bot. Gazette* **142**, 191–196 (1981).

49. H. Z. Enoch and R. G. Hurd. Effect of light intensity, carbon-dioxide concentration and leaf temperature on gas exchange of spray carnation plants. *J. Exp. Bot.* **28**, 84–95 (1977).

50. A. I. Kudish, J. Gale, and Y. Zarmi. A low cost design solar desalination unit. *Energy Conserv. Manag.* **22**, 269–274 (1983).

51. H. J. von Daunicht. Zur Entwicklung von Wüstengewächshausern. *Dropenland-wirte* **40**, 126–155 (1977).

52. A. Bassham. Increasing crop production through more controlled photosynthesis. *Science* **197**, 630–643 (1977).

53. A. San Pietro. *Biosaline Research: A Look to the Future.* Plenum, New York 1982, p. 578.

21

AN ECONOMIC ANALYSIS
OF PLANT IMPROVEMENT
STRATEGIES FOR
SALINE CONDITIONS

Charles V. Moore

National Economics Division
U.S. Department of Agriculture

and

Department of Agricultural Economics
University of California
Davis, California

Like a dark cloud, salinity has cast a shadow over every irrigation development in the arid and semiarid regions of the world since man first started to record his activities. Once man began to understand this odorless and colorless compound he was able to ameliorate its impact on his food supply and survival through management of the concentration and location of these salts within the plant root zone.

1. MAGNITUDE OF THE PROBLEM

What is the magnitude of the "salt" problem? To assess the magnitude it is necessary to make more definite this thing or compound called "salt." For purposes of this chapter, sodium chloride will be the salt of major interest, keeping in mind there can be many other salts present and widely varying in their composition.

A search of the literature plus some personal communications with researchers around the world revealed two points: first, we do not know very much about the problem, and second, what we do know or think we know is quite out of date. No detailed data of a global nature could be located which provided any indication of the *degree* of salinization. In every source located, the simple dichotomy of affected or unaffected was used. No references were found which provided detailed data on salinized soils distinguishing between irrigated and nonirrigated lands. An aggregate estimate by Epstein (1) places the proportion of all arid and semiarid lands affected by salts at about 15% or 950 × 10^6 ha. Epstein cites other estimates by Wittwer and Eckhold placing the proportion of irrigated land affected by salts in the world at about one-third or 76.7 × 10^6 ha.

The area of salt-affected lands currently under irrigation is useful as a starting point but fails to account for the additional lands that could be brought under irrigation if more salt-tolerant species or varieties were available to the cultivator. One can only speculate at the potentially irrigable lands if soil salinity or water quality were no longer a limitation to production. This would include much of the areas along the seashores in temperate parts of the world and ignoring the cost of transporting seawater, the millions of hectares of desert lands of Africa, the Middle East, Asia, and North America.

A second area where our knowledge is far from complete deals with the supply and quality of irrigation water. A search of the literature revealed virtually no data on the quantity and quality of waters available for irrigation, although Rhoades (2) provides data on water quality for a few locations in the western United States. No data were found on supplies of saline groundwater, irrigation return flows, or brackish waters from estuaries. Of course, there was no information on the cost of capturing and transporting these water supplies to potentially irrigable lands. Transportation costs loom importantly because the probability that these saline water supplies are in close geographic proximity to arable lands is quite low.

2. ROLE OF THE ECONOMIST

Economists spend their time dealing with scarce resources. Economists understand very well that when a resource is limited in supply, this resource becomes more valuable. An economist earns his salt or pay when a decision must be made to allocate a very scarce resource because, by implication, the cost of an incorrect decision can be very high and very painful.

If reservations are held by plant scientists, gathered at this conference, over the fact that an economist was given time on the program, it is understandable. A Nobel Laureate in economics, Theodore W. Schultz (3), stated the case most succinctly when he said,

> The point that is dear to economists is that the creation, maintenance and distribution of knowledge requires scarce resources. Moreover, these activities have become large and expensive. But scholars and scientists are, in general, uneasy about this point. They prefer not to have the economic calculus applied to

their activities because they fear that it will inevitably debase the true value of their contributions.

The cold facts are that this conference brings into focus two very scarce and valuable resources. First, but not necessarily the most important, is research funding. Not a single person attending this conference would concede that he could not put additional research money to very good use. Research funding, especially for the types of basic research being discussed at this conference, is a limitation to the rate of technological progress in every country in the world.

The second and probably the most scarce resource is the supply of trained researchers or in the words of Professor Schultz, "human capital." Any country or research institution can produce or have on-board only a limited number of trained scientists. Assigning and funding these resources so as to maximize their contribution to the society they serve is a monumental task and is usually done on a more or less ad hoc basis.

Regardless of how distasteful it is to you personally or professionally, someone will have to decide on the relative merits of your current line of endeavor as compared to your productivity, if engaged in a different line of inquiry.

If an economist can make any contribution to the topic of discussion this week, he should at least be able to provide some guidance to the relative benefits and costs of resources invested in alternative lines of research.

3. GOALS AND OBJECTIVES OF RESEARCH

3.1. Prior Work

Until very recently, the salt sensitivity of a given species or variety of plant was taken as given, an exogenous parameter in the planning process. The plant physiologist, agronomist, soils physicist, chemist, hydrologist, and engineer spent their time developing new techniques to manage salts found in both the irrigation water and the soil. Although no literature search was attempted to ascertain the individuals or institutions who historically first attempted to measure in a scientific manner plant response to saline environments, this writer gained a great deal of early understanding from the staff of the U.S. Salinity Laboratory at Riverside. An early discussion with Leon Bernstein, then Director of the Laboratory, was very insightful. In essence he said, "We had two choices, one, we could withdraw into our ivory towers and contemplate the way in which a plant grows or second, observing the major decision being made involving millions of dollars based on very scant data, invest our resources in more accurately quantifying plant response to saline conditions." To the benefit of all, they close the latter alternative.

Given the availability of reasonably reliable salinity response curves, it was possible for economists to develop models which could optimize management techniques and present information on the trade-offs between water quantity

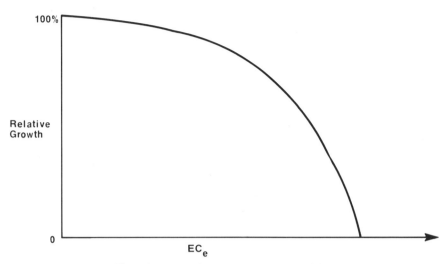

Figure 21.1 Idealized response curve to salinity.

and water quality. For an example of this type of analysis, see Moore et al. (4). Bernstein's earlier estimates of the response function to salinity were curvilinear and took the form as shown in Figure 21.1. Later, in an extensive review article, Mass and Hoffman (5) linearized these functions indicating that up to a point (as measured by EC of soil saturation extract), unique for each variety, there was no yield declination due to soil salinity. Once this critical threshold point was reached, a negative and linear response was observed to salt concentration in the root zone. More recent data (6) reveals the onset of yield declinations at very low salinity levels.

3.2. Osmoregulation Work

For the first time, it may be possible to consider plant response to salinity as an endogenous variable in the planning process, subject to manipulation. Ideally, the goal of the work discussed here is to shift the entire production response curve, as depicted by Mass and Hoffman, to the right (from a to b) as shown in Figure 21.2. If it is not possible to shift all the response function to the right, then the next best objective would be to reduce the slope of the negative portion of the response curve such as depicted by curve segment c in Figure 21.2.

4. STRATEGIES IN OSMOREGULATION RESEARCH

The author is deeply grateful to Professor D. W. Rains of the University of California, Davis, for his advice and materials supplied which were invaluable in the conceptualization of this section.

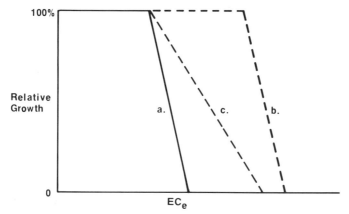

Figure 21.2   Linearized response curve to salinity.

Predicting technological change is an extremely risky business because we are attempting to predict an event that has never before occurred. We do know that the greater the number of resources applied to a problem, the greater the probability of success. Take, for example, the development of the atomic bomb during World War II or the race to place man in orbit around the earth and on the moon during the 1960s.

Before defining the possible strategies that might be analyzed, it is useful to define those elements that could not be included. It was not possible to refine the analysis to the level where the benefits of additional work on the formation of organic molecules versus accumulation and transport of inorganic ions versus the structural properties of cells could be taken into account.

There appears to be three major research thrusts or strategies in developing plants with increased tolerance to saline conditions, although they certainly are not mutually exclusive or exhaustive. The general categories used in this chapter are (i) plant breeding from the existing gene pool; (ii) cell culture with subsequent plant breeding; and (iii) genetic engineering. These three approaches will be described individually.

4.1. Plant Breeding from the Existing Gene Pool

From a layman's point of view, this approach involves utilizing existing genotypes, including wild species and subjecting them to a highly saline environment. Plants that survive and produce economic yields are considered tolerant and are used in further breeding work to develop varieties acceptable for cultivation on a commercial scale. A more detailed description is available in Ramage (7).

The major advantage of following this approach, from an economist's point of view, is that over the past 30 years agricultural colleges throughout the world have produced a relatively large number of plant geneticists and plant breeders.

Thus, with respect to one of the two major limiting resources mentioned earlier, there would appear to be a relatively abundant supply of human capital that could be invested into such a thrust. Second, based on limited work by Epstein (1), results indicate modest gains in salt tolerance could be achieved rather quickly at a fairly low investment in the second most limiting resource, monetary capital or operating funds.

This approach has one major negative factor, however; the potential increase in salt tolerance within a species is limited by the variability of the existing gene pool. Thus, one might expect fairly rapid small gains in tolerance with a rather high probability, but quantum jumps are rather unlikely even over an extended period of time.

4.2. Cell Culture

This approach utilizes cell tissues that can be subjected to mutagenic agents in order to expand the variability of the gene pool beyond that available in the existing gene pool for a species. Advantages and disadvantages are described in Rains (8). Briefly, these include increased control of environmental factors, a greatly expanded number of treatments and replications with reduced manpower, and a vastly increased potential for selection of salt-tolerant variants. Disadvantages include the difficulty of selecting for characteristics that are manifested in subsequent growth stages such as yields. Thus, the cells must still be regenerated and grown out, thereby providing additional breeding material for standard breeding and selection work.

Due to the greater variability of the gene pool, the potential for a larger more significant shift to the right in the response curve is present. The upper limit to this shift is problematical and would be difficult to estimate at this time. Because the use of cell culture adds one additional step to the process of developing plant varieties with increased salt tolerance, the initial manpower and funding requirement will also be increased over and above the standard plant breeding and selection strategy. However, the cost per unit of change in salinity tolerance may be lower. On the other hand, there appears to be some serendipity between the work on cell culture and genetic engineering, the third major thrust or strategy.

4.3. Genetic Engineering

As understood by the layman, genetic engineering involves determining the mechanism or mechanisms within the plant cell which controls the plant's response to saline environments, then locating the genetic code that controls these mechanisms, and finally transferring the DNA that controls this process onto the DNA structure of the subject genes from the cell or plant of interest.

Given there are both simple and complex halophytes available, it is easy to speculate that the potential shift in the response curve to salinity for some economic crops such as wheat or barley is virtually unlimited. The problems of achieving such a goal based on a cursory review of the literature, however, appears formidable. These problems appear as great in attempting to achieve a small increase in salt tolerance as for achieving a major shift in the response curve.

The major problems mentioned above center on the lack of knowledge about exactly how the control mechanisms within the plant cell operate. Although there appears to be some general agreement with respect to very simple plants, gaining understanding of complex plants will require the investment of many scientist years before any concrete results can be expected. Thus, due to the expensive laboratory equipment essential to this type of work and the amount of basic "pick and shovel" work that must be done, this approach will require a significant commitment of both funding and researchers.

Thus, in summary, the probability of success in achieving small increases in salinity tolerance in a major food crop, such as wheat, using standard plant breeding techniques in the next 10 years appears fairly high. The probability for major shifts in the response curve using plant breeding, on the other hand, are rather low. For cell culture techniques the probability of a significant shift in the response curve is fairly good but quantum shifts are not expected. Finally, due to the large amount of "basic" work yet to be done, the probability of success for a significant shift is the same as for even minor gains. The ability to attach even subjective probabilities to the success of different research strategies makes the problem amenable to a tool referred to by statisticians as statistical decision theory.

5. STATISTICAL DECISION THEORY

Statistical decision theory is a formalized method of dealing with problems where outcomes, costs, and/or benefits are not known with certainty.

5.1. Subjective Probabilities

Two types of uncertainty can be defined: first, objective probabilities, that is, the probability of alternative outcomes that can be established based on a large number of empirical observations; and second, subjective probabilities. Savage (9) points out, however, that both subjective and objective information can be combined in Bay's model. Technological breakthroughs in science are not based on historical occurrences and thus the parameters of the probability distribution are unknown. In forecasting technological change, therefore, the analyst must rely on subjective probabilities. According to Savage (9), subjective probability reflects the decision maker's "degree of belief" about a proposition. Compared to objective probabilities, subjective or personal probabilities allow the incorporation of intuitive knowledge and the recognition that the future may not be like the past (10). The major difficulty for most people in the acceptance of subjective probability is the seeming loss of "scientific objectivity" which occurs when two people facing the same problem with the same data make different decisions. Misplaced confidence in "scientific objectivity," however, can lead to what Schlaifer (11) called an error of the third kind: "when a statistician delivers a carefully computed solution to the wrong problem."

Any personal belief or subjective probability estimate for developing a successful plant with increased tolerance to salinity will be framed by two questions: first, how many resources, scientists, and funding will be invested, and second, what is the time span or planning horizon? To compress this three-dimensional problem into two dimensions it was assumed that the existing level of resources would continue over a 10-year planning horizon for all three strategies.

Hypothesized Prior Probability Table

A prior probability table is constructed by first defining the alternative actions or strategies for this problem which have been simplified to (i) plant breeding, (ii) cell culture, and (iii) genetic engineering as shown in Table 21.1. The left-hand side of the table (the stub) shows the possible states the system may achieve; in this problem, the threshold tolerance of the crop in question. States are defined in terms of the salinity tolerance as a fraction of the salinity of seawater. In the example problem, wheat, *Triticum aestivum*, with a threshold tolerance of $EC_e = 6.0$ or about 1/10 the salinity of seawater, is the base line against which any improvements would be measured. Thus, achieving a threshold tolerance of 2/10 would require a doubling of the tolerance.

Probabilities of achievement in Table 21.1 are based on subjective estimates and are subject to revision. The sample of three plant scientists who reviewed

Table 21.1.   Prior Probability of Achieving Specified Threshold Salinity Tolerance for Wheat, 10-Year Planning Horizon

Salinity as Fraction of Seawater	Actions (probability)		
	Plant Breeding	Cell Culture	Genetic Engineering
2/10	0.8	0.8	0.2
3/10	0.6	0.7	0.2
4/10	0.4	0.6	0.2
5/10	0.2	0.4	0.2
6/10	0	0.2	0.2
7/10		0	0.2
8/10			0.2
9/10			0.2
10/10			0.2

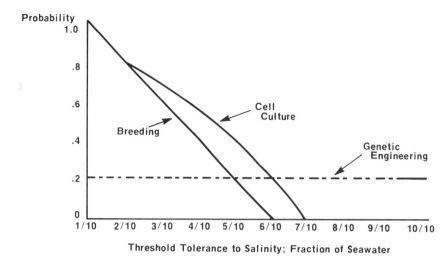

Figure 21.3 Probability of achieving specified salinity tolerance, wheat.

these estimate did not have a high degree of confidence in these or their own personal estimates. Graphically, these probabilities appear as shown in Figure 21.3. Subsequently, therefore, sensitivity analysis will be performed where the probability of success through genetic engineering is reduced to 0.1.

Pay-Off Table

Due to the paucity of data on the degree of salinization of existing irrigated lands, potentially irrigable lands and the salt concentration of potential irrigation water sources, confidence in the estimates of potential benefits if a salt-tolerant crop were to be produced is lower than in the previous probability assessments.

The standard practice of economists in estimating the net benefits of any project is to estimate the changes in yield, market value, and costs due to the project. Increases in annual net revenue are calculated over the expected economic life of the innovation and discounted back to their present value using an appropriate discount rate. Similar analyses are conducted for investment in research, development, and dissemination of new technology. Again these costs are discounted back to their present value using the same discount rate in order to determine a benefit/cost ratio or the excess present value of benefits over costs.

Using the previous estimate (1) of salt-affected irrigated land of 76. 7×10^6 ha, it becomes obvious that development of a highly saline-tolerant wheat variety would create benefits of such a magnitude that any research and development costs would be relatively insignificant. Thus, for this example, net benefits are assumed to equal gross benefits.

Given the total potential area that could be planted is unknown as well as the

Table 21.2.   Hypothesized Pay-Off Table, Present Value of Net Benefits for Wheat[a]

Tolerance to Seawater	Plant Breeding		Cell Culture		Genetic Engineering	
	Index	Probability	Index	Probability	Index	Probability
2/10	20	0.8	20	0.8	20	0.2
3/10	30	0.6	30	0.7	30	0.2
4/10	40	0.4	40	0.6	40	0.2
5/10	50	0.2	50	0.4	50	0.2
6/10			60	0.2	60	0.2
7/10					70	0.2
8/10					80	0.2
9/10					90	0.2
10/10					100	0.2
Expected value	60	—	93	—	108	—

[a]Pay-offs are calculated only for states where the probability of achievement is positive.

level of benefits per hectare, it was decided to construct an index of net benefits where the index of the present value of net benefits with a threshold tolerance of seawater equals 100.* A table based on these assumptions is presented in Table 21.2.

The bottom row of Table 21.2 provides the expected benefits if each of the three possible strategies were to be followed. An expected benefits index for the genetic engineering alternative of 108 is largest, followed by cell culture with 93 and finally plant breeding with 60. If research administrators were very confident of both the probability assessments and the pay-off table values, the greatest total benefits would be achieved by investing all research and development resources into the genetic engineering strategy. However, as discussed earlier, the level of confidence in the subjective probabilities is low and the pay-offs for each state are subject to wide variations depending on the assumptions used to estimate them. A valid question at this point is, how stable are these results? Analysis of this question will be discussed in the next section.

Sensitivity Analysis

In developing pay-off Table 21.2 a question arose concerning the shape of the benefit curve as the threshold tolerance shifted to the right. Two variables working in opposite directions come into play. First, as the threshold tolerance shifts, yields on existing salt-affected lands would be expected to increase and total production due to new lands coming under irrigation would also increase.

*The assumption of linearity in this index is somewhat simplistic; however, since the distribution of saline soils and irrigation waters is unknown, it is a reasonable assumption.

Table 21.3.   Pay-Off Table with Benefits Increasing at a Decreasing Rate for Wheat

Seawater	Plant Breeding	Cell Culture	Genetic Engineering
2/10	18.8	18.8	18.8
3/10	27.3	27.3	27.3
4/10	35.2	35.2	35.2
5/10	42.5	42.5	42.5
6/10		49.2	49.2
7/10			55.3
8/10			61.8
9/10			65.7
10/10			70.0
Expected value	54.0	82.0	85.2
Expected value if probability of success for genetic engineering = 0.1			42.6

Second, however, as total production in the world increased, world market prices and thus the net benefits to growers would be expected to decline, but benefits to consumers would probably increase.

The pay-off values in Table 21.2 assume benefits increase linearly with the threshold tolerance.* To test this assumption, two additional assumptions were made which will allow each individual to select the set of assumptions with which he feels most comfortable.

Alternative pay-off Table 21.3 is based on the assumption that benefits increase with increasing threshold tolerance but at a *decreasing* rate.† The results of this conservative set of assumptions are shown.

Changing the shape of the benefit curve or function does not change the relative ranking of the research strategies, although the position of plant breeding with respect to genetic engineering is improved. However, if, as some feel, the probability of breakthrough in genetic engineering is only 0.1, the relative position of plant breeding and genetic engineering are reversed.

Table 21.4 takes a more optimistic view of total net benefits assuming that benefits increase at an increasing rate as the threshold tolerance shifts to the right.‡

The results of the more optimistic perception of the net benefit function does not change the relative ranking of the three strategies with genetic engineering having an index of 131.0, followed by cell culture with 104.0, and finally plant

---

*Index of benefits = threshold tolerance or $y = bX$ where y = benefits and X the threshold tolerance in percent.

†Benefits = $bX - 0.003X^2$ where X is threshold tolerance expressed as a percent of seawater.

‡Benefits = $y = bX + 0.003X^2$.

Table 21.4.   Pay-Off Table with Benefits Increasing at an Increasing Rate for Wheat

Seawater	Plant Breeding	Cell Culture	Genetic Engineering
2/10	21.2	21.2	21.2
3/10	32.7	32.7	32.7
4/10	44.8	44.8	44.8
5/10	57.5	57.5	57.5
6/10		70.8	70.8
7/10			84.7
8/10			99.2
9/10			114.3
10/10			130.0
Expected value	66.0	104.0	131.0
Expected value if probability of success for genetic engineering = 0.1			65.1

breeding with an index of 66.0. However, as noted in the last line of Table 21.3, an assumption that the probability of a breakthrough in genetic engineering is really only 0.1 reduces the index of expected net benefits to being essentially the same as that derived from the plant breeding strategy, 65.1.

Data Needs for Improved Estimation

Clearly, the above analysis points out the need for additional, accurate information if these estimates are to be improved. Probabilities of a technological breakthrough will always be very subjective and one educated person's guess is about as good as another. Significant improvements, however, can be made in the estimation of the stream of net benefits. As mentioned earlier, data on the amount of salt-affected land by degree of severity would be critical. In addition, potentially irrigable lands now affected by salt but economically reclaimable would be very helpful. In regard to water supply, an inventory of actual and potentially usable irrigation water sources and volumes, including return flows and brackish sources grouped by electroconductivity, would be vital. The last step would be attach monetary costs for bringing additional lands into production, including drainage and the costs of capturing and transporting currently unused water supplies to the irrigable lands.

6.   NECESSARY AND SUFFICIENT CONDITIONS FOR A LONG-TERM IRRIGATED AGRICULTURE*

If irrigated agriculture does not learn from its history, then history is doomed to repeat itself. Societies based on an irrigated agriculture have appeared and

*This section draws heavily on a paper by Moore (12).

disappeared through the several millennia of man's history. Analogous counterparts of the process continue even today.

Salinization is not a static affliction like a broken bone or sore toe that remains isolated and constant; rather, it is like a malignancy, constantly growing and spreading, infecting new and enlarged areas of the anatomy. Thus, the larger question facing development planners and indirectly the deliberations of this conference is, are the benefits of increased salt tolerance through genetic manipulation sustainable? Without an investment in drainage and disposal facilities, will the investment in increased salinity tolerance be quickly dissipated?

6.1. Physical Conditions

The deleterious impact of salts on a soil or an irrigation water depends on variables other than salt, per se. Bernstein (13) considered three factors or conditions affecting water quality determinations: drainage, soil permeability, and salt tolerance of crops.

Soil permeability and water infiltration rates are factors in soil salinity; they influence the irrigator's ability to manage the amounts of water needed to meet both the evapotranspirational losses and the water required to carry the excess salts past the root zone. The permeability of some soils may be so low that water stands in the field for several days after irrigation. In this event, damage to the crop will often occur from poor aeration and scalding.

For an irrigation system to be successful, provision must be made for percolation of some water below the root zone to leach away the yield-depressing salts. Adequate drainage must also be available to prevent a rise in the water table, and thus forestall an upward movement of salts by capillary action. If subsurface drainage is not available naturally, it must be supplied by artificial means, such as buried drains.

Using Bernstein's (13) definition of the leaching fraction as:

$$\mathrm{LF} = \frac{D_\mathrm{d}}{D_\mathrm{i}}, \tag{1}$$

or the ratio of the depth of drainage water D_d to the total amount of irrigation water percolated into the soil D_i, then $D_\mathrm{d} = D_\mathrm{i} - D_\mathrm{e}$, where D_e, equals the amount evapotranspired.

If the drainage rate is limiting, then the difference in the quantity between infiltration and evapotranspiration cannot exceed the amount of drainage without causing waterlogging. If D is the average drainage rate per day without a rising water table, then $\mathrm{LF} = D/E + D$. Therefore, there is an upper limit to the leaching fraction which limits the salt content of usuable irrigation water.

Given the physical upper limit of LF, the suitability of an irrigation water source can be determined by the ratio of its conductivity to that of the drainage water:

$$LF = \frac{EC_i}{EC_d} \quad \text{or} \quad EC_i = LF \times EC_d \tag{2}$$

where EC_d in this analysis is the threshold EC from Mass and Hoffman (5). The important point here is that even if geneticists are able to shift the threshold tolerance quite far to the right, thus greatly increasing the permissible conductivity of the drainage water, for all practical purposes, drainage cannot be reduced to zero. Therefore, a necessary condition for a long-term irrigated agriculture is a soil permeability, either natural or artificially enhanced through mechanical means, that is compatible with quality of the irrigation water and the salt tolerance of the crop to be grown. Drainage, natural or artificial, must be sufficient to prevent waterlogging.

6.2. Economic Conditions

The soil root zone can be thought of as a stock resource, providing a flow of services each year. The soil is a reservoir for holding moisture and nutrients and a repository of precipitated salts. If placed under stress by the excess deposition of salts or inadequate drainage in relation to the quantity of water applied, the bundle of resources composing the root zone can pass a critical point or level and become irretrievably salinized.

Given the level of irrigation technology, the optimum rate of consumption of the absorptive capacity of the root zone (salt buildup) is dependent on a long-term interest rate and a positive net income in each planning period. Thus, economic survival implies that the present value of the future stream of agricultural income must always be greater than zero. In other words, if the long-term expectations of income are not suffcient to cover all costs including land reclamation, development of a water supply, production expenses, and, most important of all, a means to remove excess salts through a system of underground and collector drains, an irrigation project is doomed to failure. Thus, although the physical parameters of irrigation and salinity management define the necessary conditions, economic parameters define the sufficient conditions for a long-term irrigated agriculture.

6.3. Genetic Research

Osmoregulation research has an important bearing on the above analysis in three ways. First, by shifting the threshold tolerance level to the right, the maximum permissible EC of drainage water is increased and therefore the maximum permissible EC of irrigation water is also increased. For potentially new irrigated lands, shifting the threshold tolerance to the right will allow resources with no current economic value, such as saline return flows, brackish waters, and heavily salt-affected lands, to be brought into production. On existing irrigated lands in steady-state equilibrium, increased salt tolerance in plants would allow a higher equilibrium level of salts in the root zone without deleterious effects, thus allowing the use of lower quality water sources.

For lands where the salt level in the root zone is currently *not* in equilibrium (progressive salinization), osmoregulation research merely postpones the time when the investment in drainage and disposal must be made. The unknown factor is: How much time can be purchased by investing in this new technology? If adequate time for landowners to accumulate sufficient capital to make the drainage investment is provided by such an innovation, the benefit stream from osmoregulation research would be sustainable. In the short run for the individual grower, income would be higher with the salt-tolerant varieties than without.

The second major impact of osmoregulation research, if directed toward historically more salt-sensitive crops such as fresh vegetables and other speciality crops, would be to increase the profitability of irrigated agriculture if production did not expand so rapidly as to severely depress market prices. Higher threshold tolerances for high-valued but salt-sensitive crops would allow a shift in the crop pattern of a region to crops that could not previously be grown due to saline conditions. Increased profitability could allow sufficient capital to be accumulated for investment in drainage and disposal facilities which could be used to bring root zone salts into a steady-state equilibrium at an acceptable level. One note of caution, however, is that shifting a large area to nonstorable speciality crops could have a significant depressing effect on their market prices.

A third affect may be less obvious. After the "Green Revolution," based on short stature wheat and rice varieties, was well under way, criticisms were raised based on the distributional impacts of the new technology. The primary complaint was that most of the benefits accrued to wealthy landowners in the most productive regions of a country. My conjecture with respect to the distribution of benefits from osmoregulation research would be as follows:

1. Benefits would accrue to existing landowners but with one significant difference. Landowners of the most salt-affected lands would be the major beneficiaries not the owners of the best quality lands.
2. The potential to bring new lands into production, assuming these lands are presently in the public sector, provides a unique opportunity for governments to distribute these previously useless lands to landless people, thereby broadening the distribution of benefits.

7. SUMMARY, CONCLUSIONS, AND RECOMMENDATIONS

An estimated one-third of the world's irrigated lands are affected by salts to some degree. In addition, there is an unestimated area potentially or reclaimable using highly saline water supplies such as irrigation return flows, brackish waters, saline groundwater, and possibly seawater.

The role of the economist is to assist research administrators in allocating a scarce supply of scientists and research funds in order to maximize the benefits to the societies they serve. Statistical decision theory was used to formalize the subjective probabilities of achieving specified threshold salinity tolerances under three alternative research thrusts or strategies: (i) plant breeding, (ii) cell culture, and (iii) genetic engineering. An hypothesized pay-off or benefit table was constructed based on a set of assumed yield and income increases for wheat with varying levels of threshold tolerance to salt.

Under all alternative assumptions of the probability of success and benefits, the genetic engineering strategy always ranked highest when a probability of success of 0.2 was assumed. Plant breeding and genetic engineering generated indexes of total benefits of about the same magnitude when the probability of success for genetic engineering was assumed to be 0.1.

Genetic manipulation to shift the threshold tolerance to the right is not a universal panacea. To sustain a long-term irrigated agriculture the physical parameters of water quality, percolation, drainage, and crop salt tolerance become a necessary but not a sufficient condition. The sufficient conditions are defined by a positive, present value of the future stream of net agricultural income where costs include land reclamation, water supply, production expenses, and a means of removing and disposing of excess salts from the root zone.

Three major conclusions can be drawn from this analysis. First and foremost, the potential economic and social benefits of osmoregulation research are so large as to dwarf the research and development investment that will be required. Second, given the great uncertainty surrounding any technological break-through, to recommend a single course of action or research approach would be unwise. Third, even though the genetic engineering approach always ranked highest in the analysis, a prudent economist would only recommend that additional resources be allocated to this approach with funds and human capital continuing to be invested in plant breeding and cell culture.

Osmoregulation research to increase salt tolerance in plants can "buy time" for cultivators to accumulate capital to invest in improved drainage and disposal of salts from the plant root zone. In most irrigated areas of the world where root zone salts are not at a steady state equilibrium level, progressive salinization will continue to spread like the cancer that it is and the full potential from this research may never be realized. Thus, osmoregulation research must be accompanied by research in improved methods of drainage and saline management techniques to prolong and enhance the economic life of the new technology.

Finally, it was hypothesized that the benefits of osmoregulation research would be distributed toward the low income farmers operating the lowest quality lands and to the landless who could take up newly reclaimed lands that were previously useless.

REFERENCES

1. E. Epstein. Responses of plants to saline environments. In *Genetic Engineering of Osmoregulation*. Plenum, New York, 1980, pp. 7–21.

2. J. D. Rhoades. Potential for using saline agricultural drainage waters for irrigation. In *Proceedings of the Water Management for Irrigation and Drainage, ASCE*, Reno, Nevada, July 20–22, 1977, pp. 85–115.

3. T. W. Schultz. *Knowledge Activities of Universities: A Critical View*, Paper No. 82-1, University of Arizona, January 19, 1981.

4. C. V. Moore, J. H. Snyder, and P. Sun. Effects of Colorado River water quality and supply on irrigated agriculture. *Water Resour. Res.* **10**(2), 137–144 (1974).

5. E. V. Mass and G. J. Hoffman. Crop salt tolerance—Current assessment, *J. Irrig. Drainage Div. ASCE*, pp. 115–134 (June 1977).

6. H. Greenway and R. Munns. Mechanism of salt tolerance in nonhalophytes. *Annu Rev. Plant Physiol.* **31**, 49–90, (1980).

7. R. T. Ramage. Genetic methods to breed salt tolerance in crops. *Genetic Engineering of Osmoregulation*. Plenum, New York, 1980, pp. 311–318.

8. D. W. Rains. Salt tolerance—New developments. *Advances in Food Producing Systems for Arid and Semiarid Lands*. Academic Press, New York, 1981, pp. 431–455.

9. L. J. Savage. *Foundations of Statistics*. Wiley, New York, 1954.

10. W. R. Lin. *Decisions under Uncertainty in an Empirical Application and Test of Decision Theory in Agriculture*. Unpublished Ph.D. dissertation, University of California, Davis, 1973, p. 218.

11. R. Schlaifer. *Probability and Statistics for Business Decisions*. McGraw-Hill, New York, 1959.

12. C. V. Moore. On the necessary and sufficient conditions for a long-term irrigated agriculture. *Water Resour. Bull.* **8**(4), 802–812 (1972).

13. L. Bernstein. Quantitative assessment of irrigation water quality. *Water Quality Criteria*. American Society for Testing and Materials, Special Tech. Bull. No. 416, 1967.

22

REVIEW OF THE WORLD FOOD
SITUATION AND THE ROLE
OF SALT—TOLERANT PLANTS

Gary H. Toenniessen

Agricultural Sciences
The Rockefeller Foundation
New York, New York

1. ANALYSIS OF FOOD NEEDS AND DEMANDS

1.1. Accelerating Demand for Food

The world is faced today with an accelerating growth in the demand for food. The most obvious cause is the continuing growth in the world's human population. Despite promising indications that fertility is declining and that the rate of population growth may have peaked, the momentum built into the current population structure will carry the world to a population of roughly 6 billion people by the year 2000. This represents a rate of increase of 70–90 million people per year and over 90% of the increase will occur in the less-developed countries. As indicated in Table 22.1, by the year 2000, about 80% of the world's population will live in less-developed regions with China alone having over 20% and India 16%. Hence, population growth will cause about a 2%/year increase in demand for food.

The other major contributor toward driving up world food demand is increasing affluence and the desire for higher quality foods, such as meat and poultry. The centrally planned economies, particularly the Soviet Union, are already purchasing large quantities of grain imports. In addition, there are a growing number of developing countries, some with large populations, where the success of development is causing a rapid growth in per capita incomes. As people move up from a subsistence level economy, more food must be produced

Table 22.1.   Population Projections for World, Major Regions, and Selected Countries[a]

	Population (millions)		Percent Increase by 2000	Average Annual Percent Increase	Percent of World Population in 2000
	1975	2000			
World	4090	6351	55	1.8	100
More developed regions	1131	1323	17	0.6	21
Less developed regions	2959	5028	70	2.1	79
Major regions					
Africa	399	814	104	2.9	13
Asia and Oceania	2274	3630	60	1.9	57
Latin America	325	637	96	2.7	10
U.S.S.R. and Eastern Europe	384	460	20	0.7	7
North America, Western Europe, Japan, Australia, and New Zealand	708	809	14	0.5	13
Selected countries and regions					
People's Republic of China	935	1329	14	1.4	21
India	618	1021	65	2.0	16
Indonesia	135	226	68	2.1	4
Bangladesh	79	159	100	2.8	2
Pakistan	71	149	111	3.0	2
Philippines	43	73	71	2.1	1
Thailand	42	75	77	2.3	1
South Korea	37	57	55	1.7	1
Egypt	37	65	77	2.3	1
Nigeria	63	135	114	3.0	2
Brazil	109	226	108	2.9	4
Mexico	60	131	119	3.1	2
United States	214	248	16	0.6	4
U.S.S.R.	254	309	21	0.8	5
Japan	112	133	19	0.7	2
Eastern Europe	130	152	17	0.6	2
Western Europe	344	378	10	0.4	6

[a] From ref. 1. The Global 2000 Report to the President of the U.S. (1981). Reprinted with permission from Pergamon Press, Ltd.

to feed each individual. There is greater variability in the rate of increase in food demand caused by increasing affluence, but it is generally estimated to average 1–2%/year.

Hence most estimates of the overall rate of increase in world food demand during the remainder of this century are in the 3–4%/year range and much of this will occur in developing countries.

1.2. Regional Differences

Asia, Africa, and Latin America differ with respect to their food problems and their potential for dealing with them. In terms of sheer numbers of undernourished and malnourished people, the situation is worst in Asia. However, food production gains have been impressive in the vast irrigated areas in many of these countries. In addition, it appears that adequate research, training, and investment will be applied to providing for further gains. A comparable effort is urgently needed to improve rainfed agriculture.

Although there may be fewer undernourished people, the food situation in sub-Saharan Africa is bad and growing worse. As shown in Figure 22.1, over the past two decades per capita food production declined in most of the 45 countries of this region (2). Many African agricultural development projects have failed, and even when successful they are offset by production declines in other sectors of the food economy.

Although these constraints to development are numerous and complex it should also be noted that most African countries have been independent of colonial powers for about 20 years or less. What is more, at independence, many were left with little more than their boundaries and even these had been

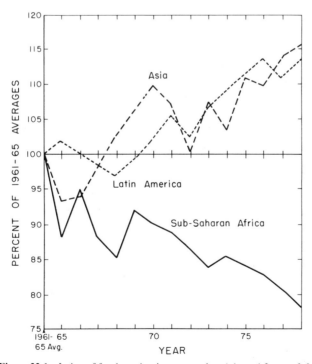

Figure 22.1   Index of food production per capita. Adapted from ref. 2.

established by Europeans without much concern for African interests. Conse-quently, most countries of sub-Saharan Africa are simply at a much earlier stage of development than the countries of Asia. Much more building of human, institutional, and governmental capability will need to occur in Africa before these countries are able to take effective advantage of the technologies and capital resources available from external sources.

Hunger and malnutrition are also serious problems in Latin America but they are primarily a function of extreme inequalities in the distribution of income and food. As with the world as a whole, per capita food production in Latin America is adequate, but unequal distribution of purchasing power results in large numbers of poor people being underfed.

Production and distribution issues are obviously important in all three regions. However, in terms of emphasis, the need in Asia is to maintain or improve on the reasonably satisfactory production growth of the past decade which resulted from the use of improved seed and fertilizer in the irrigated areas. With these relatively easy gains already made and the enormity of the population, the task ahead in Asia is formidable. In Africa, the need is to invigorate a stagnant agriculture; in Latin America the need is to improve the distribution of reasonably adequate food supplies through raising the ability of the poor to purchase food.

2.   ANALYSIS OF FOOD PRODUCING RESOURCES

Based on the above projections, it can be concluded that agricultural production in the developing countries will need to increase in the range of 3–4%/year over the next half century. Since the earth is finite and there are limits to the land, water, energy, and human resources that can be committed to agriculture and food production, two questions immediately arise: are there adequate physical resources, and if so, can they be developed and managed in order to achieve this goal.

2.1.   Land

The earth's land resources vary significantly with regard to agricultural potential. Of the approximately 140 million km^2 of land surface worldwide, well over half have essentially no agricultural potential due to ice cover, lack of soil, poor climate, or other factors. At the other extreme, regions of recent geologic origin (glacial, alluvial, volcanic) often have deep soils of high native fertility. When combined with a moist warm climate and relatively flat topography they provide ideal conditions for cultivated crop production. Large continuous areas of such high quality arable land exist in North America, Europe, the subcontinent of Asia, and China. Elsewhere they occur more sporadically, tending to be patchy, rather than continuous. In addition, the fertility of soil can be built up by good management, as well as destroyed by poor management. The productivity of many soils in northern Europe has been increased over two centuries of steadily improving farming techniques.

Until the late nineteenth century, increased food demand was met almost entirely by expanding the area devoted to crop production and grazing. During the last century this has been supplemented and for the last 30 years overshadowed by increases in crop yield per unit of land area. Still, on a worldwide basis, the expansion of agriculture has continued. Over the past decade, 6–8 million ha of additional cultivated land have been brought into production each year primarily in the developing world (3). For any given locality, the best agricultural lands are developed first and essentially all higher quality land is now under cultivation or has been converted to other societal uses. The tens of millions of additional hectares that will need to be developed for agriculture over the next few decades will have to come from poorer quality land. It is likely that greater inputs and improved management will be required and that the real cost of the food produced in these areas will be higher.

Approximately 1,500 million ha, or about 11% of the world's land surface is currently under cultivation. The amount of land presently not cultivated but potentially available for crop production is difficult to estimate. It will be determined not only by the physical nature of the land but by factors that are not easily predicted, such as the future level of agricultural technology. However, assuming existing technology, the Food and Agriculture Organization and the World Bank have estimated that about 1,000 million additional ha are available (3, 4). About 800 million ha are in the humid tropics and suffer from low fertility soils. Table 22.2 presents the FAO estimate broken down on a regional basis.

In Latin America, the physical resources in most countries are adequate to permit a substantial increase in food output over the next several decades. High fertility lands are relatively limited, and some are still inefficiently used due to land tenure inequities. Consequently, agriculture is expanding in the great areas of relatively infertile but potentially arable land that dominate much of the region. In particular, the Cerrado of Brazil, the Llanos of Colombia and Venezuela, and the forested Amazon Basin are being developed for agriculture, despite concern that many of the agronomic practices presently employed are nonsustainable.

Sub-Saharan Africa also has vast underutilized lands potentially available for agricultural development. There are even river valleys with fertile alluvial soils that remain relatively unused due to insect transmitted diseases such as onchocerciasis. Tsetse fly, the vector for trypanosomiasis, limits development of the vast savanna areas of Africa. Much of the remaining potentially arable land in Africa is classified as semiarid, with inadequate rainfall and saline soils and groundwater being important limitations. Where rainfall is adequate, as in Zaire and Gabon, low fertility soils predominate.

Table 22.2.   Land Under Cultivation in Developing Countries and Potential for Expansion by Major Region[a]

Region	Land in Annual or Permanent Crops (1000 ha)			Potential for Increase (1000 ha)	% of World Potential for Increase
	1966	1976	% Increase		
Latin America	123,558	143,568	16.19	442,432	44.0
Africa	173,375	185,610	7.06	280,390	27.92
Asia centrally planned	132,432	141,266	6.67	62,734	6.25
Near East	76,827	81,062	5.51	30,938	3.08
Far East	251,538	266,329	5.88	5,671	0.56
Total developing countries[b]	758,743	818,943	7.93	823,057	81.96
World total	1,418,210	1,491,800	5.19	1,004,200	100.

[a]From ref. 3.
[b]Includes island countries not considered part of major regions.

Asia has large regions of highly productive agricultural land. They have long been intensively farmed and support dense human populations. Due to population pressure, most adjacent poorer quality lands have also been developed. Future food needs in this region will have to be met through even more intensive use of existing farm land. The most notable exceptions are the vast outer islands of Indonesia and Malaysia. Though dominated by low fertility soils typical of the humid tropics, they are being developed for agriculture and have the potential of significantly contributing to food production in the region.

About 20% of the earth's land surface (2800 million ha) is in permanent pasture and meadows. In developing countries, a majority of the grazing land has a very low stocking rate. The livestock produced are of great importance to the herdsman involved, but the total contribution to food production is modest compared to that of arable agriculture. There is great potential to increase the efficiency of livestock production from such land.

Clearly, resources are available for bringing additional land under cultivation in some developing countries, particularly those of Latin America and Africa. For such cultivation to be stable over the long term, improved agronomic practices will be necessary. Yields are still likely to be lower than those from land presently in production. The current annual rate of agricultural expansion in the developing world is about 0.8% (see Table 22.2), and it has been suggested that this rate could be increased to 1%/yr (5). If adequate research is committed to the development of sustainable production systems on marginal lands, it is reasonable to expect that agricultural expansion can contribute one-forth to one-third of the 3–4% annual increase in food production that is needed to meet the aggregate needs of developing countries over the next few decades. Since underutilized lands are not evenly distributed, their greater use will make the major contribution in some countries, but an almost negligible contribution in others.

2.2 Water

On both existing farms and virgin soils, under irrigated as well as rainfed conditions, the ability of land of produce food is highly dependent on the supply, efficient use, and conservation of water resources. Water is the most important factor limiting crop yields, as well as the productivity of grazing lands. In almost all areas of the world, improved water management can lead to increased food production be it through supplemental irrigation, drainage of excess water, channeling of runoff for crop production, or other techniques, all of which usually involve land management as much as water management.

About 85% percent of all cultivated land is dependent for moisture on rainfall, and a majority of the poorer people in developing countries are dependent for their food on rain-fed agriculture. The magnitude of the water resource available for rain-fed farming is variable and dependent on the amount and timing of rain that falls on any given region. It is one of the least manageable and most risky aspects of farming. Rainfall stimulation has had very limited success, and then only in areas where, and at times when, clouds appropriate for seeding already exist. The weather always has been, and for the foreseeable future will continue to be, the most important variable affecting rainfed crop production. The best a farmer can do is to maximize the efficiency with which rainwater is used by appropriate selection of crops and through agronomic techniques that control runoff and increase the moisture retention capacity of soils.

Irrigation is one of the oldest and most successful means by which man has managed his environment to support food production. It can significantly reduce the risk of drought induced crop failure, lead to substantial increases in yield per crop, allow for the harvest of two or even three crops per year from the same field in tropical regions, and facilitate the use of other technologies.

Worldwide, only about 4% of total river runoff is presently used for irrigation but most of the runoff occurs in areas where and at times when it is not needed. Much of the runoff in semiarid regions is already being used. Most developing countries, however, have considerable potential for new irrigation development. Worldwide, about 210 million ha of cultivated land were under irrigation in 1979, with about 75% of this in developing countries (6). As indicated in Table 22.3, China and India together account for about 90 million ha. During the decade of the 1980s, an additional 50 million ha are scheduled to come under irrigation, and a majority of this will also be in Asia (7). Kovda (8) estimates that worldwide the area of land suitable for irrigation is 1 billion ha.

Unfortunately, irrigation systems in many developing countries are currently operated very inefficiently. In some cases, salinization and waterlogging occur due to inadequate drainage and are destroying the productive potential of the

Table 22.3. Developing Countries with over 1 Million Hectares Irrigated[a]

Country	Irrigated Area (million ha)
China	49.2
India	39.1
Pakistan	14.5
Iran	5.9
Indonesia	5.4
Mexico	5.1
Egypt	2.9
Thailand	2.6
Afghanistan	2.5
Turkey	2.1
Brazil	1.7
Sudan	1.7
Iraq	1.7
Vietnam	1.7
Argentina	1.6
Bangladesh	1.5
Philippines	1.3
Chile	1.3
Peru	1.2
Burma	1.0
Korea Republic	1.0
Korea DPR	1.0

[a]From ref. 6.

land. Severe salinization has occurred in over 50% of the Euphrates valley and over 80% of the irrigated land in Pakistan is affected to some degree. Improved management combined with rehabilitation and modernization of existing irrigation facilities is needed and can provide for considerable increases in food production at a lower cost than building new facilities. About 50 million ha of presently irrigated land are scheduled for such rehabilitation during the 1980s (7).

Groundwater resources exist in many developing countries, but the magnitude of the resource is not well known. Successful development of groundwater irrigation in the tropics has occurred primarily in the Asian subcontinent. Many groundwaters, particularly those in semiarid and coastal regions, have salinity problems.

Irrigation expansion and rehabilitation are both proceeding at an average rate of about 5 million ha per year. The International Food Policy Research Institute has estimated that this could generate about 40% of the additional food needed to meet demand through 1990 (7). The great majority of this would be produced in Asian and Middle Eastern countries. It is clear that more than sufficient water resources are available to make this a doable task. Access to the large sums of investment capital necessary will probably be a more limiting factor than access to water. Past experience indicates, however, that much improvement will be needed in the management of irrigation systems if problems such as salinization and waterlogging are to be avoided. This will require development of new management techniques and substantially enhanced management capacity at the national and local level.

2.3. Energy

The energy resources used in agriculture can be divided into three basic types: solar, commercial (oil, gas, coal, electricity), and traditional (wood, dung, animals, crop waste, and human labor).

The photosynthetic conversion of carbon dioxide to organic materials using solar energy is the basis of essentially all crop production. The amount of solar radiation reaching a given area of the earth's surface varies with time of day, season, and atmospheric conditions, but on an annual basis is essentially constant. Man is not able to alter the incident radiation from the sun but he is able to make some improvements in the efficiency with which it is used for food production. Some plants have higher photosynthetic efficiencies and are inherently more productive. Increasing crop density and use of mixed cropping can provide a plant canopy that captures and uses productively a greater amount of solar radiation.

Commercial energy has become a fundamental requirement of modern agriculture. It is primarily used for fertilizer, irrigation, mechanical power, and crop drying. Agriculture must compete for commercial energy with a wide variety of other economic activities. Petroleum, for example, is a highly movable universal resource. For the world, only about 3.5% of total commercial energy is used for agricultural production (9). In most national energy budgets agriculture plays a relatively minor role. Consequently, though data exist on known and potential fossil fuel reserves, they are not very useful in assessing whether or not adequate energy inputs will be available for agriculture in developing countries. It is safe to say that the total fossil fuel energy resources that will be available over the next half century far exceed agriculture's needs. What is more, a major portion of the remaining and still largely unexploited energy resources are in developing countries. Even over the long term, dependence on diminishing fossil fuel resources is likely to be a less critical issue in agriculture than to other segments of the economy where there are few alternatives available. This, however, does not mean that farmers in developing countries will have timely access to and be able to afford energy based inputs in the future, anymore than they can today. Many of these farmers use little or no such inputs. The determining factors are and will continue to be government policies, manufacturing capacity, pricing stucture, and adequacy of infrastructure.

Fertilizers are the most important form of commercial energy input used by agriculture, accounting for about 45% of the total. About 95% of the N in fertilizers is derived from snythetic ammonia, and about 70% of this is produced from natural gas. Several developing countries have, or are in the process of establishing N fertilizer manufacturing facilities as means of converting their natural gas resources into a more useful and marketable product. For the next decade, manufacturing capacity should be sufficient to meet projected needs. The price of N fertilizer is likely to increase or decrease in accordance with the price of energy, but not to the same degree as with petroleum. The price of natural gas is considerably below parity with petroleum, particularly in developing countries where there is limited infrastructure for its use. In addition, there are major capital costs associated with fertilizer manufacturing which reduce fluctuations in the cost of the final product, and which provide incentives for the continued operation of such facilities once they are established. In many countries the real price of fertilizer today is about equal to what it was in the mid-1960s. As more rural roads, marketing facilities, and credit systems are established in developing countries, a greater number of farmers should have access to fertilizer. There is considerable research under way which should lead to significant increases in the efficiency with which fertilizer is used for crop production. Much potential exists in this regard, since only about 5–10% of the N currently applied actually get used to promote plant growth.

At any price, commercial fertilizer will still be expensive to poor farmers with very limited cash. Hence, alternative and complementary means of providing plant nutrients need to be developed. Progress is being made in expanding the use of leguminous plants, in using crop wastes and manures more efficiently, and in crop-associated cultivation of nitrogen-fixing organisms such as blue-green algae and the azolla water fern with its symbiotic bacteria.

The most important forms of traditional energy used in agriculture are animal traction and human labor. There is considerable potential to improve the efficiency and expand the use of animal traction. New techniques, such as no-till agriculture, and equipment designed for use by farmers with small holdings will improve the efficiency of both animal and human energy devoted to agriculture.

Energy resources should be sufficient to meet agriculture's needs in the coming decades. Significant effort is needed, however, to improve the efficiency of their use, to develop lower cost sources of and alternatives to commercial inputs, and to make both commercial and traditional energy inputs more broadly and reliably available to farmers in developing countries. It should also be noted that agriculture has the potential of producing energy resources, as well as consuming them.

2.4. New Varieties and Technologies

The most significant change in developing country agriculture in the last 20 years has been the introduction of improved varieties that have the potential for much higher yields under better management. High yielding varieties of rice and wheat are now grown on about 30% and about 45%, respectively, of the total area planted to these crops in developing countries, and superior varieties are continually being developed. For example, IR36, a rice variety developed by the International Rice Research Institute, has high yield potential, early maturity, multiple pest resistance, and can withstand moisture stress. Under both irrigated and rain-fed conditions, it made a major contribution to the Philippines becoming a rice exporting country. It is now providing the basis for similar progress in Indonesia. Continued improvement in the productivity of these and other food crops can be expected to result from research currently underway. In addition, the area planted to higher yielding varieties should continue to increase as greater emphasis is placed on development of varieties that are better suited for the many small scale farmers who work under a broad range of less favorable soil, water, and climate conditions, and who often do not have access to or knowledge of other factors involved in better management. Through plant breeding, varieties of most major crops are now available that have a higher degree of salinity tolerance.

There is every reason to believe that traditional plant breeding will continue to make incremental improvements in major crops during the coming decades. Moreover, it is important to remember that modern plant breeding has been focused on a very small number of food crops. Similar efforts focused on other crops certainly have the potential of leading to dramatic breakthroughs such as those that occurred with wheat and rice in the 1960s. Scientists involved in plant improvement have only begun to scratch the surface of this much larger germ plasm reservoir. These "neglected" crops can play a particularly important role in the development of sustainable agriculture on marginal lands. Expanded use of halophytic plants for food and feed production has considerable potential but its realization will require a much greater research effort.

Recent advances in molecular genetics and cell culture techniques offer great potential to expand the options available for plant improvement well beyond those of traditional plant breeding. Some cellular techniques such as embryo culture and pollen culture should provide practical results in the short term. Protoplast fusion and application of recombinant DNA techniques to plant improvement will probably be a longer-term process, but greatly expand horizons concerning what is potentially feasible. At present, advances are occurring in the basic fields of genetics much more rapidly than the results are being applied to plant improvement. New institutional and organizational arrangements are needed that can more effectively link the basic scientists with the plant breeders and agronomists who can use their results, and who better understand the types of genetic recombinations and modifications that can have practical value. A recent survey conducted by economists at the University of Minnesota indicates that agricultural scientists in the United States expect emerging biotechnologies such as plant genetic engineering to make a major contribution toward increased food production beginning in the 1990s (10). It appears very promising that basic and applied research can provide for the continual development of improved varieties of food plants (and other plants and animals) through the remainder of this century and probably beyond.

2.5.   Institutions and Professional Manpower

There are numerous agencies and institutions that contribute to agricultural development in poor countries. National agricultural research agencies and extension services and the international agricultural research institutes have this as their principal mission. In others, such as national planning, multilateral lending, and bilateral assistance agencies, agriculture is just one segment of a more comprehensive set of development objectives. Some institutions, such as basic research laboratories in the developed world, contribute indirectly.

The research centers and laboratories supported by the Consultative Group for International Agricultural Research provide a setting where interdisciplinary teams of highly qualified agricultural scientists from throughout the world address specific problems that limit food production and consumption in developing countries. Of the nine centers located in the tropics, seven have a plant commodity orientation. They focus on development of better crop varieties and improved production practices that can increase yields and solve major problems associated with these crops over a broad region. The other two centers are concerned with improving livestock production. The strength of the CGIAR centers is that they are able to bring together the critical mass of talent and effort necessary to make real progress in reaching sharply focused goals. With adequate support, the research conducted at these centers will continue to make major contributions toward increased food production. Funding shortages have already occurred, however, and important research programs are being terminated.

It should also be realized that of the hundreds of plant and animal species that are important sources of human nutrition in developing countries, the CGIAR centers are concerned with only about 20 plant species, and just a few animal species. Several international research centers concerned with certain other food sources have been established but they are not part of the CGIAR system and are far from adequately funded. Almost no international research effort is devoted to many other potentially important food crops with tropical perennials being a largely neglected group.

At the national level there are four general types of agencies and organizations that contribute to agricultural development and improved nutrition: those that set policies, those that provide inputs and services such as credit, those that conduct research, and those that transfer research results to the farmers. In order for a country to make rapid progress in agricultural development, all four need to work in concert and they need to have access to the financing, new technologies, and technical know-how available internationally (11).

Without a strong government commitment to agricultural development little progress toward such development can be made. If there is a strong commitment, it needs to be reflected in government policies that provide farmers with adequate incentives to produce more and better crops, and that ensure that the necessary inputs, credits, and marketing systems are available. Many developing country governments are committed to agricultural development, but they lack the capacity to formulate and implement policies that are effective in attaining their goals. Where the commitment exists but the policies do not, development of improved capacity for policy formation can make a major contribution to increased food production.

The weakness of national agricultural research agencies and extension services has been and continues to be one of the major constraints to agricultural development. Many national scientists are not well supported with facilities and technicians, and their salaries are often far below those of other professionals. Africa is particularly short of indigenous research capability and of the educational institutions that are needed to train researchers. Fortunately, the World Bank and other major lenders are now willing to provide support to train people and to build research facilities and progress is being made in selected countries.

2.6. Summary

It can be concluded that there are adequate land, water, energy, and technological resources available to meet the world's food production needs during the coming decades. Much human intellect and effort will be needed, however, to provide for their sound management and sustainable use. Agencies and institutions capable of bringing together the necessary human and physical resources for expansion of food production do exist, but they will need to be significantly expanded and strengthened if they are to meet the challenges that lie ahead.

3. ROLE OF RESEARCH ON SALT—TOLERANT PLANTS

The high-yielding wheat, rice, and maize varieties that are available today and that permitted much of the production increases of the past few decades are often appropriate only under favorable physical, economic, and biological conditions. The challenge for plant breeders today is to produce crop varieties that can yield well and reliably under adverse conditions without costly inputs. We must learn how to overcome the numerous problems associated with poor soil conditions, diseases, pests, water deficiencies or excesses, water impurities, and other production constraints that have kept food production for millions of poor farmers in the developing countries at or below a subsistence level. One of the more important of these constraints is salinity in soil and irrigation waters.

Salinization of cropland primarily results from poorly drained irrigation. On irrigated fields evaporation and transpiration extract almost pure water, leaving behind salts that have previously been dissolved in the water. If the salt is not flushed and drained from the ground by rainfall or additional irrigation, it can concentrate in and on the surface of the soil. Since most of today's conventional plants are harmed by salinities of 3000 ppm and most cannot survive more than 5000 ppm, continued salt buildup will ultimately destroy the land's productivity. It is estimated that about one-third of the world's irrigated land is affected to some degree by excess salinity. The best and only long-term solution to salinization where irrigation is practiced is to provide adequate drainage.

Unfortunately, solutions that are best and long-term are also usually expensive and this is indeed the case with providing adequate drainage. In fact, an adequate drainage system often costs more than the associated water delivery system. The reality is therefore that many irrigation systems, particularly in developing countries, lack adequate drainage and many crops consequently now suffer and will continue to suffer from salinity problems. It is a particularly serious situation in Pakistan, India, Iraq, Egypt, and Peru. Developing crop plants tolerant to saline irrigation water has the potential of making an important contribution to food production in many countries. However, it should not be viewed as a substitute for providing adequate drainage.

Some lands or waters have naturally occurring salinity and salt-tolerant crop plants may provide a better or perhaps the only means of utilizing these resources for food production. In many countries, for example, rice and other crops are grown in estaurine and coastal swamp areas where seawater intrusion can cause serious salinity problems. Arid and semiarid lands comprise about 40% of the world's land surface and most groundwaters and some surface waters in these regions are characterized by a high degree of natural salinity. Under such conditions development of salt-tolerant crops may provide for food production in regions where no food is currently produced. Moreover, salt-tolerant plants may allow good quality irrigation water to be used at least twice and perhaps three times before it becomes too brackish for further use. This could significantly increase food production in areas where quality water for irrigation is available but limited.

Some progress has already been made toward developing salt-tolerant crop plants and there appears to be potential for further genetic improvement (12). However, development of crop varieties that combine high yield potential and high salinity tolerance will not be an easy task. Salinity tolerance is a complex whole plant characteristic with physiological and biochemical functions controlled by numerous genes. Moreover, environmental and soil factors heavily influence its expression. Salt-tolerant plants use energy to exclude or sequester salt, thereby maintaining a reasonably low salt concentration in the cytoplasm of their cells. Use of energy for this purpose will ultimately cost the plant in the form of lower yields. Hence, alternatives to the use of saline soil and irrigation waters should always be considered. Where few or no alternatives exist salt-tolerant crop varieties will be needed and can make an important contribution toward meeting the world's growing food demand.

REFERENCES

1. Anonymous. *The Global 2000 Report to the President of the U.S.*, Vol. 1. Pergamon Press, New York, 1981.
2. USDA *Food Problems and Prospects in Sub-Saharan Africa*, Foreign Agricultural Research Report No. 166, 1981.
3. FAO. How much good land is left. *FAO Review on Agriculture and Development.* July/August 1978.
4. World Bank, *Agricultural Land Settlement.* A World Bank Issues Paper, January 1978.
5. W. H. Pawley. World picture—present and future. In A. N. Duckham, J. G. W. Jones, and E. H. Robert, eds., *Food Production and Consumption.* North-Holland, Amsterdam, 1976.
6. FAO. *1980 FAO Production Yearbook*, Vol. 34. Rome, 1981.
7. P. Oram, J. Zapata, G. Alibaruho, and S. Roy. Investment and input requirements for accelerating food production and low-income countries by 1990. International Food Policy Research Institute Research Report No. 10. Washington, D.C., September 1979.
8. V. A. Kovda. Arid land irrigation and soil fertility: Problems of salinity, alkalinity, compaction. In E. B. Worthington, ed., *Arid Land Irrigation in Developing Countries.* Pergamon Press, Oxford, 1977.
9. M. S. Mudahar, and T. P. Hignett. *Energy and Fertilizer Policy Implications and Options for Developing Countries.* International Fertilizer Development Center Technical Bulletin No. 20, 1982.
10. W. B. Sundquist. K. M. Menz, and C. F. Neumeyer. *A Technology Assessment of Commercial Corn Production in the United States.* University of Minnesota Agricultural Experiment Station, Bulletin 546, 1982.
11. S. Wortman and R. W. Cummings, Jr. *To Feed This World: The Challenge and the Strategy.* Johns Hopkins University Press, Baltimore, Maryland, 1978.
12. D. W. Rains, R. C. Valentine, and A. Hollaender, eds. *Genetic Engineering of Osmoregulation.* Plenum Press, New York, 1980.

AUTHOR INDEX

SUBJECT INDEX

Abscisic acid (ABA):
  effect on sodium distribution, 52–53
  tomato and potato mutations affecting, 311,
    312
Absorption zone of root, 5
Acholeplasma laidlawii, sterol effect on cell
  membranes of, 79
Aegilops searsii, salt-stress effects on, 193, 194,
  200
Aegilops squarrosa, see Triticum tauschii
Aegopodium, asymmetric hybrids of, 345
Africa, land use in, 403
Agar gels, salinized, use in breeding studies,
  244
Agrobacterium tumefaciens, tumorigenic
  transformation by, 336
Agropyron, salt tolerance in, breed for, 234–235
Agropyron pungens, salt response of, 218
Agropyrum:
  growth habitat and genetics of, 191
  potassium-sodium exchange at cellular
    membranes in, 38
  salt tolerance in, 195–198, 200
Agropyrum curvifolium:
  glycinebetaine accumulation in, 199
  salt tolerance in, 198, 201
Agropyrum elongatum:
  glycinebetaine accumulation in, 199
  salt tolerance of, 192, 196, 197, 236
Agropyrum intermedium:
  glycinebetaine accumulation in, 199
  salt tolerance in, 197, 198
Agropyrum junceum:
  glycinebetaine accumulation in, 199
  hybrids with *Triticum* sp., 190
  salt tolerance in, 197, 198, 200
Agropyrum scirpeum:
  glycinebetaine accumulation in, 199
  salt tolerance in, 198, 200, 201
Agrostis stolonifera:
  genetic variation in, 215
  salt response of, 218, 236
Aizoaceae, salinity response in, 126

Alfalfa:
  controlled environment agriculture of, 376
  salt tolerance in:
    glycolipids and, 83
    tissue-culture studies of, 306
  see also Medicago sativa
Algae, nitrogen fixation by, 222
Allium cepa, potassium-sodium exchange at
  cellular membranes in, 38, 40, 53
Aloë spp., succulent tissue cells of, 134
Aluminum toxicity, screening for, 245
Ammonium chloride, membrane leakiness from,
  70, 71, 73
Antimonate, as ion precipitating agent, 19
Apium graveolens, gene mutations of mineral
  nutrition in, 309
Apoplast, transport in, 107–109
Apoplastic pathway, blockage of, 7
Arabidopsis-Brassica, heterokaryon-derived
  plants from, 343, 344
Arabidopsis thaliana, mutation affecting
  osmoregulation in, 311
Arachis hypogea, 171
  salt resistance of, 172
Armeria maritima, growth of, 221
Asia, land use in, 404
Asperella, growth habitat and genetics of, 191
Aster, salinity effects on humidity and, 118
Aster tripolium:
  potassium and sodium transport in, 102, 105–
    106, 112–114
  salt response of, 218, 219, 220
ATPases:
  role in potassium-sodium exchange at cell
    membranes, 40, 304
    genetics of, 313
  salinity effects on, 78, 79, 85–87
  in tonoplast transport, 133
Atriplex:
  photosynthesis in, stress effects on, 137–138
  potassium-sodium exchange in, 42, 49, 60,
    97, 108
  potassium and sodium selectivity in, 102